The Cosmic
Perspective
Fundamentals

Jeffrey Bennett | University of Colorado at Boulder

Megan Donahue | Michigan State University

Nicholas Schneider | University of Colorado at Boulder

Mark Voit | Michigan State University

Addison-Wesley

Boston Columbus Indianapolis New York San Francisco Upper Saddle River
Amsterdam Cape Town Dubai London Madrid Milan Munich Paris Montréal Toronto
Delhi Mexico City São Paulo Sydney Hong Kong Seoul Singapore Taipei Tokyo

Publisher: James Smith
Executive Editor: Nancy Whilton
Director of Development: Michael Gillespie
Associate Development Editor: Ashley Eklund
Senior Media Producer: Deb Greco
Media Producer: Kate Brayton
Senior Marketing Manager: Scott Dustan
Associate Director of Production: Erin Gregg
Managing Editor: Corinne Benson
Production Supervisor: Mary O'Connell
Production Service: Lifland et al., Bookmakers
Composition: Progressive Information Technologies
Text and Cover Design: Derek Bacchus
Manufacturing Buyer: Jeffrey Sargent
Photo Research: David Chavez
Director, Image Resource Center: Melinda Patelli
Manager, Rights and Permissions: Zina Arabia
Manager, Cover Visual Research & Permissions: Karen Sanatar
Image Permission Coordinator: Elaine Soares
Printer and Binder: Quebecor Versailles
Cover Printer: Phoenix Color
Cover Photo Credits: Noctilucent Clouds, Morton Ross; Two Tails of Comet Lulin, Richard D. Richins

Library of Congress Cataloging-in-Publication Data

The cosmic perspective fundamentals/Jeffrey Bennett ... [et.al.].
    p.   cm.
Includes index.
ISBN 978-0-321-56704-8
1. Astronomy—Textbooks.   I. Bennett, Jeffrey O.

QB43.3.C683 2010
520—dc22

                                        2009022610

ISBN-10: 0-321-56704-8; ISBN-13: 978-0-321-56704-8

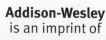

Addison-Wesley
is an imprint of

1 2 3 4 5 6 7 8 9 10—QWV—15 14 13 12 11 10 09

Dedication

To all who have ever wondered about the mysteries of the universe. We hope that this book will answer some of your questions—and that it will also raise new questions in your mind that will keep you curious and interested in the ongoing human adventure of astronomy.

And, especially, to Michaela, Emily, Sebastian, Grant, Nathan, Brooke, and Angela. The study of the universe begins at birth, and we hope that you will grow up in a world with far less poverty, hatred, and war so that all people will have the opportunity to contemplate the mysteries of the universe into which they are born.

Brief Contents

Detailed Contents

Preface

WE HUMANS HAVE GAZED into the sky for countless generations. We have wondered how our lives are connected to the Sun, Moon, planets, and stars that adorn the heavens. Today, through the science of astronomy, we know that these connections go far deeper than our ancestors ever imagined. This book focuses on the story of modern astronomy and the new perspective—*the cosmic perspective*—that astronomy gives us on ourselves and our planet.

Who Is This Book For?

The Cosmic Perspective Fundamentals is designed to serve as a textbook for one-term college courses in introductory astronomy. Our other textbooks, *The Cosmic Perspective* and *The Essential Cosmic Perspective*, are fairly complete and authoritative references, from which instructors can pick and choose the material that they will emphasize in their individual courses. In recent years, however, some instructors have adopted a new approach to teaching in which they rely more heavily on Web-based materials and active or collaborative learning techniques and less on textbook-based lectures. This new approach inspired us to create a much shorter book that would retain the pedagogy and flexibility of our longer texts but focus only on the most essential concepts of modern astronomy. *The Cosmic Perspective Fundamentals* is the result.

Topical Selection

A dramatically shorter textbook must necessarily cover fewer topics. We have carefully selected those topics using the following four criteria:

- *Importance.* We surveyed a large number of professors to identify the topics considered of greatest importance in a college-level astronomy course, in order to ensure that the most fundamental concepts are covered in this text.

- *Engagement.* Most students in a college astronomy course are there to satisfy a general education requirement, but the subject is sufficiently interesting that it should be possible to choose topics that students will find highly engaging—and that they will therefore be willing to work hard to learn.

- *Process of science.* We believe that the primary purpose of a general education requirement in science is to ensure that students learn about science itself. Throughout the book, we have chosen topics that illustrate important aspects of the process of science, and each chapter concludes with a section called **The Process of Science in Action,** which presents a case study of how the process of science has helped (or is currently helping) to provide greater insight into key topics in astronomy.

- *Active learning.* Educational research has shown that students learn scientific concepts best by actively solving conceptual problems, both individually and in collaboration with other students. We have emphasized topics that are well suited to active learning, and each chapter includes **Think About It** critical thinking questions for in-class discussion and **See It for Yourself** hands-on activities to further promote active learning. These in-text features are reinforced by a variety of active learning resources on the Companion Web Site.

We recognize that most astronomy courses follow a similar structure, beginning with topics such as the scale of the universe, seasons, and phases of the Moon and then progressing to study of the planets, stars, galaxies, and cosmology. Our selected topics have been organized in a similar fashion. The fifteen chapters are designed so that they can be covered in a typical semester at a rate of approximately one chapter per week.

Book Structure

To facilitate student learning, we have created a simple pedagogical structure used in each of the book's fifteen chapters:

- Each chapter begins with an opening page that includes a brief overview of the chapter content and a clear set of **Learning Goals** associated with the chapter; each Learning Goal is phrased as a question to engage students as they read.

- Each chapter consists of three sections. The first two sections focus on the key topics of the chapter; the third section builds on the ideas from the first two sections, but focuses on **The Process of Science in Action.**

- Each section is written to address the Learning Goal questions from the chapter opening page.

- Each chapter concludes with a **visual summary** that provides a concise review of the answers to the Learning Goal questions.

The summary is followed by a 12-question **Quick Quiz** (with answers provided in Appendix D) and a set of **short-answer, essay, and quantitative questions.**

Additional features of the book include the following:

- **Tools of Science** boxes, which present a brief overview of key tools that astronomers use, including theories, equations, observational techniques, and technology. Each chapter includes one Tools of Science box related to the chapter content.

- **Common Misconceptions** boxes, which address popularly held but incorrect ideas about topics in the text.

- **Annotated Figures and Photos**, which act like the voice of an instructor, walking students through the key ideas presented in complex figures, photos, and graphs.

- **Cosmic Context Figures**, which combine text and illustrations into accessible and coherent two-page visual summaries that will help improve student understanding of essential topics.

Companion Web Site

www.pearsonhighered.com/bennett

This password-protected Web Site provides one gradable quiz per chapter, ten Interactive Figures and Photos based on figures from the book, four Self-Guided Tutorials that interactively teach about and test comprehension of key topics, four movies, and an interactive glossary, plus a set of activities written by the authors for use in lab or as homework.

Acknowledgments

A textbook may carry author names, but it is the result of hard work by a long list of committed individuals, as well as many reviewers. We could not possibly list everyone who has helped, but we would like to thank in particular our current editorial and design team: Jim Smith, Nancy Whilton, Ashley Eklund, Michael Gillespie, Sally Lifland, Mary O'Connell, Derek Bacchus, and Mark Ong. In addition, we thank the following professors who have reviewed this new book during its development:

Philip Blanco, *Grossmont College*

Jeff Bodart, *Chipola Junior College*

Debra Burris, *University of Central Arkansas*

Scott Calvin, *Sarah Lawrence College*

David Falk, *Los Angeles Valley College*

Martin Gaskell, *University of Texas, Austin*

Harold Geller, *George Mason University*

Javier Hasbun, *University of West Georgia*

David Herrick, *Maysville Community College*

Tracy Hodge, *Berea College*

William Keel, *University of Alabama*

Arthur Kosowsky, *University of Pittsburgh*

David Lamp, *Texas Technical University*

Michael LoPresto, *Henry Ford Community College*

Ihor Luhach, *Valencia Community College*

Lauren Novatne, *Reedley College*

Richard Olenick, *University of Dallas*

Jessica Smay, *San Jose City College*

Brent Sorenson, *Southern Utah University*

Adriane Steinacker, *University of California, Santa Cruz*

Peter Stine, *Bloomsburg University of Pennsylvania*

Irina Struganova, *Valencia Community College*

Grant Wilson, *University of Massachusetts Amherst*

Tim Young, *University of North Dakota*

Finally, we thank the many people who have greatly influenced our outlook on education and our perspective on the universe over the years, including Tom Ayres, Fran Bagenal, Forrest Boley, Robert A. Brown, George Dulk, Erica Ellingson, Katy Garmany, Jeff Goldstein, David Grinspoon, Robin Heyden, Don Hunten, Joan Marsh, Catherine McCord, Dick McCray, Dee Mook, Cherilynn Morrow, Charlie Pellerin, Carl Sagan, Mike Shull, John Spencer, and John Stocke.

Jeff Bennett

Megan Donahue

Nick Schneider

Mark Voit

About the Authors

Jeffrey Bennett holds a B.A. (1981) in biophysics from the University of California, San Diego, and an M.S. and Ph.D. (1987) in astrophysics from the University of Colorado, Boulder. He has taught at every level from preschool through graduate school, including more than 50 college classes in astronomy, physics, mathematics, and education. He served 2 years as a visiting senior scientist at NASA headquarters, where he created NASA's "IDEAS" program, started a program to fly teachers aboard NASA's airborne observatories (including the hopefully soon-to-be-flying SOFIA), and worked on numerous educational programs for the Hubble Space Telescope and other space science missions. He also proposed the idea for and helped develop both the Colorado Scale Model Solar System on the CU-Boulder campus and the Voyage Scale Model Solar System on the National Mall in Washington, D.C. (He is pictured here with the model Sun.) In addition to this astronomy textbook, he has written college-level textbooks in astrobiology, mathematics, and statistics; two books for the general public, *On the Cosmic Horizon* (Pearson Addison-Wesley, 2001) and *Beyond UFOs* (Princeton University Press, 2008); and an award-winning series of children's books that includes *Max Goes to the Moon, Max Goes to Mars, Max Goes to Jupiter, and Max's Ice Age Adventure*. When not working, he enjoys participating in masters swimming and in the daily adventures of life with his wife, Lisa; his children, Grant and Brooke; and his dog, Cosmo. His personal Web site is www.jeffreybennett.com.

Megan Donahue is a professor in the Department of Physics and Astronomy at Michigan State University. Her current research is mainly on clusters of galaxies: their contents—dark matter, hot gas, galaxies, active galactic nuclei—and what they reveal about the contents of the universe and how galaxies form and evolve. She grew up on a farm in Nebraska and received a B.A. in physics from MIT, where she began her research career as an X-ray astronomer. She has a Ph.D. in astrophysics from the University of Colorado, for a thesis on theory and optical observations of intergalactic and intracluster gas. That thesis won the 1993 Trumpler Award from the Astronomical Society for the Pacific for an outstanding astrophysics doctoral dissertation in North America. She continued postdoctoral research in optical and X-ray observations as a Carnegie Fellow at Carnegie Observatories in Pasadena, California, and later as an STScI Institute Fellow at Space Telescope. Megan was a staff astronomer at the Space Telescope Science Institute until 2003, when she joined the MSU faculty. Megan is married to Mark Voit, and they collaborate on many projects, including this textbook and the raising of their children, Michaela, Sebastian, and Angela. Between the births of Sebastian and Angela, Megan qualified for and ran the 2000 Boston Marathon. These days, Megan runs, orienteers, and plays piano and bass guitar whenever her children allow it.

Nicholas Schneider is an associate professor in the Department of Astrophysical and Planetary Sciences at the University of Colorado and a researcher in the Laboratory for Atmospheric and Space Physics. He received his B.A. in physics and astronomy from Dartmouth College in 1979 and his Ph.D. in planetary science from the University of Arizona in 1988. In 1991, he received the National Science Foundation's Presidential Young Investigator Award. His research interests include planetary atmospheres and planetary astronomy, with a focus on the odd case of Jupiter's moon Io. He enjoys teaching at all levels and is active in efforts to improve undergraduate astronomy education. Off the job, he enjoys exploring the outdoors with his family and figuring out how things work.

Mark Voit is a professor in the Department of Physics and Astronomy at Michigan State University. He earned his B.A. in astrophysical sciences at Princeton University and his Ph.D. in astrophysics at the University of Colorado in 1990. He continued his studies at the California Institute of Technology, where he was a research fellow in theoretical astrophysics, and then moved on to Johns Hopkins University as a Hubble Fellow. Before going to Michigan State, Mark worked in the Office of Public Outreach at the Space Telescope, where he developed museum exhibitions about the Hubble Space Telescope and was the scientist behind NASA's HubbleSite. His research interests range from interstellar processes in our own galaxy to the clustering of galaxies in the early universe. He is married to coauthor Megan Donahue, and they try to play outdoors with their three children whenever possible, enjoying hiking, camping, running, and orienteering. Mark is also author of the popular book *Hubble Space Telescope: New Views of the Universe.*

How to Succeed in Your Astronomy Course

Using This Book

Each chapter in this book is designed to make it easy for you to study effectively and efficiently. To get the most out of each chapter, you might wish to use the following study plan:

- A textbook is not a novel, and you'll learn best by reading the elements of this text in the following order:

  1. Start by reading the Learning Goals and the introductory paragraphs at the beginning of the chapter so that you'll know what you are trying to learn.

  2. Next, get an overview of the key concepts by studying the illustrations and reading their captions. The illustrations highlight almost all of the major concepts, so this "illustrations first" strategy gives you an opportunity to survey the concepts before you read about them in depth. You will find the *Cosmic Context* figures to be especially useful.

  3. Read the chapter narrative, but save the boxed features (Common Misconceptions, Tools of Science) to read later. As you read, make notes on the pages to remind yourself of ideas you'll want to review later. Avoid using a highlight pen; underlining with pen or pencil is far more effective, because it forces you to take greater care and therefore helps keep you alert as you study. Be careful to underline selectively—it won't help you later if you've underlined everything.

  4. After reading the chapter once, go back through and read the Common Misconceptions and Tools of Science.

  5. Then turn your attention to the Summary of Key Concepts. The best way to use the summary is to try to answer the Learning Goal questions for yourself before reading the short answers given in the summary.

- After completing the reading as described above, start testing your understanding with the end-of-chapter exercises. A good way to begin is to make sure you can answer all of the Quick Quiz questions; if you don't know an answer, look back through the chapter until you figure it out. Then test your understanding a little more deeply by trying the Short Answer/Essay questions.

- You can further check your understanding and get feedback on difficulties by trying the online quizzes at www.wps.aw.com/aw_bennett_tcpf_1.

- Don't stop there; visit the site again and make use of other resources that will help you further build your understanding. These resources have been developed specifically to help you learn the most important ideas in your astronomy course, and they have been extensively tested to make sure they are effective. They really do work, and the only way you'll gain their benefits is by going on the Web site and using them.

The Key to Success: Study Time

The single most important key to success in any college course is to spend enough time studying. A general rule of thumb for college classes is that you should expect to study about 2 to 3 hours per week *outside* of class for each unit of credit. For example, based on this rule of thumb, a student taking 15 credit hours should expect to spend 30 to 45 hours each week studying outside of class. Combined with time in class, this works out to a total of 45 to 60 hours spent on academic work—not much more than the time a typical job requires, and you get to choose your own hours. Of course, if you are working while you attend school, you will need to budget your time carefully.

As a rough guideline, your studying time in astronomy might be divided as shown in the table at the bottom of this page. If you find that you are spending fewer hours than these guidelines suggest, you can probably improve your grade by studying longer. If you are spending more hours than these guidelines suggest, you may be studying inefficiently; in that case, you should talk to your instructor about how to study more effectively.

continued

If Your Course Is	Times for Reading the Assigned Text (per week)	Times for Homework Assignments (per week)	Times for Review and Test Preparation (average per week)	Total Study Time (per week)
3 credits	2 to 4 hours	2 to 3 hours	2 hours	6 to 9 hours
4 credits	3 to 5 hours	2 to 4 hours	3 hours	8 to 12 hours
5 credits	3 to 5 hours	3 to 6 hours	4 hours	10 to 15 hours

General Strategies for Studying

- Don't miss class. Listening to lectures and participating in discussions is much more effective than reading someone else's notes. Active participation will help you retain what you are learning.

- Take advantage of the resources offered by your professor, whether it be e-mail, office hours, review sessions, online chats, or simply opportunities to talk to and get to know your professor. Most professors will go out of their way to help you learn in any way they can.

- Budget your time effectively. Studying 1 or 2 hours each day is more effective, and far less painful, than studying all night before homework is due or before exams.

- If a concept gives you trouble, do additional reading or studying beyond what has been assigned. And if you still have trouble, ask for help: You surely can find friends, peers, or teachers who will be glad to help you learn.

- Working together with friends can be valuable in helping you understand difficult concepts. However, be sure that you learn *with* your friends and do not become dependent on them.

- Be sure that any work you turn in is of *collegiate quality:* neat and easy to read, well organized, and demonstrating mastery of the subject matter. Although it takes extra effort to make your work look this good, the effort will help you solidify your learning and is also good practice for the expectations that future professors and employers will have.

Preparing for Exams

- Study the Review Questions, and rework problems and other assignments; try additional questions to be sure you understand the concepts. Study your performance on assignments, quizzes, or exams from earlier in the term.

- Study the relevant online tutorials and chapter quizzes available at www.pearsonhighered.com/bennett.

- Study your notes from lectures and discussions. Pay attention to what your instructor expects you to know for an exam.

- Reread the relevant sections in the textbook, paying special attention to notes you have made on the pages.

- Study individually *before* joining a study group with friends. Study groups are effective only if every individual comes prepared to contribute.

- Don't stay up too late before an exam. Don't eat a big meal within an hour of the exam (thinking is more difficult when blood is being diverted to the digestive system).

- Try to relax before and during the exam. If you have studied effectively, you are capable of doing well. Staying relaxed will help you think clearly.

1

A Modern View of the Universe

Learning Goals

1.1 Our Place in the Universe
- What is our place in the universe?
- How big is the universe?

1.2 A Brief History of the Universe
- How did we come to be?
- How do our lifetimes compare to the age of the universe?

✷ THE PROCESS OF SCIENCE IN ACTION

1.3 Defining Planets
- What is a planet?

This Hubble Space Telescope photo shows a piece of the sky so small that you could cover it with a grain of sand at arm's length. Yet it is filled with galaxies, each containing billions of stars much like our Sun, perhaps orbited by their own families of planets. The picture shows more than just the vastness of space, however. It also shows the depths of time: Some of the smaller smudges are galaxies so far away that their light has taken more than 12 billion years to reach us. A major goal of this book is to help you understand what you see in this photograph. We'll begin with a brief survey of our modern, scientific view of the universe.

1.1 Our Place in the Universe

Our ancestors imagined the universe to be relatively small. They placed Earth at the center, surrounded by circles or spheres that carried the Sun, Moon, and planets around us each day. Beyond the planets, they imagined the boundary of the universe to be a sphere filled with stars. These ideas made sense at the time, because they agreed with everyday experience: The Sun, Moon, planets, and stars all *appear* to circle around us each day, and we cannot feel the constant motion of Earth as it rotates on its axis and orbits the Sun. But today we know that these appearances are deceiving, that Earth is *not* the center of the universe, and that our universe is far larger and filled with far greater wonders than our ancestors ever imagined.

What is our place in the universe?

Before we can discuss the universe and its great wonders, we first need to develop a general sense of our place within it. We can do this by thinking about what we might call our "cosmic address," illustrated in **Figure 1.1.**

Earth is a planet in our **solar system,** which consists of the Sun, the planets and their moons, and countless smaller objects that include rocky *asteroids* and icy *comets*. Keep in mind that our Sun is a *star,* just like the stars we see in our night sky.

Our solar system belongs to the huge, disk-shaped collection of stars called the **Milky Way Galaxy.** A **galaxy** is a great island of stars in space, containing from a few hundred million to a trillion or more stars. The Milky Way is a relatively large galaxy, containing more than 100 billion stars. Our solar system is located a little over halfway from the galactic center to the edge of the galactic disk.

Billions of other galaxies are scattered throughout space. Some galaxies are fairly isolated, but many others are found in groups. Our Milky Way, for example, is one of the two largest among about 40 galaxies in the **Local Group.** Groups of galaxies with more than a few dozen members are often called **galaxy clusters.**

On a very large scale, observations show that galaxies and galaxy clusters appear to be arranged in giant chains and sheets with huge voids between them; the background of Figure 1.1 shows this large-scale structure. The regions in which galaxies and galaxy clusters are most tightly packed are called **superclusters,** which are essentially clusters of galaxy clusters. Our Local Group is located in the outskirts of the **Local Supercluster.**

Together, all these structures make up our **universe.** In other words, the universe is the sum total of all matter and energy, encompassing the superclusters and voids and everything within them.

 Think about it Some people think that our tiny physical size in the vast universe makes us insignificant. Others think that our ability to learn about the wonders of the universe gives us significance despite our small size. What do *you* think?

Astronomical Distance Measurements   Notice that Figure 1.1 is labeled with an approximate size for each structure in kilometers. You can convert from kilometers into miles by multiplying by 0.6 (because 1 kilometer ≈ 0.6 mile); for example, 10,000 (10^4) kilometers is about 6000 miles. In astronomy, many distances are so large that kilometers are not the most convenient unit. We will therefore make frequent use of two other distance units:

- One **astronomical unit (AU)** is Earth's average distance from the Sun, which is about 150 million kilometers.

- One **light-year (ly)** is the *distance* that light can travel in 1 year, which is about 10 trillion kilometers. Note that you can find this distance by multiplying the speed of light—300,000 kilometers per second—by the number of seconds in one year (see Tools of Science, p. 9).

Common Misconceptions

THE MEANING OF A LIGHT-YEAR

You've probably heard people say things like "It will take me light-years to finish this homework!" But a statement like this one doesn't make sense, because light-years are a unit of *distance,* not time. If you are unsure whether the term *light-year* is being used correctly, try testing the statement by using the fact that 1 light-year is about 10 trillion kilometers, or 6 trillion miles. The statement then reads "It will take me 6 trillion miles to finish this homework," which clearly does not make sense.

Figure 1.1

Our Cosmic Address

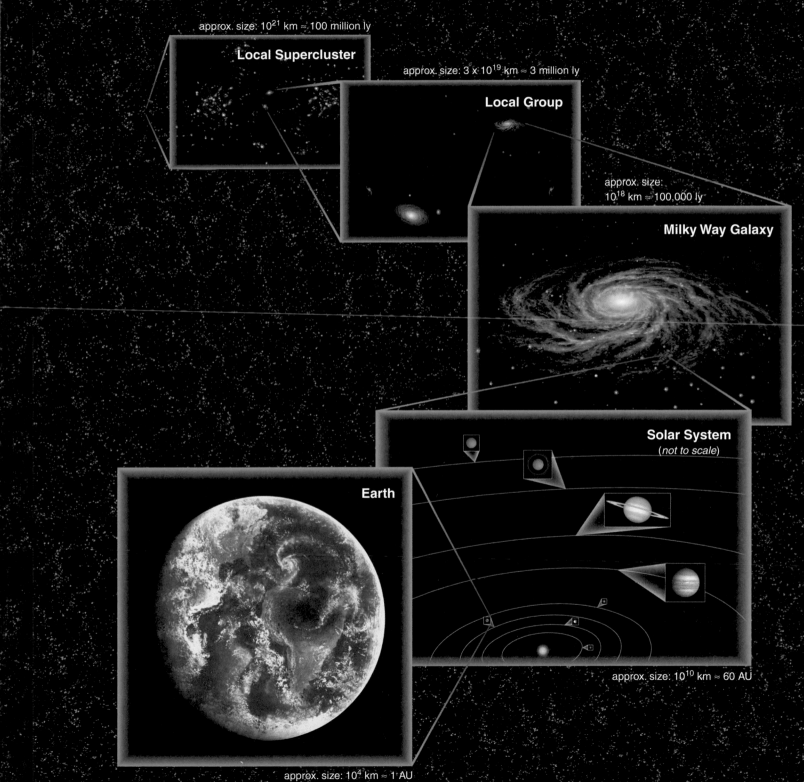

Universe

approx. size: 10^{21} km $\approx$ 100 million ly

Local Supercluster

approx. size: 3×10^{19} km $\approx$ 3 million ly

Local Group

approx. size:
10^{18} km $\approx$ 100,000 ly

Milky Way Galaxy

Solar System
(*not to scale*)

Earth

approx. size: 10^{10} km $\approx$ 60 AU

approx. size: 10^4 km $\approx$ 1 AU

Figure 1.2 | The Andromeda Galaxy (M31). When we look at this galaxy, we see light that traveled through space for 2.5 million years. The inset shows its location in the constellation Andromeda.

Measurements in AU are useful for describing distances in our solar system, while light-years are more useful for describing the distances of stars and galaxies.

Looking Back in Time   Light-years are a unit of distance, but they are related to the time it takes light to travel through space. Consider Sirius, the brightest star in the night sky, which is located about 8 light-years away. Because it takes light 8 years to travel this distance, we see Sirius not as it is today, but rather as it was 8 years ago. The star Betelgeuse, a bright red star in the constellation Orion, is 427 light-years away, which means we see it as it was 427 years ago. If Betelgeuse exploded in the past 427 years (a possibility we'll discuss in Chapter 10), we would not yet know it, because the light from the explosion could not yet have reached us.

The general idea that light takes time to travel through space leads to a remarkable fact: **The farther away we look in distance, the further back we look in time.** The effect is dramatic for large distances. The Andromeda Galaxy (**Figure 1.2**) is about 2.5 million light-years away, which means we see it as it looked about 2.5 million years ago. We see more distant galaxies as they were even further in the past.

It's also amazing to realize that any "snapshot" of a distant galaxy is a picture of both space and time. For example, because the Andromeda Galaxy is about 100,000 light-years in diameter, the light we see from the far side of the galaxy must have left on its journey to us some 100,000 years before the light from the near side. Figure 1.2 therefore shows different parts of the galaxy spread over a time period of 100,000 years. When we study the universe, it is impossible to separate space and time.

See it for yourself   You can see the Andromeda Galaxy for yourself, because it is faintly visible to the naked eye. You'll need a dark site and a star chart to find it. When you have this opportunity, remember that you are seeing light that spent 2.5 million years in space before reaching your eyes. If students on a planet in the Andromeda Galaxy were looking at the Milky Way Galaxy, what would they see? Could they know that we exist here on Earth?

The Observable Universe   As we'll discuss in Section 1.2, astronomers estimate that the universe is about 14 billion years old. This fact, combined with the fact that looking deep into space means looking far back in time, places a limit on the portion of the universe that we can see, even in principle.

Figure 1.3 shows the idea. If we look at a galaxy that is 7 billion light-years away, we see it as it looked 7 billion years ago—which means we see it as it was when the universe was half its current age. If we look at a galaxy that is 12 billion light-years away—like the most distant ones in the Hubble Space Telescope

Figure 1.3 | The farther away we look in space, the further back we look in time. The age of the universe therefore puts a limit on the size of the *observable* universe—the portion of the entire universe that we could observe in principle.

Far: We see a galaxy 7 billion light-years away as it was 7 billion years ago—when the universe was half its current age of 14 billion years.

Farther: We see a galaxy 12 billion light-years away as it was 12 billion years ago—when the universe was only about 2 billion years old.

The limit of our observable universe: Light from nearly 14 billion light-years away shows the universe as it looked shortly after the Big Bang, before galaxies existed.

Beyond the observable universe: We cannot see anything farther than 14 billion light-years away, because light has not had enough time to reach us.

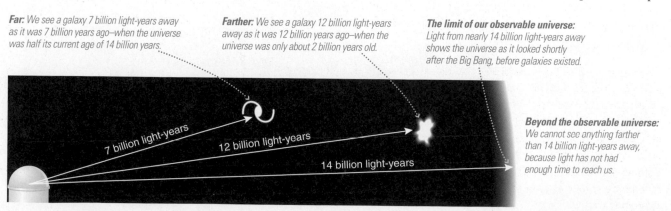

7 billion light-years

12 billion light-years

14 billion light-years

photo on page 1—we see it as it was 12 billion years ago, when the universe was only 2 billion years old. And if we tried to look beyond 14 billion light-years, we'd be trying to look to a time more than 14 billion years ago—which is before the universe existed. We cannot see anything more than 14 billion light-years away, because light from such distant objects has not yet had time to reach us. This distance of 14 billion light-years therefore marks the boundary of our **observable universe**—the portion of the entire universe that we can potentially observe. Note that this fact does not put any limit on the size of the *entire* universe, which may be far larger than our observable universe. We simply have no hope of seeing or studying anything beyond the bounds of our observable universe.

How big is the universe?

Figure 1.1 put numbers on the sizes of different structures in the universe, but these numbers have little meaning for most people—after all, they are literally astronomical. Let's try to put these numbers into perspective.

The Scale of the Solar System Illustrations and photo montages often make our solar system look as if it were crowded with planets and moons, but the reality is far different. One of the best ways to develop perspective on cosmic sizes and distances is to imagine our solar system shrunk down

Basic Astronomical Objects, Units, and Motions

This box summarizes key definitions used throughout this book.

Basic Astronomical Objects

star A large, glowing ball of gas that generates heat and light through nuclear fusion in its core. Our Sun is a star.

planet A moderately large object that orbits a star and shines primarily by reflecting light from its star. According to a definition adopted in 2006, an object can be considered a planet only if it (1) orbits a star; (2) is large enough for its own gravity to make it round; and (3) has cleared most other objects from its orbital path. An object that meets the first two criteria but not the third, like Pluto, is designated a *dwarf planet.*

moon (or satellite) An object that orbits a planet. The term *satellite* is also used more generally to refer to any object orbiting another object.

asteroid A relatively small and rocky object that orbits a star.

comet A relatively small and ice-rich object that orbits a star.

small solar system body An asteroid, comet, or other object that orbits a star but is too small to qualify as a planet or dwarf planet.

Collections of Astronomical Objects

solar system The Sun and all the material that orbits it, including the planets, dwarf planets, and small solar system bodies. Although the term *solar system* technically refers only to our own star system (*solar* means "of the Sun"), it is often applied to other star systems as well.

star system A star (sometimes more than one star) and any planets and other materials that orbit it.

galaxy A great island of stars in space, containing from a few hundred million to a trillion or more stars, all held together by gravity and orbiting a common center.

cluster (or group) of galaxies A collection of galaxies bound together by gravity. Small collections (up to a few dozen galaxies) are generally called *groups*, while larger collections are called *clusters*.

supercluster A gigantic region of space where many individual galaxies and many groups and clusters of galaxies are packed more closely together than elsewhere in the universe.

universe (or cosmos) The sum total of all matter and energy—that is, all galaxies and everything between them.

observable universe The portion of the entire universe that can be seen from Earth, at least in principle. The observable universe is probably only a tiny portion of the entire universe.

Astronomical Distance Units

astronomical unit (AU) The average distance between Earth and the Sun, which is about 150 million kilometers. More technically, 1 AU is the length of the semimajor axis of Earth's orbit.

light-year The distance that light can travel in 1 year, which is about 9.46 trillion kilometers.

Terms Relating to Motion

rotation The spinning of an object around its axis, such as Earth's daily rotation around the axis. For example, Earth rotates once each day around its axis, which is an imaginary line connecting the North and South Poles.

orbit (revolution) The orbital motion of one object around another due to gravity. For example, Earth orbits around the Sun once each year.

expansion (of the universe) The increase in the average distance between galaxies as time progresses.

Figure 1.4 | This photo shows the pedestals housing the Sun (the gold sphere on the nearest pedestal) and the inner planets in the Voyage scale model solar system (Washington, D.C.). The model planets are encased in the sidewalk-facing disks visible at about eye level on the planet pedestals. The building at the left is the National Air and Space Museum.

to a scale that would allow you to walk through it. The Voyage scale model solar system in Washington, D.C., makes such a walk possible (**Figure 1.4**). This model shows the Sun and the planets, and the distances between them, at *one ten-billionth* of their actual sizes and distances.

Figure 1.5a shows the Sun and planets at their correct sizes (but not distances) on the Voyage scale: The model Sun is about the size of a large grapefruit, Jupiter is about the size of a marble, and Earth is about the size of the ball point in a pen. You can immediately see some key facts about our solar system. For example, the Sun is far larger than any of the planets; in mass, the Sun outweighs all the planets combined by a factor of more than 1000. The planets also vary considerably in size: The storm on Jupiter known as the Great Red Spot (visible near Jupiter's lower left in the painting) could swallow up the entire Earth.

The scale of the solar system is even more remarkable when you combine the sizes shown in Figure 1.5a with the distances illustrated by the map of the Voyage model in **Figure 1.5b.** For example, the ball-point-sized Earth is located about 15 meters (16.5 yards) from the grapefruit-sized Sun, which means you can picture Earth's orbit as a circle of radius 15 meters around a grapefruit.

Perhaps the most striking feature of our solar system when we view it to scale is its emptiness. The Voyage model shows the planets along a straight path, so we'd need to draw each planet's orbit around the model Sun to show the full extent of our planetary system. Fitting all these orbits would require an area measuring more than a kilometer on a side—an area equivalent to more than 300 football fields arranged in a grid. Spread over this large area, only the grapefruit-size Sun, the planets, and a few moons would be big enough to notice with your eyes. The rest of it would look virtually empty (that's why we call it *space*!).

 Think about it Earth is the only place in our solar system—and the only place we yet know of in the universe—with conditions suitable for human life. How does visualizing Earth to scale affect your perspective on human existence?

a This painting shows the scaled sizes (but not distances) of the Sun, the planets, and the two largest known dwarf planets.

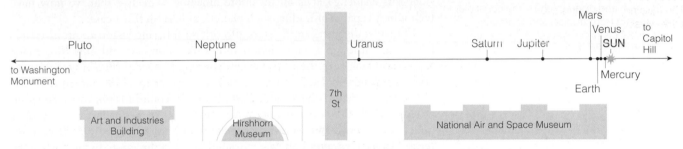

b This map shows the locations of the Sun and planets in the Voyage model; the distance from the Sun to Pluto is about 600 meters (1/3 mile). Planets are lined up in the model, but in reality each planet orbits the Sun independently and a perfect alignment never occurs.

Figure 1.5 | The Voyage scale model represents the solar system at *one ten-billionth* of its actual size. Pluto is included in the Voyage model, which was built before the International Astronomical Union reclassified Pluto as a dwarf planet.

Seeing our solar system to scale also helps put space exploration into perspective. The Moon, the only other world on which humans have ever stepped (**Figure 1.6**), lies only about 4 centimeters ($1\frac{1}{2}$ inches) from Earth in the Voyage model. On this scale, the palm of your hand can cover the entire region of the universe in which humans have so far traveled. The trip to Mars is more than 150 times as far as the trip to the Moon, even when Mars is on the same side of its orbit as Earth. And while you can walk from the Sun to Pluto in just a few minutes on the Voyage scale, the *New Horizons* spacecraft that is making the real journey will have been in space nearly a decade when it finally flies past Pluto in 2015.

Figure 1.6 | This famous photograph from the first Moon landing (*Apollo 11* in July 1969) shows astronaut Buzz Aldrin, with Neil Armstrong reflected in his visor. Armstrong was the first to step onto the Moon's surface, saying, "That's one small step for a man, one giant leap for mankind."

Figure 1.7 | This photograph and diagram show the constellation Centaurus, which is visible from tropical and southern latitudes. Alpha Centauri's real distance is about 4.4 light-years, which is 4400 kilometers on the 1-to-10-billion Voyage scale.

Distance to Stars If you visit the Voyage model in Washington, D.C., you can walk the roughly 600-meter distance from the Sun to Pluto in just a few minutes. But how far would you have to walk to reach the next star on this scale?

Amazingly, you would need to walk to California. If this answer seems hard to believe, you can check it for yourself. A light-year is about 10 trillion kilometers, which becomes 1000 kilometers on the 1-to-10-billion scale (because 10 trillion ÷ 10 billion = 1000). The nearest star system to our own, a three-star system called Alpha Centauri (**Figure 1.7**), is about 4.4 light-years away. That distance is about 4400 kilometers (2700 miles) on the 1-to-10-billion scale, roughly equivalent to the distance across the United States.

The tremendous distances to the stars give us some perspective on the technological challenge of astronomy. For example, because the largest star of the Alpha Centauri system is roughly the same size and brightness as our Sun, viewing it in the night sky is somewhat like being in Washington, D.C., and seeing a very bright grapefruit in San Francisco (neglecting the problems introduced by the curvature of the Earth). It may seem remarkable that we can see this star at all, but the blackness of the night sky allows the naked eye to see it as a faint dot of light. It looks much brighter through powerful telescopes, but we still cannot see features of the star's surface.

Now, consider the difficulty of detecting *planets* orbiting nearby stars, which is equivalent to looking from Washington, D.C., and trying to find ball points or marbles orbiting grapefruits in California or beyond. When you consider this challenge, it is all the more amazing to realize that we now have technology capable of finding such planets, at least in some cases.

The vast distances to the stars also offer a sobering lesson about interstellar travel. Although science fiction shows like *Star Trek* and *Star Wars* make such travel look easy, the reality is far different. Consider the *Voyager 2* spacecraft. Launched in 1977, *Voyager 2* flew by Jupiter in 1979, Saturn in 1981, Uranus in 1986, and Neptune in 1989. It is now bound for the stars at a speed of close to 50,000 kilometers per hour—about 100 times as fast as a speeding bullet. But even at this speed, *Voyager 2* would take about 100,000 years to reach Alpha Centauri if it were headed in that direction (which it's not). Convenient interstellar travel remains well beyond our present technology.

The Size of the Milky Way Galaxy We must change our scale to visualize the galaxy, because most stars are so far away that they would not even fit on Earth with the 1-to-10-billion scale we used to visualize the solar system. Let's therefore reduce our scale by another factor of 1 billion (making it a scale of 1 to 10^{19}).

On this new scale, each light-year becomes 1 millimeter, and the 100,000-light-year diameter of the Milky Way Galaxy becomes 100 meters, or about the length of a football field. Visualize a football field with a scale model of our galaxy centered over midfield. Our entire solar system is a microscopic dot located around the 20-yard line. The 4.4-light-year separation between our solar system and Alpha Centauri becomes just 4.4 millimeters on this scale—smaller than the width of your little finger. If you stood at the position of our solar system in this model, millions of star systems would lie within reach of your arms.

Another way to put the galaxy into perspective is to consider its number of stars—more than 100 billion. Imagine that tonight you are having difficulty falling asleep (perhaps because you are contemplating the scale of the universe). Instead of counting sheep, you decide to count stars. If you are able to count about one star each second, how long would it take you to count 100 billion stars in the Milky Way? Clearly, the answer is 100 billion (10^{11}) seconds, but how long is that? Amazingly, 100 billion seconds turns out to be more than 3000 years. (You can confirm this by dividing 100 billion by the number of seconds in 1 year.) You would need thousands of years just to *count* the stars in the Milky Way Galaxy, and this assumes you never take a break—no sleeping, no eating, and absolutely no dying!

Common Misconceptions

CONFUSING VERY DIFFERENT THINGS

Many people mix up the terms *solar system* and *galaxy*, but they refer to very different things. Our solar system is a single star system, while our galaxy is a vast collection of more than 100 billion star systems. In fact, if you look at the sizes in Figure 1.1, you'll see that our galaxy is about 100 million times larger in diameter than our solar system. So be careful with these terms: Mixing up *solar system* and *galaxy* is a very big mistake!

Stars in the Observable Universe   As incredible as the scale of our galaxy may seem, the Milky Way is only one of roughly 100 billion galaxies in the observable universe. Just as it would take thousands of years to count the stars in the Milky Way, it would take thousands of years to count all the galaxies.

Think for a moment about the total number of stars in all these galaxies. If we assume 100 billion stars per galaxy, the total number of stars in the observable universe is roughly 100 billion × 100 billion, or 10,000,000,000,000,000,000,000 (10^{22}). How big is this number? Visit a beach. Run your hands through the fine-grained sand. Imagine counting each tiny grain of sand as it slips through your fingers. Then imagine counting every grain of sand on the beach and continuing to count *every* grain of dry sand on *every* beach on Earth. If you could actually complete this task, you would find that the number of grains of sand was comparable to the number of stars in the observable universe (**Figure 1.8**).

Figure 1.8 | The number of stars in the observable universe is comparable to the number of grains of dry sand on all the beaches on Earth.

Think about it   Contemplate the fact that there may be as many stars in the observable universe as grains of sand on all the beaches on Earth. How does this affect your view of the possibility that other intelligent civilizations exist?

Tools of Science: Doing the Math

Mathematics is one of the most important tools of science, because it allows scientists to make precise, numerical predictions that can be tested through observations or experiments. These types of tests make it possible for us to gain confidence in scientific ideas. That is why the development of science and mathematics has often gone hand in hand. For example, Sir Isaac Newton developed the mathematics of calculus so that he could do the calculations necessary to test his theory of gravity, and Einstein used new mathematical ideas to work out the details of his general theory of relativity. Fortunately, you don't have to be a Newton or an Einstein to benefit from mathematics in science. Calculations using only multiplication and division can still provide important insights into scientific ideas. Let's look at a few examples.

Example 1:   How far is a light-year?

Solution: A light-year (ly) is the distance that light can travel in one year; recall that light travels at the *speed of light*, which is 300,000 km/s. Just as we can find the distance that a car travels in two hours by multiplying the car's speed by two hours, we can find a light-year by multiplying the speed of light by one year. Because we are given the speed of light in kilometers per second, we must carry out the multiplication while converting 1 year into seconds. See Appendix C if you need a review of unit conversions; here, we show the result for this case:

$$1 \text{ ly} = \left(300{,}000 \ \frac{\text{km}}{\text{s}}\right) \times (1 \text{ yr})$$

$$= \left(300{,}000 \ \frac{\text{km}}{\text{s}}\right) \times \left(1 \ \text{yr} \times 365 \ \frac{\text{day}}{\text{yr}} \times 24 \ \frac{\text{hr}}{\text{day}} \times 60 \ \frac{\text{min}}{\text{hr}} \times 60 \ \frac{\text{s}}{\text{min}}\right)$$

$$= 9{,}460{,}000{,}000{,}000 \text{ km}$$

That is, 1 light-year is equivalent to 9.46 trillion kilometers, which is easier to remember as almost 10 trillion kilometers.

Example 2:   How big is the Sun on the 1-to-10-billion scale?

Solution: The Sun's actual radius is 695,000 km, which we express in scientific notation as 6.95×10^5 km. (See Appendix C if you need to review

powers of 10 and scientific notation.) To find the Sun's radius on the 1-to-10-billion scale, we divide this actual radius by 10 billion, or 10^{10}:

$$\text{scaled radius} = \frac{\text{actual radius}}{10^{10}}$$

$$= \frac{6.95 \times 10^5 \text{ km}}{10^{10}}$$

$$= 6.95 \times 10^{(5-10)} \text{ km}$$

$$= 6.95 \times 10^{-5} \text{ km}$$

This answer is easier to interpret if we convert it to centimeters, which we can do by recalling that there are 1000 ($= 10^3$) meters in a kilometer and 100 ($= 10^2$) centimeters in a meter:

$$6.95 \times 10^{-5} \text{ km} \times \frac{10^3 \text{ m}}{1 \text{ km}} \times \frac{10^2 \text{ cm}}{1 \text{ m}} = 6.95 \text{ cm}$$

On the 1-to-10-billion scale, the Sun is just under 7 centimeters in radius, or 14 centimeters in diameter.

Example 3:   How fast is Earth orbiting the Sun?

Solution: Earth completes one orbit in one year, so we can find its average orbital speed by dividing the circumference of its orbit by one year. Earth's orbit is nearly circular with radius of 1 AU ($= 1.5 \times 10^8$ km); the circumference of a circle is given by the formula $2\pi \times$ radius. If we want the speed to come out in units of km/hr, we divide this circumference by 1 year converted to hours, as follows:

$$\text{orbital speed} = \frac{\text{orbital circumference}}{1 \text{ yr}}$$

$$= \frac{2 \times \pi \times (1.5 \times 10^8 \text{ km})}{1 \text{ yr} \times \frac{365 \text{ day}}{\text{yr}} \times \frac{24 \text{ hr}}{\text{day}}}$$

$$\approx 107{,}000 \text{ km/hr}$$

Earth's average speed as it orbits the Sun is more than 100,000 km/hr.

1.2 A Brief History of the Universe

Our universe is vast not only in space, but also in time. Scientific measurements show that we live in a universe that is about 14 billion years old. To gain some perspective on this age, let's first explore our modern view of how we came to exist at this point in time, then consider how our human lifetimes compare to the age of the universe.

How did we come to be?

Figure 1.9 summarizes the history of the universe according to modern science. Follow the figure as you read through this section.

The Big Bang and the Expanding Universe Telescopic observations of distant galaxies show that the entire universe is **expanding,** meaning that average distances between galaxies are increasing with time. This fact implies that galaxies must have been closer together in the past, and if we go back far enough, we must reach the point at which the expansion began. We call this beginning the **Big Bang,** and scientists use the observed rate of expansion to calculate that it occurred about 14 billion years ago. The three cubes in the upper left corner of Figure 1.9 represent the expansion of a small piece of the entire universe through time.

The universe as a whole has continued to expand ever since the Big Bang, but on smaller scales the force of gravity has drawn matter together. Structures such as galaxies and galaxy clusters occupy regions where gravity has won out against the overall expansion. That is, while the universe as a whole continues to expand, individual galaxies and galaxy clusters (and objects within them such as stars and planets) do *not* expand. The three cubes in Figure 1.9 illustrate this idea. Notice that as the cube as a whole grew larger, the matter within it clumped into galaxies and galaxy clusters. Most galaxies, including our own Milky Way, formed within a few billion years after the Big Bang.

Stellar Lives and Galactic Recycling Within galaxies like the Milky Way, gravity drives the collapse of clouds of gas and dust to form stars and planets. Stars are not living organisms, but they nonetheless go through "life cycles." A star is born when gravity compresses the material in a cloud until the center becomes dense enough and hot enough to generate energy by **nuclear fusion,** the process in which lightweight atomic nuclei smash together and stick (or fuse) to make heavier nuclei. The star "lives" as long as it can shine with energy from fusion, and "dies" when it exhausts its usable fuel.

In its final death throes, a star blows much of its content back out into space. In particular, massive stars die in titanic explosions called *supernovae.* The returned matter mixes with other matter floating between the stars in the galaxy, eventually becoming part of new clouds of gas and dust from which new generations of stars can be born. Galaxies therefore function as cosmic recycling plants, recycling material expelled from dying stars into new generations of stars and planets. This cycle is illustrated in the lower right of Figure 1.9. Our own solar system is a product of many generations of such recycling.

Element Creation in Stars The recycling of stellar material is connected to our existence in an even deeper way. By studying stars of different ages, we have learned that the early universe contained only the simplest chemical elements: hydrogen and helium (and a trace of lithium). We and

Figure 1.9

Our Cosmic Origins

Birth of the Universe: The expansion of the universe began with the hot and dense Big Bang. The cubes show how one region of the universe has expanded with time. The universe continues to expand, but on smaller scales gravity has pulled matter together to make galaxies.

Galaxies as Cosmic Recycling Plants: The early universe contained only two chemical elements: hydrogen and helium. All other elements were made by stars and recycled from one stellar generation to the next within galaxies like our Milky Way.

Stars are born in clouds of gas and dust; planets may form in surrounding disks.

Massive stars explode when they die, scattering the elements they've produced into space.

Stars shine with energy released by nuclear fusion, which ultimately manufactures all elements heavier than hydrogen and helium.

Earth and Life: By the time our solar system was born, 4½ billion years ago, about 2% of the original hydrogen and helium had been converted into heavier elements. We are therefore "star stuff," because we and our planet are made from elements manufactured in stars that lived and died long ago.

Life Cycles of Stars: Many generations of stars have lived and died in the Milky Way.

Earth are made primarily of other elements, such as carbon, nitrogen, oxygen, and iron. Where did these other elements come from? Evidence shows that elements besides hydrogen and helium were manufactured by stars, some through the nuclear fusion that makes stars shine and others through nuclear reactions accompanying the explosions that end stellar lives.

By the time our solar system formed, about $4\frac{1}{2}$ billion years ago, earlier generations of stars had already converted about 2% of our galaxy's original hydrogen and helium into heavier elements. Therefore, the cloud that gave birth to our solar system was made of about 98% hydrogen and helium and 2% other elements. This 2% may sound small, but it was more than enough to make the small rocky planets of our solar system, including Earth. On Earth, some of these elements became the raw ingredients of life, which ultimately blossomed into the great diversity of life on Earth today.

In summary, most of the material from which we and our planet are made was created inside stars that lived and died before the birth of our Sun. As astronomer Carl Sagan said, we are "star stuff."

How do our lifetimes compare to the age of the universe?

We can put the 14-billion-year age of the universe into perspective by imagining this time compressed into a single year, so that each month represents a little more than 1 billion years. On this *cosmic calendar*, the Big Bang occurs at the first instant of January 1 and the present is the stroke of midnight on December 31 (**Figure 1.10**).

On this time scale, the Milky Way Galaxy probably formed in February. Many generations of stars lived and died in the subsequent cosmic months, enriching the galaxy with the "star stuff" from which we and our planet are made.

Our solar system and our planet did not form until early September on this scale, or $4\frac{1}{2}$ billion years ago in real time. By late September, life on Earth was flourishing. However, for most of Earth's history, living organisms remained relatively primitive and microscopic. On the scale of the cosmic calendar, recognizable animals became prominent only in mid-December. Early dinosaurs appeared on the day after Christmas. Then, in a cosmic instant, the dinosaurs disappeared forever—probably because of the impact of an asteroid

Figure 1.10 | The cosmic calendar compresses the 14-billion-year history of the universe into 1 year, so that each month represents a little more than 1 billion years. (This cosmic calendar is adapted from a version created by Carl Sagan.)

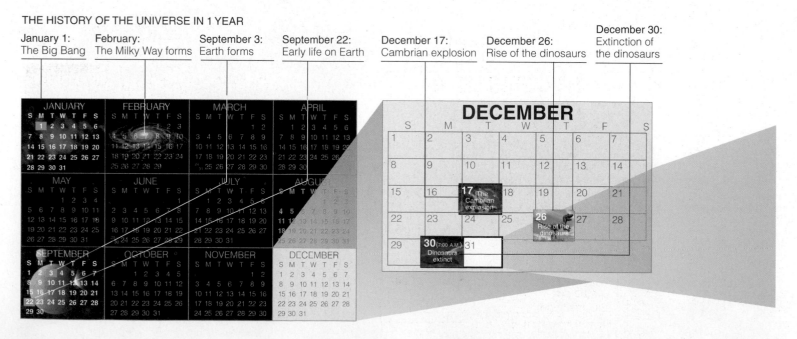

THE HISTORY OF THE UNIVERSE IN 1 YEAR

January 1: The Big Bang

February: The Milky Way forms

September 3: Earth forms

September 22: Early life on Earth

December 17: Cambrian explosion

December 26: Rise of the dinosaurs

December 30: Extinction of the dinosaurs

or a comet [Section 6.3]. In real time the death of the dinosaurs occurred some 65 million years ago, but on the cosmic calendar it was only yesterday. With the dinosaurs gone, small furry mammals inherited Earth. Some 60 million years later, or around 9 P.M. on December 31 of the cosmic calendar, early hominids (human ancestors) began to walk upright.

Perhaps the most astonishing thing about the cosmic calendar is that the entire history of human civilization falls into just the last half-minute. The ancient Egyptians built the pyramids only about 11 seconds ago on this scale. About 1 second ago, Kepler and Galileo proved that Earth orbits the Sun rather than vice versa. The average college student was born about 0.05 second ago, around 11:59:59.95 P.M. on the cosmic calendar. On the scale of cosmic time, the human species is the youngest of infants, and a human lifetime is a mere blink of an eye.

✳ THE PROCESS OF SCIENCE IN ACTION

1.3 Defining Planets

One of the goals of this book is to help you learn more about science in general as you study the science of astronomy. We will therefore conclude each chapter with a case study that illustrates the process of science in action. Here, we look at the process of scientific classification, a topic that made the news with the 2006 demotion of Pluto from *planet* to *dwarf planet*.

Science begins with observations of the world around us, and after observing we often try to classify the objects we find. Scientific classification helps us organize our thinking and provides a common language for discussion. Consider living things, which were long classified only as either plants or animals. This classification is clearly helpful, since "plant" immediately brings up a different mental picture than "animal." But it also has limitations: We now know that most living things are neither plants nor animals, but instead are members of diverse groups of microscopic organisms.

In this chapter, we have already classified objects into categories such as planets, stars, and galaxies. The box on page 5 provides basic definitions for these categories but does not explain *how* we came to classify objects in this way. You may notice that the definition of the term *planet* is particularly

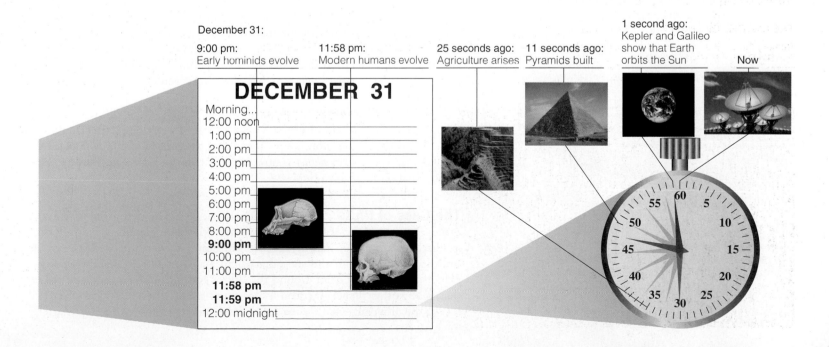

December 31:

9:00 pm: Early hominids evolve

11:58 pm: Modern humans evolve

25 seconds ago: Agriculture arises

11 seconds ago: Pyramids built

1 second ago: Kepler and Galileo show that Earth orbits the Sun

Now

DECEMBER 31

Morning...
12:00 noon
1:00 pm
2:00 pm
3:00 pm
4:00 pm
5:00 pm
6:00 pm
7:00 pm
8:00 pm
9:00 pm
10:00 pm
11:00 pm
11:58 pm
11:59 pm
12:00 midnight

complicated. The story behind this definition provides an excellent example of how scientists classify objects and how scientific classification must adapt to new discoveries.

What is a planet?

The difference between a star and a planet is not obvious from a casual glance at the night sky. In fact, the term *star* historically applied to almost any shining object in the night sky, including the planets and even the brief flashes of light known as "shooting stars" (or *meteors*), which we now know to be caused by comet dust entering Earth's atmosphere. To the naked eye, the difference between stars and planets becomes clear only if you observe the sky over a period of many days or weeks: Stars remain fixed in the patterns of the constellations, while planets appear to move slowly among the constellations of stars [Section 2.3].

Planets as Wanderers The word *planet* comes from the Greek for "wanderer," and in ancient times it applied to all objects that appear to move, or wander, among the constellations. The Sun and the Moon were counted as planets, because they move steadily through the constellations. Earth did *not* count as a planet, since it is not something we see in the sky and it was presumed to be stationary at the center of the universe. Ancient observers therefore recognized seven objects as planets: the Sun, the Moon, and the five planets that are easily visible to the naked eye (Mercury, Venus, Mars, Jupiter, and Saturn). The special status of these seven objects is still enshrined in the names of the seven days of the week. In English, only Sunday, Moonday, and Saturday are obvious, but if you know a romance language like Spanish you'll be able to figure out the rest: Tuesday is Mars day (martes), Wednesday is Mercury day (miércoles), Thursday is Jupiter day (jueves), and Friday is Venus day (viernes).

This original definition of *planet* began to change about 400 years ago, when we learned that Earth is *not* the center of the universe but rather one of the objects that orbit the Sun. The term *planet* came to mean any object that orbits the Sun, which added Earth to the list of planets and removed the Sun and Moon (because the Moon orbits Earth). This definition successfully accommodated the planets Uranus and Neptune after their discoveries in 1781 and 1846, respectively.

A weakness of this definition became apparent as scientists began to discover asteroids, starting with the discovery of Ceres in 1801. Ceres was initially hailed as a new "planet," but as the number of known asteroids grew—and as we realized that asteroids were all much smaller than the traditional planets—scientists decided that these relatively small worlds should count only as "minor planets."

This division between "minor planets" and "planets" worked fine for more than a century, but recent discoveries have forced further changes in classification. For example, we now know that other stars have their own planets [Section 7.1], so the term *planet* is now used for these objects as well as for the planets that orbit our own Sun.

The Case of Pluto The most recent change in the definition of *planet* comes from the story of Pluto. Pluto was quickly given planetary status upon its discovery in 1930, mainly because astronomers overestimated its mass. Still, it was clear from the start that Pluto was a misfit among the known planets. Its 248-year orbit around the Sun is more elongated in shape than that of any other planet. Its orbit is also significantly tilted relative to the orbits of the other planets (**Figure 1.11**). It became even more of a misfit after

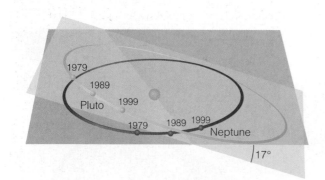

Figure 1.11 | Pluto's orbit is significantly elongated and tilted with respect to those of the other planets. It even comes closer to the Sun than Neptune for 20 years in each 248-year orbit, as was the case between 1979 and 1999. There's no danger of a collision, however, because Neptune completes exactly three orbits for every two of Pluto's orbits.

astronomers pinned down its mass and composition: Pluto is only about $\frac{1}{25}$ as massive as Mercury, smallest of the first eight planets, and its ice-rich composition is more similar to that of a comet than to that of any of the other planets.

Pluto's low mass, unusual orbit, and comet-like composition eventually began to cause controversy over its status as a planet. Scientists had known since the 1950s that many of the comets that we see in the inner solar system come from the same region of the solar system in which Pluto orbits (called the *Kuiper belt* [Section 4.1]). By the 1990s, telescope technology had reached the point where scientists could begin to observe this region of the solar system directly. Many Pluto-like objects were soon discovered, though for a while these objects were all smaller than Pluto. But in 2005, Caltech astronomer Mike Brown announced the discovery of an object that is slightly larger than Pluto, now known as Eris. Astronomers were forced to consider the question of whether Eris should count as a planet, and this led to reconsideration of the status of Pluto and of other objects that are close to Pluto in size.

At least three general ways of defining *planet* have been considered since Eris's discovery. The first would set Pluto's size as the minimum for a planet, thereby preserving Pluto's cultural status as the ninth planet. Eris would become the tenth planet, and objects discovered in the future would join the list only if they proved to be larger than Pluto. However, many scientists object that this definition is too arbitrary, as there is nothing special about Pluto's size.

A second possible definition would make planetary status depend solely on an object's physical characteristics. Several ways of doing this have been proposed; the most popular would define a planet as an object that is large enough for its own gravity to make it round. This definition would probably make dozens of Pluto-like objects count as planets (along with one or two asteroids), so our solar system might have 40 or more planets. Moreover, because there undoubtedly would be borderline cases, we might never be able to state the precise number of planets in our solar system. This definition also raises the question of whether large moons—including Earth's Moon—would count as planets because they are round.

The third idea defines *planet* so that neither Pluto nor Eris counts, leaving only eight planets in our solar system. This is the option chosen in 2006 by the International Astronomical Union (IAU), the organization responsible for astronomical names and definitions. The approved definition has some obvious quirks (such as defining a planet as something that orbits our Sun, which implies that no other star can have planets), but the underlying idea defines a planet as an object that (1) orbits a star (but is not itself a star); (2) is massive enough for its own gravity to make it round; and (3) dominates its orbital region. Objects that meet the first two criteria but not the third are designated *dwarf planets;* these include Pluto and Eris, which are round but share their orbital region with many similarly sized objects.

Does this definition make scientific sense? For the time being, the answer seems to be yes. **Figure 1.12** contrasts the sizes of various objects in our solar system. The eight planets clearly divide into two groups, while Pluto and Eris clearly belong to a different group. Nevertheless, we can envision future discoveries that could cause problems for the new definition. For example, what if

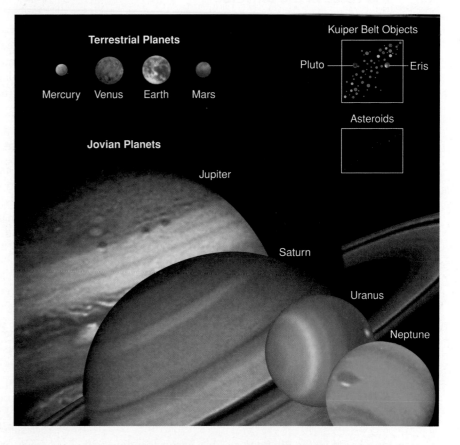

Figure 1.12 | Relative sizes of various objects in the solar system. Notice that the eight planets divide clearly into two groups, known as the *jovian planets* (Jupiter, Saturn, Uranus, and Neptune) and the *terrestrial planets* (Mercury, Venus, Earth, and Mars). Pluto and Eris clearly belong to a group of much smaller but more numerous objects.

we were to discover an Earth-size object orbiting in Pluto's region of the solar system; would its location mean it would count as a dwarf planet, even though it would be larger than Venus and Mars?

Perhaps the most important scientific idea to take away from this debate is that nature need not obey the classification systems we propose. Whether we call Pluto a planet, a dwarf planet, or a large comet may affect the way we think about it, but it does not change what it actually is. A key part of science is learning to adapt our own notions of organization to the underlying reality of nature. As we discover new things, we must sometimes change our definitions. Future astronomers may define *planet* differently than we do now.

 Think about it The 2009 IAU meeting will occur shortly after this book goes to press. Look for news reports on the meeting; was the definition of *planet* reconsidered? What's your opinion of the current definition?

Summary of Key Concepts

 ## 1.1 Our Place in the Universe

What is our place in the universe?

Earth is a planet orbiting the Sun. Our Sun is one of more than 100 billion stars in the **Milky Way Galaxy.** Our galaxy is one of about 40 galaxies in the **Local Group.** The Local Group is one small part of the **Local Supercluster,** which is one small part of the **universe.**

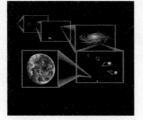

How big is the universe?

On a 1-to-10-billion scale, the Sun is the size of a grapefruit, Earth is a ball point about 15 meters away, and the nearest stars are thousands of kilometers away. Our galaxy has so many stars that it would take thousands of years just to count them. The **observable universe** contains some 100 billion galaxies, with a total number of stars comparable to the number of grains of dry sand on all the beaches on Earth.

 ## 1.2 A Brief History of the Universe

How did we come to be?

The universe began in the **Big Bang** and has been expanding ever since, except in localized regions where gravity has caused matter to collapse into galaxies and stars. The Big Bang essentially pro-

duced only two chemical elements: hydrogen and helium. All other elements have been produced by stars and recycled within galaxies from one generation of stars to the next. We and our planet are made of this recycled "star stuff."

How do our lifetimes compare to the age of the universe?

On a cosmic calendar that compresses the history of the universe into 1 year, human civilization is just a few seconds old, and a human lifetime lasts only a fraction of a second.

 ✹ THE PROCESS OF SCIENCE IN ACTION

1.3 Defining Planets

What is a planet?

The definition of the term *planet* has changed over time. It originally referred to objects that wandered among the constellations, but now applies to Earth and seven other objects in our solar system, while Pluto, Eris, and similar objects are classified as *dwarf planets*. The definition may yet change again, showing how scientific classification must adapt to new discoveries.

Investigations

Quick Quiz

Choose the best answer to each of the following; answers are in Appendix D. Explain your reasoning with one or more complete sentences.

1. Which of the following correctly lists our "cosmic address" from smallest to largest? (a) Earth, solar system, Milky Way Galaxy, Local Group, Local Supercluster, universe (b) Earth, solar system, Local Group, Local Supercluster, Milky Way Galaxy, universe (c) Earth, Milky Way Galaxy, solar system, Local Group, Local Supercluster, universe

2. An *astronomical unit* is (a) any planet's average distance from the Sun. (b) Earth's average distance from the Sun. (c) any large astronomical distance.

3. A *light-year* is (a) about 10 trillion kilometers. (b) the time it takes light to reach the nearest star. (c) the time it takes light to travel around the Sun.

4. The star Antares is 604 light-years away. If it explodes tonight, (a) we'll know because it will be brighter than the full Moon. (b) we'll know because debris from the explosion will rain down on us from space. (c) we won't know about it for another 604 years.

5. Could we see a galaxy that is 20 billion light-years away? (a) Yes, if we had a big enough telescope. (b) No, because it would be beyond the bounds of our observable universe. (c) No, because a galaxy could not possibly be that far away.

6. If we represent the solar system on a scale that allows us to walk from the Sun to Pluto in a few minutes, then (a) the planets are the size of basketballs and the nearest stars are a few miles away. (b) the planets are marble size or smaller and the nearest stars are thousands of miles away. (c) the planets are microscopic and the stars are millions of miles away.

7. The number of stars in the Milky Way Galaxy is roughly (a) 100,000. (b) 100 million. (c) 100 billion.

8. When we say the universe is *expanding,* we mean that (a) everything in the universe is growing in size. (b) the average distance between galaxies is growing with time. (c) the number of stars in the universe is growing with time.

9. The *Big Bang* is the name astronomers give to (a) the explosion that occurs when a star dies. (b) the largest explosion ever observed. (c) the birth of the universe.

10. We are "star stuff" in the sense that (a) we are made of elements that were produced in stars. (b) our bodies have the same chemical composition as stars. (c) we are born, live, and die, just like stars.

11. The age of our solar system is about (a) $\frac{1}{3}$ of the age of the universe. (b) $\frac{3}{4}$ of the age of the universe. (c) the same as the age of the universe.

12. The event that triggered the change in Pluto's status from planet to dwarf planet was the discovery that (a) it is smaller than the planet Mercury. (b) it has a comet-like composition of ice and rock. (c) it is not the largest object in its region of the solar system.

Short-Answer/Essay Questions

Explain all answers clearly, using complete sentences and proper essay structure if needed. An asterisk (*) designates a quantitative problem, for which you should show all your work.

13. *Our Cosmic Origins.* Write one to three paragraphs summarizing why we could not be here if the universe did not contain both stars and galaxies.

14. *Alien Technology.* Some people believe that Earth is being visited by aliens from other star systems. If this is true, how would the alien technology compare to our own? Using ideas of scale introduced in this chapter, write one to two paragraphs to give a sense of the technological difference.

15. *Looking for Evidence.* This chapter discussed the scientific story of the universe but not the evidence that backs it up. Choose one idea from this chapter—such as the idea that there are billions of galaxies, that the universe was born in the Big Bang, or that we are "star stuff"—and briefly discuss the type of evidence you would like to see before accepting the idea. (Hint: You can look ahead in the book to see the evidence presented in later chapters.)

16. *The Value of Classification.* Section 1.3 discussed difficulties that can arise with attempts to define scientific classifications precisely, such as those that occur with the term *planet.* Make a bullet list of pros and cons (at least three of each) to having classification schemes. Then write a one-paragraph summary stating your opinion of the value (or lack of value) of scientific classification.

17. *The Cosmic Perspective.* Write a one-page essay describing how the ideas presented in this chapter affect your perspectives on your own life and on human civilization.

*18. *Light-Minute.* Just as a light-year is the distance that light can travel in 1 year, a light-minute is the distance that light can travel in 1 minute. What is a light-minute in kilometers?

*19. *Sunlight.* Use the speed of light and the Earth–Sun distance of 1 AU to calculate how long it takes light to travel from the Sun to Earth.

*20. *Cosmic Calendar.* The cosmic calendar condenses the 14-billion-year history of the universe into 1 year. How long does 1 second represent on the cosmic calendar?

*21. *Saturn vs. the Milky Way.* Photos of Saturn with its rings can look so similar to photos of galaxies that children often think they are similar objects, but of course galaxies are far larger. About how many times larger in diameter is the Milky Way Galaxy than Saturn's rings? (*Data:* Saturn's rings are 270,000 km in diameter; the Milky Way is 100,000 light-years in diameter.)

*22. *Galactic Rotation.* Our solar system is located about 28,000 light-years from the galactic center and orbits the center once every 230 million years. How fast are we traveling around the galaxy, in km/hr?

2 Understanding the Sky

This time-exposure photo, taken at Arches National Park in Utah, shows how the entire sky seems to circle daily around a point above Earth's North Pole (or South Pole, for the Southern Hemisphere). Stars far to the north, like those visible within the arch, complete their entire circles above the northern horizon. Other stars— along with the Sun, Moon, and planets—follow circles that cross the horizon, which is why they rise in the east and set in the west each day. This daily circling of the sky explains why most of our ancestors assumed that the universe revolved around Earth. Today, we know that it is actually Earth's daily rotation that makes the sky appear to turn. How did we learn this fact? Through careful study of other, more subtle patterns of change in the sky, including the patterns we will study in this chapter.

2.1 Understanding the Seasons

We're all familiar with seasonal changes, such as longer, warmer days in summer and shorter, cooler days in winter. But do you know why the seasons occur? In this section, we'll examine the cause of the seasons and the seasonal changes that we can observe in the sky.

What causes the seasons?

Seasons are caused by the tilt of Earth's axis, which in turn causes the Sun's path through our sky and the intensity of sunlight to vary over the course of each year. To understand exactly how this works, we'll first look at how Earth's rotation produces the apparent daily motion of the Sun through the sky, then examine how Earth's axis tilt causes the Sun's daily path through the sky to change as Earth orbits the Sun.

The Sun's Daily Path   The Sun's daily path through the sky is fairly simple: It rises somewhere in the east, sets somewhere in the west, and in between reaches its highest point around noon. Ancient people assumed that the Sun actually circled Earth each day, but we now know that its daily motion arises from Earth's rotation. Just as the world seems to circle around you if you spin in place, the Sun seems to go around us on our rotating planet. However, while the Sun always follows the same general daily pattern, its precise path through the sky varies throughout the year.

To describe the Sun's daily path more clearly, we need to define some key reference points in the sky (**Figure 2.1**). The boundary between Earth and sky defines the **horizon.** The point directly overhead is the **zenith.** The **meridian** is an imaginary half-circle stretching from the horizon due south, through the zenith, to the horizon due north. We can pinpoint the position of any object in the sky by stating its *direction* along the horizon (sometimes expressed as *azimuth*) and its *altitude* above the horizon.

Figure 2.2 shows how the Sun's daily path differs at different times of years for typical locations in the Northern Hemisphere. Notice that the Sun's path is long and high in summer, with the Sun rising well north of due east and setting well north of due west. In the winter, the Sun's path is short and low as it rises well south of due east and sets well south of due west. Notice also that the Sun *never* passes directly overhead at this latitude. The Sun can reach the zenith only for locations within the *tropics*, meaning locations on Earth that are between latitudes of $23\frac{1}{2}°$N and $23\frac{1}{2}°$S.

The Reason for Seasons   What causes these seasonal changes in the Sun's path through the sky? The answer is Earth's axis tilt, and you can understand why by studying **Figure 2.3**. Step 1 illustrates how Earth's rotation axis is tilted with respect to its orbit. Notice that the axis remains pointed in the same direction in space (toward the North Star, Polaris) throughout the year. As a result, the orientation of the axis *relative to the Sun* changes over the course of each orbit: The Northern Hemisphere is tipped toward the Sun in June and away from the Sun in December, while the reverse is true for the Southern Hemisphere. That is why the two hemispheres experience opposite seasons. The rest of the figure shows how the changing angle of sunlight on the two hemispheres leads directly to seasons.

Step 2 shows Earth in June, when the axis tilt causes sunlight to strike the Northern Hemisphere at a steeper angle and the Southern Hemisphere at a shallower angle. The steeper sunlight angle makes it summer in the Northern Hemisphere for two reasons. First, as shown in the zoom-out, the steeper angle means more concentrated sunlight, which tends to make it warmer. Second, if you visualize what happens as Earth rotates each day, you'll see that the

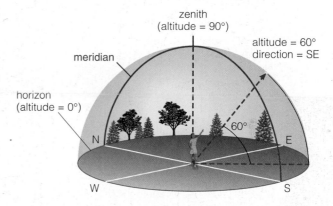

Figure 2.1 | From any place on Earth, the local sky looks like a dome (hemisphere). This diagram shows key reference points in the local sky. It also shows how we can describe any position in the local sky by its altitude and direction.

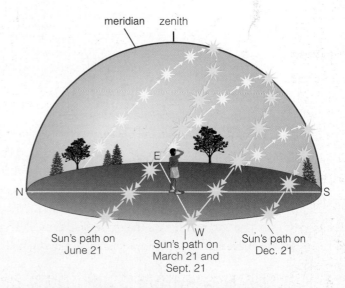

Figure 2.2 | This diagram shows the Sun's path in different seasons for a Northern Hemisphere location (latitude 40°N). The precise paths are different for other latitudes.

Earth's seasons are caused by the tilt of its rotation axis, which is why the seasons are opposite in the two hemispheres. The seasons do *not* depend on Earth's distance from the Sun, which varies only slightly throughout the year.

① Axis Tilt: Earth's axis points in the same direction throughout the year, which causes changes in Earth's orientation *relative to the Sun.*

② Northern Summer/Southern Winter: In June, sunlight falls more directly on the Northern Hemisphere, which makes it summer there because solar energy is more concentrated and the Sun follows a longer and higher path through the sky. The Southern Hemisphere receives less direct sunlight, making it winter.

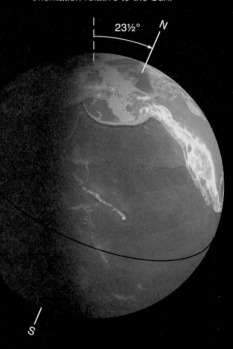

23½° N

S

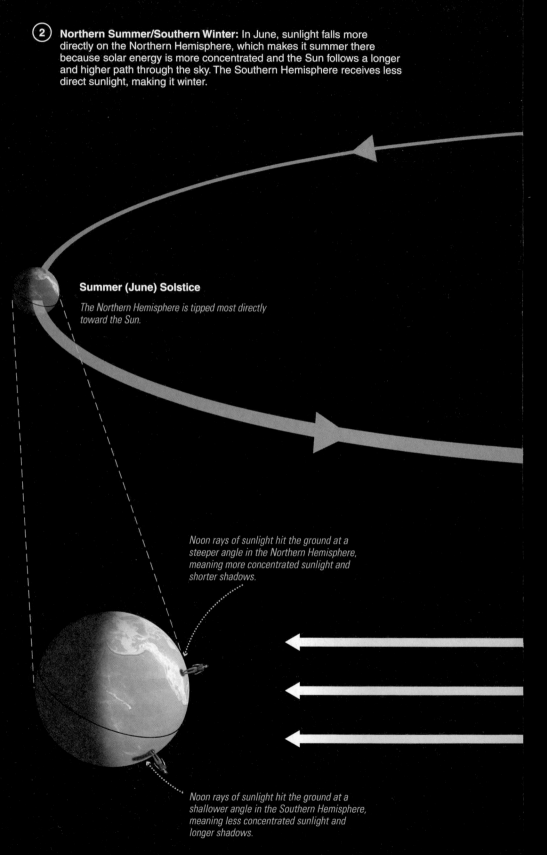

Summer (June) Solstice

The Northern Hemisphere is tipped most directly toward the Sun.

Noon rays of sunlight hit the ground at a steeper angle in the Northern Hemisphere, meaning more concentrated sunlight and shorter shadows.

Noon rays of sunlight hit the ground at a shallower angle in the Southern Hemisphere, meaning less concentrated sunlight and longer shadows.

Interpreting the Diagram

To interpret the seasons diagram properly, keep in mind:

1. Earth's size relative to its orbit would be microscopic on this scale, meaning that both hemispheres are at essentially the same distance from the Sun.

2. The diagram is a side view of Earth's orbit. A top-down view (below) shows that Earth orbits in a nearly perfect circle and comes closest to the Sun in January.

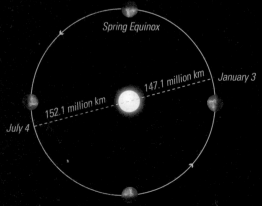

Spring Equinox

147.1 million km   *January 3*

152.1 million km

July 4

Fall Equinox

3 **Spring/Fall:** Spring and fall begin when sunlight falls equally on both hemispheres, which happens twice a year: In March, when spring begins in the Northern Hemisphere and fall in the Southern Hemisphere; and in September, when fall begins in the Northern Hemisphere and spring in the Southern Hemisphere.

4 **Northern Winter/Southern Summer:** In December, sunlight falls less directly on the Northern Hemisphere, which makes it winter because solar energy is less concentrated and the Sun follows a shorter and lower path through the sky. The Southern Hemisphere receives more direct sunlight, making it summer.

Spring (March) Equinox

The Sun shines equally on both hemispheres.

The variation in Earth's orientation relative to the Sun means that the seasons are linked to four special points in Earth's orbit:

Solstices *are the two points at which sunlight becomes most extreme for the two hemispheres.*

Equinoxes *are the two points at which the hemispheres are equally illuminated.*

Winter (December) Solstice

The Southern Hemisphere is tipped most directly toward the Sun.

Fall (September) Equinox

The Sun shines equally on both hemispheres.

Noon rays of sunlight hit the ground at a shallower angle in the Northern Hemisphere, meaning less concentrated sunlight and longer shadows.

Noon rays of sunlight hit the ground at a steeper angle in the Southern Hemisphere, meaning more concentrated sunlight and shorter shadows.

Common Misconceptions

THE CAUSE OF SEASONS

Many people guess that seasons are caused by variations in Earth's distance from the Sun. But if this were true, the whole Earth would have summer or winter at the same time, and it doesn't: The seasons are opposite in the Northern and Southern Hemispheres. In fact, Earth's slightly varying orbital distance has virtually no effect on the weather. The real cause of seasons is Earth's axis tilt, which causes the two hemispheres to take turns being tipped toward the Sun over the course of each year.

steeper angle also means the Sun follows a longer and higher path through the sky (the path shown for June in Figure 2.2), giving the Northern Hemisphere more hours of daylight during which it is warmed by the Sun. The opposite is true for the Southern Hemisphere at this time: The shallower sunlight angle makes it winter there, because sunlight is less concentrated and the Sun follows a shorter, lower path through the sky.

The sunlight angle gradually changes as Earth moves along its orbit around the Sun. At the opposite side of Earth's orbit, Step 4 shows that it has become winter for the Northern Hemisphere and summer for the Southern Hemisphere. In between these two extremes, Step 3 shows that both hemispheres are illuminated equally in March and September. It is therefore spring for the hemisphere that is on the way from winter to summer, and fall for the hemisphere on the way from summer to winter.

The key point is this: Seasons occur because of the combination of Earth's axis tilt and its orbit around the Sun. If Earth did not have an axis tilt, we would not have seasons.

Think about it Jupiter has an axis tilt of about 3°, small enough to be insignificant. Saturn has an axis tilt of about 27°, slightly greater than that of Earth. Both planets have nearly circular orbits around the Sun. Do you expect Jupiter to have seasons? Do you expect Saturn to have seasons? Explain.

Solstices and Equinoxes To help us mark the changing seasons, we define four special moments in the year, each of which corresponds to one of the four special positions in Earth's orbit shown in Figure 2.3.

- The **summer (June) solstice,** which occurs around June 21, is the moment when the Northern Hemisphere is tipped most directly toward the Sun and receives the most direct sunlight.

- The **winter (December) solstice,** which occurs around December 21, is the moment when the Northern Hemisphere receives the least direct sunlight.

- The **spring (March) equinox,** which occurs around March 21, is the moment when the Northern Hemisphere goes from being tipped slightly away from the Sun to being tipped slightly toward the Sun.

- The **fall (September) equinox,** which occurs around September 22, is the moment when the Northern Hemisphere first starts to be tipped away from the Sun.

The exact dates and times of the solstices and equinoxes vary from year to year but stay within a couple of days of the dates given here. In fact, our modern calendar includes leap years in a pattern specifically designed to keep the solstices and equinoxes around the same dates: We usually add a day (February 29) for leap year every fourth year, but skip leap year when a century changes, unless the century year is divisible by 400. This pattern makes the average length of the calendar year match the true length of the year, which is just slightly short of $365\frac{1}{4}$ days.

First Days of Seasons We usually say that each equinox and solstice marks the first day of a season. For example, the day of the summer solstice is called the "first day of summer." Notice, however, that the summer solstice occurs when the Northern Hemisphere has its *maximum* tilt toward the Sun. You might then wonder why we consider the summer solstice to be the beginning rather than the midpoint of summer.

Choosing the summer solstice to be the first day of summer makes sense in at least two ways. First, it was much easier for ancient people to identify the days on which the Sun reached extreme positions in the sky—such as

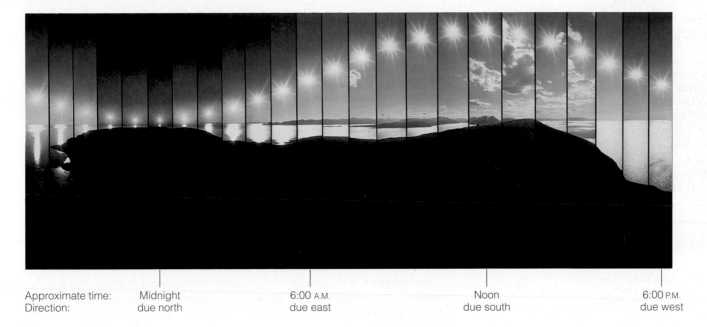

Approximate time:	Midnight	6:00 A.M.	Noon	6:00 P.M.
Direction:	due north	due east	due south	due west

when it reached its highest point on the summer solstice—than other days in between. Second, we usually think of the seasons in terms of weather, and the solstices and equinoxes correspond quite well with the beginnings of seasonal weather patterns. The Sun's path through the Northern Hemisphere sky is longest and highest on the day of the summer solstice, and the warmest days tend to come 1 to 2 months later. To understand why the warmest days come after the solstice, think about what happens when you heat a pot of cold soup. Even though you may have the stove turned on high from the start, it takes a while for the soup to warm up. In the same way, it takes some time for sunlight to heat the ground and oceans from the cold of winter to the warmth of summer. "Midsummer" in terms of weather therefore comes in late July and early August, making the summer solstice a pretty good choice for the "first day of summer." For similar reasons, the winter solstice is a good choice for the first day of winter, and the spring and fall equinoxes are good choices for the first days of those seasons.

Seasons Around the World   Notice that the names of the solstices and equinoxes reflect the northern seasons, and therefore sound backward to people who live in the Southern Hemisphere. For example, Southern Hemisphere winter begins when Earth is at the orbital point usually called the *summer* solstice. This apparent injustice to people in the Southern Hemisphere arose because the solstices and equinoxes were named long ago by people living in the Northern Hemisphere. A similar injustice is inflicted on people living in equatorial regions. If you study Figure 2.3 carefully, you'll see that Earth's equator gets its most direct sunlight on the two equinoxes and its least direct sunlight on the solstices. People living near the equator therefore don't experience four seasons in the same way as people living at mid-latitudes. Instead, equatorial regions generally have rainy and dry seasons, with the rainy season coming when the Sun is higher in the sky.

In addition, seasonal variations around the times of the solstices are more extreme at high latitudes. For example, Vermont has much longer summer days and much longer winter nights than Florida. At the Arctic Circle (latitude $66\frac{1}{2}°$), the Sun remains above the horizon all day long on the summer solstice (**Figure 2.4**), and never rises on the winter solstice. The most extreme variations occur at the North and South Poles, where the Sun remains above the horizon for 6 months in summer and below the horizon for 6 months in winter.

Figure 2.4 | This sequence of photos shows the progression of the Sun all the way around the horizon on the summer solstice at the Arctic Circle. Notice that the Sun does not set but instead skims the northern horizon at midnight. It then gradually rises higher, reaching its highest point at noon, when it appears due south.

Common Misconceptions

HIGH NOON

When is the Sun directly overhead in your sky? Many people answer "at noon." It's true that the Sun reaches its *highest* point each day when it crosses the meridian, giving us the term "high noon" (though the meridian crossing is rarely at precisely 12:00 and occurs closer to 1 p.m. during daylight saving time). However, unless you live in the tropics (between latitudes 23.5°S and 23.5°N), the Sun is *never* directly overhead. In fact, any time you can see the Sun as you walk around, you can be sure it is *not* at your zenith. Unless you are lying down, seeing an object at the zenith requires tilting your head back into a very uncomfortable position.

Figure 2.5 | The constellation Orion is prominent in the evening sky during Northern Hemisphere winter. With binoculars or a telescope, you can see that one of its fainter "stars" is actually an interstellar cloud, called the Orion Nebula, in which many stars are being born.

Why do the constellations we see depend on the time of year?

We have seen how and why we have seasonal changes in the weather and in the Sun's daily path through the sky. But you can also track the seasons by observing the nighttime sky. For example, if you go outside on a January evening, during Northern Hemisphere winter, you'll see the constellation Orion shining prominently in your sky (**Figure 2.5**); if you go out on a July evening, you won't see Orion at all. To understand why we see different constellations during different seasons, we must think more about how the night sky looks from different vantage points in Earth's orbit around the Sun. That, in turn, requires that we first understand the general appearance of the night sky.

Constellations and the Celestial Sphere   People of nearly every culture gave names to patterns they saw in the sky. We usually refer to such patterns as constellations, but to astronomers the term has a more precise meaning: A **constellation** is a *region* of the sky with well-defined borders; the familiar patterns of stars merely help us locate these constellations. For example, while we identify Orion by recognizing the pattern of stars outlined in Figure 2.5, even the seemingly empty spots in that region of the sky are considered part of the constellation Orion. The idea is analogous to the way we identify and define states on a map of the United States: We may identify the states by looking for major cities, rivers, or mountains, but every spot of land on the map belongs to some state. In the same way, every spot in the sky belongs to some constellation.

Together, all the constellations seem to fill a great **celestial sphere** surrounding Earth (**Figure 2.6**). Of course, the entire celestial sphere is an illusion: The sky looks like a great sphere only because the stars are so far away that we have no depth perception when we look into space, and Earth seems to be in the center of the celestial sphere only because it is where we are located as we look into space. Nevertheless, the celestial sphere is a useful illusion, because it allows us to map the sky as seen from Earth. For reference, we identify four special points and circles on the celestial sphere:

- The **north celestial pole** is the point directly over Earth's North Pole.

- The **south celestial pole** is the point directly over Earth's South Pole.

- The **celestial equator,** which is a projection of Earth's equator into space, makes a complete circle around the celestial sphere.

- The **ecliptic** is the yearly path of the Sun around the celestial sphere. (Note that this is the Sun's *annual* path with respect to the constellations, *not* the daily path through the local sky.)

Daily Paths of Stars Through the Sky   Stars appear to move through the sky during the night for the same reason the Sun appears to move through the sky during the day: Earth's daily west-to-east rotation causes everything on the celestial sphere to appear to move around us from east to west (**Figure 2.7**). If we could stop Earth from rotating, the stars would appear motionless. They are so far away that, even though they move at fairly high speeds by earthly standards, their motions would be noticeable to the naked eye only if you could watch the sky for thousands of years. The constellations therefore appear to remain in fixed positions on the celestial sphere, and the celestial sphere looks essentially the same today as it did to the ancient Egyptians.

Now let's visualize how the sky appears from a location in the United States (**Figure 2.8**). You can see only half the celestial sphere at any one time, because the ground blocks your view of the other half. Notice that, because your location means you are standing at an angle to Earth's equator, your horizon appears to slice through the celestial sphere at an angle to the celestial equator. This angle

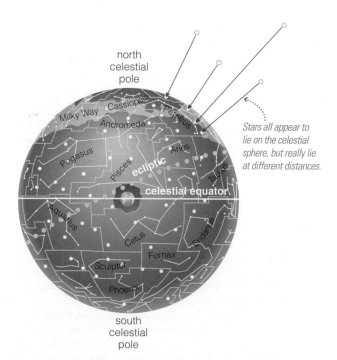

Stars all appear to lie on the celestial sphere, but really lie at different distances.

Figure 2.6 | The constellations appear to fill a great celestial sphere that surrounds Earth.

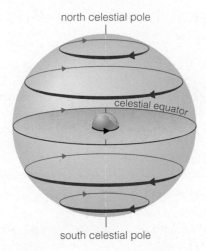

north celestial pole

celestial equator

south celestial pole

Figure 2.7 | Earth rotates from west to east (black arrow), making the celestial sphere *appear* to rotate around us from east to west (red arrows).

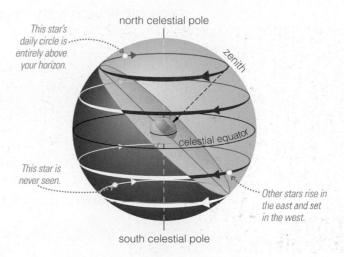

This star's daily circle is entirely above your horizon.

north celestial pole

zenith

celestial equator

This star is never seen.

Other stars rise in the east and set in the west.

south celestial pole

Figure 2.8 | The local sky as viewed from a location in the United States (latitude 40°N). The horizon slices through the celestial sphere at an angle to the equator, causing the daily circles of stars to appear tilted in the local sky. Note: It is easier to follow the star paths in the local sky if you rotate the page so that the zenith points up.

causes the daily circles of the stars on the celestial sphere to appear tilted in your local sky. As a result, stars that are near the north celestial pole will make daily circles above your horizon, like those within the arch in the photo on page 18. Stars that are near the south celestial pole will never be visible in your sky at all. In between, all other stars—along with the Sun, Moon, and planets—will appear to rise up from your eastern horizon and set into your western horizon.

Think about it Do distant galaxies also rise and set like the stars in our sky? Why or why not?

The same general ideas apply to all locations on Earth, but the details vary with latitude. Because latitude describes the angle that your zenith makes with Earth's equator, it also determines the angle at which your horizon slices through the celestial sphere. The portion of the celestial sphere that you can see and the daily paths of stars through the sky therefore both depend on your latitude; that is why the night sky looks quite different in Australia than it does in the United States.

Seasonal Changes in the Night Sky Look back at Figure 2.6 and notice the yellow dots marking the path we call the *ecliptic*. These dots show that the Sun appears to move gradually around the celestial sphere, completing one full circuit in one year. This is a direct consequence of Earth's yearly orbit around the Sun.

Figure 2.9 shows how this works. As we orbit the Sun over the course of a year, the Sun *appears* to move against the background of constellations on the celestial sphere. We don't see the Sun and the stars at the same time, but if we could we'd notice the Sun gradually moving eastward through the constellations, tracing out the path of the ecliptic. The constellations along the ecliptic are called the **zodiac.** (Tradition places 12 constellations along the zodiac, but the official borders include a wide swath of a thirteenth constellation, Ophiuchus.)

The Sun's apparent location along the ecliptic determines which constellations we see at night. For example, Figure 2.9 shows that the Sun appears to be in Leo in late August. We therefore cannot see Leo in late August, because it moves with the Sun through the daytime sky. However, we can see Aquarius

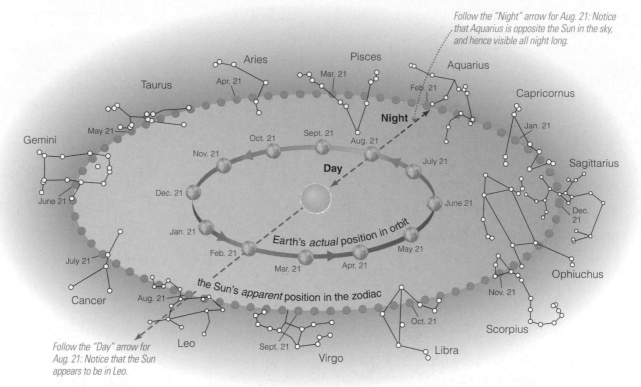

Follow the "Night" arrow for Aug. 21: Notice that Aquarius is opposite the Sun in the sky, and hence visible all night long.

Follow the "Day" arrow for Aug. 21: Notice that the Sun appears to be in Leo.

Figure 2.9 | The Sun appears to move steadily eastward along the ecliptic as Earth orbits the Sun, so we see the Sun against the background of different zodiac constellations at different times of year. For example, on August 21 the Sun appears to be in Leo, because it is between us and the much more distant stars that make up Leo.

all night long, because it is opposite Leo on the celestial sphere. Six months later, in February, we see Leo at night while Aquarius is above the horizon only in the daytime. To summarize, the Sun's apparent position among the constellations changes as Earth orbits around the Sun, and we therefore see different constellations at different times of year.

2.2 Understanding the Moon

Aside from the seasons and the daily circling of the sky, the most familiar pattern of change in the sky is that of the changing phases of the Moon. We will explore these changes in this section—along with the rarer changes that occur with eclipses—and see that they are consequences of the Moon's orbit around Earth.

Why do we see phases of the Moon?

The easiest way to understand lunar phases is with the simple demonstration illustrated in **Figure 2.10.** Take a ball outside on a sunny day. (If it's dark or cloudy, you can use a flashlight instead of the Sun; put the flashlight on a table a few meters away and shine it toward you.) Hold the ball at arm's length to represent the Moon while your head represents Earth. Slowly spin around (counterclockwise) so that the ball goes around you just as the Moon orbits Earth. As you turn, you'll see the ball go through phases just like the Moon. If you think about what's happening, you'll realize that the phases of the ball result from just two basic facts:

1. Half the ball always faces the Sun (or flashlight) and therefore is bright, while the other half faces away from the Sun and therefore is dark.

2. As you look at the ball at different positions in its "orbit" around your head, you see different combinations of its bright and dark faces.

Common Misconceptions

SHADOWS AND THE MOON

Many people guess that the Moon's phases are caused by shadows from Earth, but this is not the case. As we've seen, the Moon's phases are caused by the fact that we see different portions of its daylight and night sides at different times as it orbits around Earth. The only time that Earth's shadow falls on the Moon is during lunar eclipses, which are relatively rare (occurring about twice each year) and which last only a few hours.

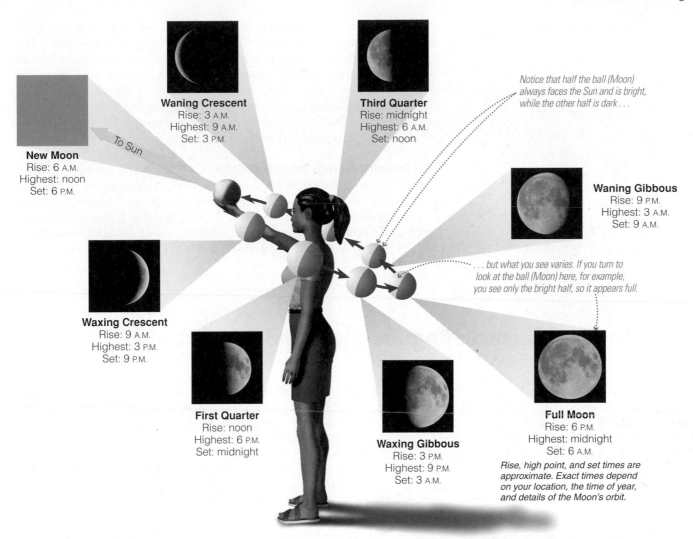

New Moon
Rise: 6 A.M.
Highest: noon
Set: 6 P.M.

To Sun

Waning Crescent
Rise: 3 A.M.
Highest: 9 A.M.
Set: 3 P.M.

Third Quarter
Rise: midnight
Highest: 6 A.M.
Set: noon

Notice that half the ball (Moon) always faces the Sun and is bright, while the other half is dark . . .

Waning Gibbous
Rise: 9 P.M.
Highest: 3 A.M.
Set: 9 A.M.

Waxing Crescent
Rise: 9 A.M.
Highest: 3 P.M.
Set: 9 P.M.

. . . but what you see varies. If you turn to look at the ball (Moon) here, for example, you see only the bright half, so it appears full.

First Quarter
Rise: noon
Highest: 6 P.M.
Set: midnight

Waxing Gibbous
Rise: 3 P.M.
Highest: 9 P.M.
Set: 3 A.M.

Full Moon
Rise: 6 P.M.
Highest: midnight
Set: 6 A.M.

Rise, high point, and set times are approximate. Exact times depend on your location, the time of year, and details of the Moon's orbit.

Figure 2.10 | A simple demonstration illustrates the phases of the Moon. The half of the ball (Moon) facing the Sun is always illuminated while the half facing away is always dark, but you see the ball go through phases as it orbits around your head (Earth). (The new moon photo shows blue sky, because a new moon is always close to the Sun in the sky and hence hidden from view by the bright light of the Sun.)

For example, when you hold the ball directly opposite the Sun, you see only the bright portion of the ball, which represents the "full" phase. When you hold the ball at its "first-quarter" position, half the face you see is dark and the other half is bright.

We see lunar phases for the same reason. Half the Moon is always illuminated by the Sun, but the portion of this illuminated half visible from Earth depends on the Moon's position in its orbit. The photographs in Figure 2.10 show how the phases look. Each complete cycle of phases, from one new moon to the next, takes about $29\frac{1}{2}$ days—hence the origin of the word *month* (think "moonth").

The Moon's phase is directly related to the time it rises, reaches its highest point in the sky, and sets. For example, the full moon must rise around sunset, because it occurs when the Moon is opposite the Sun in the sky. It therefore reaches its highest point in the sky at midnight and sets around sunrise. Similarly, a first-quarter moon must rise around noon, reach its highest point around 6 P.M., and set around midnight, because it occurs when the Moon is about 90° east of the Sun in our sky. Figure 2.10 lists the approximate rise, highest point, and set times for each phase.

 Think about it Suppose you go outside in the morning and notice that the visible face of the Moon is half light and half dark. Is this a first-quarter or third-quarter moon? How do you know?

Common Misconceptions

MOON IN THE DAYTIME

The Moon is so closely associated with night in traditions and stories that many people mistakenly believe that the Moon is visible only in the nighttime sky. In fact, the Moon is above the horizon as often in the daytime as at night, though it is easily visible only when its light is not drowned out by sunlight. For example, a first-quarter moon is easy to spot in the late afternoon as it rises through the eastern sky, and a third-quarter moon is visible in the morning as it heads toward the western horizon.

Notice that the phases from new to full are said to be *waxing,* which means "increasing." Phases from full to new are *waning,* or "decreasing." Also notice that no phase is called a "half moon." Instead, we see half the Moon's face at first-quarter and third-quarter phases; these phases mark the times when the Moon is one-quarter or three-quarters of the way through its monthly cycle (which begins at new moon). The phases just before and after new moon are called *crescent,* while those just before and after full moon are called *gibbous* (pronounced with a hard *g* as in "gift").

The Moon's Synchronous Rotation   Although we see many *phases* of the Moon, we do not see many *faces.* From Earth, we always see (nearly) the same face of the Moon. This happens because the Moon rotates on its axis in exactly the same amount of time that it takes to orbit Earth, a trait called **synchronous rotation.** A simple demonstration shows this idea (**Figure 2.11**). Place a ball on a table to represent Earth while you represent the Moon.

Tools of Science: Angular Sizes and Distances

Our lack of depth perception on the celestial sphere means we have no way to judge the true sizes or separations of the objects we see in the sky. We can directly measure only *angular* sizes or separations.

Figure 1 shows the basic idea behind angular measurement. The **angular size** of an object is the angle it appears to span in your field of view. For example, the angular sizes of the Sun and Moon are each about $\frac{1}{2}$° (**Figure 1a**). The **angular distance** between a pair of objects in the sky is

the angle that appears to separate them. For example, the angular distance between the "pointer stars" at the end of the Big Dipper's bowl is about 5° (**Figure 1b**). Note that you can use your outstretched hand to estimate angles in the sky (**Figure 1c**).

For more precise astronomical measurements, we subdivide each degree into 60 **arcminutes** and subdivide each arcminute into 60 **arcseconds** (**Figure 2**). We abbreviate arcminutes with the symbol ′ and arcseconds with the symbol ″. For example, we read 35°27′15″ as "35 degrees, 27 arcminutes, 15 arcseconds." Note that one arcminute is approximately the thickness of a fingernail at arm's length, and one arcsecond is approximately the thickness of someone else's fingernail viewed from across a football field.

Angular measurements are very useful in astronomy. We use angles to describe the locations of objects in the sky or on the celestial sphere, much as we describe the location of an object on Earth with latitude and longitude. Even more important, angular sizes can help us to calculate physical sizes if we also know an object's true distance, or vice versa. For example, we can use the Moon's distance and angular size to calculate its physical size, or we can use a planet's angular size and physical size to calculate its distance.

a The angular sizes of the Sun and the Moon are about 1/2°.

b The angular distance between the two "pointer stars" of the Big Dipper is about 5°.

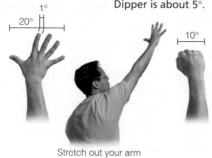

Stretch out your arm as shown here.

c You can estimate angular sizes or distances with your outstretched hand.

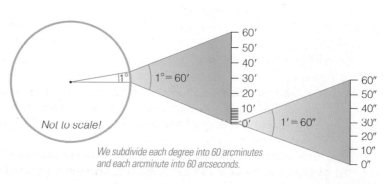

We subdivide each degree into 60 arcminutes and each arcminute into 60 arcseconds.

Figure 1   **Figure 2**

a If you do not rotate while walking around the model, you will not always face it.

b You will face the model at all times only if you rotate exactly once during each orbit.

The only way you can face the ball at all times is by completing exactly one rotation while you complete one orbit. (The Moon's synchronous rotation is not a coincidence; rather, it is a result of Earth's gravity affecting the Moon in much the same way that the Moon's gravity causes tides on Earth.)

The View from the Moon   A good way to solidify your understanding of lunar phases is to imagine that you live on the side of the Moon that faces Earth. What would you see as you looked at Earth when people on Earth saw a new moon? By remembering that a new moon occurs when the Moon is between the Sun and Earth, you'll realize that from the Moon you'd be looking at Earth's daytime side and hence would see a *full Earth*. Similarly, at full moon you would be facing the night side of Earth and would see a *new Earth*. In general, you'd always see Earth in a phase opposite the phase of the Moon being seen by people on Earth at the same time.

Think about it   About how long would each day and night last if you lived on the Moon? Explain. (*Hint*: Recall that a full cycle of phases lasts about a month.)

What causes eclipses?

Occasionally, the Moon's orbit around Earth causes events much more dramatic than lunar phases. The Moon and Earth both cast shadows in sunlight, and these shadows can create **eclipses** when the Sun, Earth, and Moon fall into a straight line. Eclipses come in two basic types:

- A **lunar eclipse** occurs when Earth lies directly between the Sun and Moon, so Earth's shadow falls on the Moon.

- A **solar eclipse** occurs when the Moon lies directly between the Sun and Earth, so the Moon's shadow falls on Earth. People living within the area covered by the Moon's shadow will see the Sun blocked or partially blocked from view.

Conditions for Eclipses   Look again at Figure 2.10. The figure makes it look as if the Sun, Earth, and Moon line up with every new and full moon. If this figure told the whole story of the Moon's orbit, we would have both a lunar and a solar eclipse every month—but we don't.

The missing piece of the story in Figure 2.10 is that the Moon's orbit is slightly inclined (by about 5°) to the plane of Earth's orbit around the Sun, called the **ecliptic plane.** To visualize this inclination, imagine the ecliptic

Common Misconceptions

THE "DARK SIDE" OF THE MOON

The phrase *dark side of the Moon* really should be used to mean the night side—that is, the side facing away from the Sun. Unfortunately, *dark side* traditionally meant what would better be called the *far side*—the face that never can be seen from Earth. Many people still refer to the far side as the "dark side," even though this side is not necessarily dark. For example, during new moon the far side faces the Sun and hence is completely sunlit. The only time the far side is completely dark is at full moon, when it faces away from both the Sun and Earth.

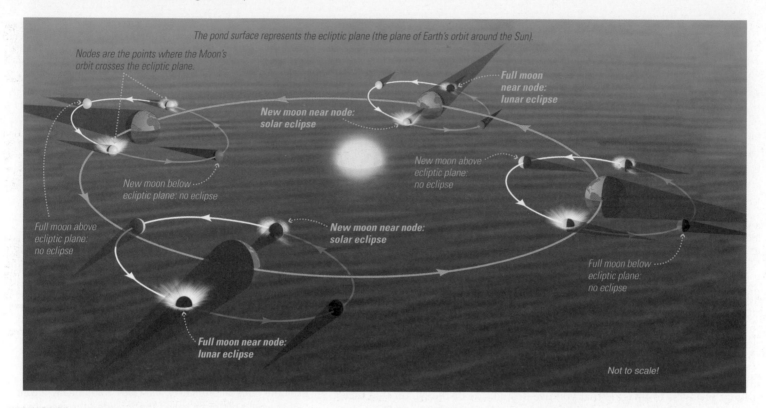

The pond surface represents the ecliptic plane (the plane of Earth's orbit around the Sun).

Nodes are the points where the Moon's orbit crosses the ecliptic plane.

Full moon near node: lunar eclipse

New moon near node: solar eclipse

New moon below ecliptic plane: no eclipse

New moon above ecliptic plane: no eclipse

Full moon above ecliptic plane: no eclipse

New moon near node: solar eclipse

Full moon below ecliptic plane: no eclipse

Full moon near node: lunar eclipse

Not to scale!

Figure 2.12 | This illustration represents the ecliptic plane as the surface of a pond. The Moon's orbit is tilted by about 5° to the ecliptic plane, so the Moon spends half of each orbit above the plane (the pond surface) and half below it. Eclipses occur only when the Moon is at both a node (passing through the pond surface) *and* a phase of either new moon (for a solar eclipse) or full moon (for a lunar eclipse)—as is the case with the lower left and top right orbits shown.

plane as the surface of a pond, as shown in **Figure 2.12**. Because of the inclination of its orbit, the Moon spends most of its time either above or below this surface. It crosses *through* this surface only twice during each orbit: once coming out and once going back in. The two points in each orbit at which the Moon crosses the surface are called the **nodes** of the Moon's orbit.

Notice that the nodes are aligned approximately the same way (diagonally on the page in Figure 2.12) throughout the year, which means they lie along a nearly straight line with the Sun and Earth about twice each year. Eclipses can occur only during these periods (sometimes called *eclipse seasons*) when the nodes line up with the Sun and Earth. A lunar eclipse occurs when a full moon occurs at one of the nodes. A solar eclipse occurs when a new moon occurs at one of the nodes.

Variations on Eclipses Although there are only two basic types of eclipse—lunar and solar—each of these types can look different depending on precisely how the shadows fall. The shadow of the Moon or Earth consists of two distinct regions: a central **umbra**, where sunlight is completely blocked, and a surrounding **penumbra**, where sunlight is only partially blocked (**Figure 2.13**).

A lunar eclipse begins at the moment when the Moon's orbit first carries it into Earth's penumbra. After that, we will see one of three types of lunar eclipse (**Figure 2.14**). If the Sun, Earth, and Moon are nearly perfectly aligned, the Moon passes through Earth's umbra and we see a **total lunar eclipse.** If the alignment is somewhat less perfect, only part of the full moon passes through the umbra (with the rest in the penumbra), and we see a **partial lunar eclipse.** If the Moon passes through *only* Earth's penumbra, we see a **penumbral lunar eclipse.** Penumbral eclipses are the most common, but they are the least visually impressive because the full moon darkens only slightly. Total lunar eclipses are the most spectacular; the Moon becomes dark and eerily red during *totality,* when the Moon is entirely engulfed in the umbra, because Earth's atmosphere bends some of the red light from the Sun toward the Moon.

We can also see three types of solar eclipse (**Figure 2.15**). Because the Moon is much smaller than Earth, its shadows are also smaller than Earth: The umbra can touch at most only a small area of Earth's surface (no more than

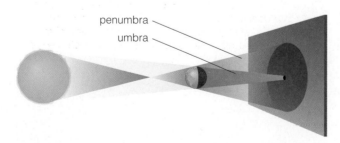

penumbra
umbra

Figure 2.13 | The shadow cast by an object in sunlight. Sunlight is fully blocked in the umbra and partially blocked in the penumbra.

about 270 kilometers in diameter), while the penumbra shadows an area with a diameter a little more than one-fourth that of Earth. You can see a **total solar eclipse,** in which the Moon blocks the full disk of the Sun, only if you are within the umbra; totality lasts no more than a few minutes, because the combination of Earth's rotation and the Moon's orbital motion causes the Moon's shadows to race across the face of Earth at a speed of about 1700 kilometers per hour. People within the penumbral shadow see a **partial solar eclipse,** in which only part of the Sun is blocked from view. Sometimes a solar eclipse occurs when the Moon is relatively far from Earth in its orbit; in that case, the umbra may not reach Earth's surface at all, and people directly behind the umbra will see an **annular solar eclipse,** in which a ring of sunlight surrounds the disk of the Moon.

 Think about it The Moon's distance from Earth varies along its orbit from about 356,000 km to about 407,000 km. When is an annular eclipse more likely: when the Moon is near its minimum or maximum distance? Explain.

Predicting Eclipses Few phenomena have so inspired and humbled humans throughout the ages as eclipses. For many cultures, eclipses were mystical events associated with fate or the gods, and countless stories and legends surround them. Much of the mystery of eclipses probably stems from the relative difficulty of predicting them.

To understand this difficulty, look again at Figure 2.12, which shows two periods (the eclipse seasons) during the year in which the nodes of the Moon's orbit are closely aligned with the Sun. If this were the end of the story, these periods would always occur 6 months apart and predicting eclipses would be easy. For example, if these periods occurred in January and July, we'd always have a solar eclipse at new moon in those months and a lunar eclipse at full moon. Actual eclipse prediction is more difficult than this because of something the figure does not show: The nodes slowly move around the Moon's orbit, so the alignments actually occur slightly less than 6 months apart (about 173 days apart). As a result, instead of recurring with an annual pattern, eclipses recur in a pattern that repeats about every 18 years $11\frac{1}{3}$ days (often called the *saros cycle*).

Some ancient civilizations learned to recognize this 18-year cycle of eclipses and therefore were able to predict when eclipses would occur. Even then, however, they could not accurately predict the precise location from which an eclipse would be visible or the type of eclipse (such as partial versus total) that would occur. Such detailed predictions are possible only with a full understanding of the orbit of Earth around the Sun and of the Moon around Earth—and hence could be accomplished only after we'd learned that Earth is a planet orbiting the Sun, along with the laws that govern orbital motion.

☀ THE PROCESS OF SCIENCE IN ACTION
2.3 The Puzzle of Planetary Motion

In this chapter we have seen how Earth's rotation leads to the daily rise and set of the Sun, Moon, and stars and how Earth's orbit makes the Sun appear to move through the constellations. Ancient people could also explain these phenomena, though their explanations differed from the modern ones because they believed in an Earth-centered universe.

Planetary motion posed a much bigger puzzle in ancient times, and we will see in Chapter 3 that the quest to understand this motion ultimately led to the realization about four centuries ago that Earth is a planet in

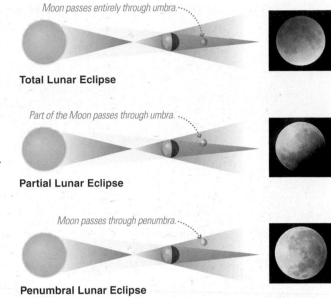

Figure 2.14 | The three types of lunar eclipse: The diagrams show how they occur, and the photos show how the Moon appears during each type of eclipse.

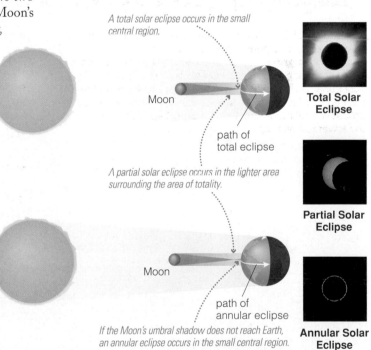

Figure 2.15 | The three types of solar eclipse. The diagrams show the Moon's shadow falling on Earth; note the dark central umbra surrounded by the much lighter penumbra. The photos show how the Sun appears during each type of eclipse.

orbit around the Sun. But it was not the first time this idea had been raised: The ancient Greeks also considered the possibility that Earth moves around the Sun. In fact, the Greeks took the idea quite seriously, and rejected it only because they could not find sufficient evidence to support it with the observations possible at the time. The story of how the Greeks considered but ultimately rejected the correct explanation for planetary motion provides an early example of scientific thinking in action, and demonstrates that ancient civilizations thought more deeply about nature than many people realize.

Why did the ancient Greeks reject the real explanation for planetary motion?

Over the course of a single night, planets behave like all the other objects in the sky—Earth's rotation makes them appear to rise in the east and set in the west. But if you continue to watch the planets night after night, you will notice that their movements among the constellations become quite complex. Instead of moving steadily eastward relative to the stars, like the Sun and Moon, the planets vary substantially in both speed and brightness. Moreover, while the planets *usually* move eastward through the constellations, they occasionally reverse course, moving westward rather than eastward through the zodiac (**Figure 2.16**). These periods of **apparent retrograde motion** (*retrograde* means "backward") last from a few weeks to a few months, depending on the planet.

Apparent retrograde motion is very difficult to explain if you believe in an Earth-centered universe; after all, what could make planets sometimes turn around and go backward if everything moves in circles around Earth? The ancient Greeks nevertheless came up with some very clever ways to explain it, but their explanations (which we'll study in Chapter 3) were quite complex. In contrast, the explanation for apparent retrograde motion is quite simple in a Sun-centered solar system. With the help of a friend, you can see for yourself how it works (**Figure 2.17a**). Pick a spot in an open area to represent the Sun. You can represent Earth by walking counterclockwise around the Sun, while your friend represents a more distant planet (such as Mars or Jupiter) by walking in the same direction around the Sun at a greater distance. Your friend should walk more slowly than you, because more distant planets orbit the Sun more slowly. As you walk, watch how your friend appears to move relative to buildings or trees in the distance. Although both of you always walk the same way around the Sun, your friend will appear to move backward against the background during the part of your "orbit" at which you catch up to and pass him or her. **Figure 2.17b** shows how the same idea applies to Mars. (To understand the apparent retrograde motions of Mercury and Venus, which are closer to the Sun than Earth, simply switch places with your friend and repeat the demonstration.)

The ancient Greeks were aware of how simply a Sun-centered system could explain apparent retrograde motion, and this may have been one reason that the Greek astronomer Aristarchus proposed in about 260 B.C. that Earth goes around the Sun. However, Aristarchus's contemporaries rejected his idea, and the Sun-centered solar system did not gain wide acceptance until almost 2000 years later.

Although there were many reasons why the Greeks were reluctant to abandon the idea of an Earth-centered universe, one of the most important

Figure 2.16 | This composite of 29 individual photos shows Mars from June through November 2003. Notice that Mars usually moves eastward (left) relative to the stars, but reverses course during its apparent retrograde motion. Note also that Mars is biggest and brightest in the middle of the retrograde loop, because that is where it is closest to Earth in its orbit. (The line of white dots just right of center shows the planet Uranus, which happened to be in the same part of the sky.)

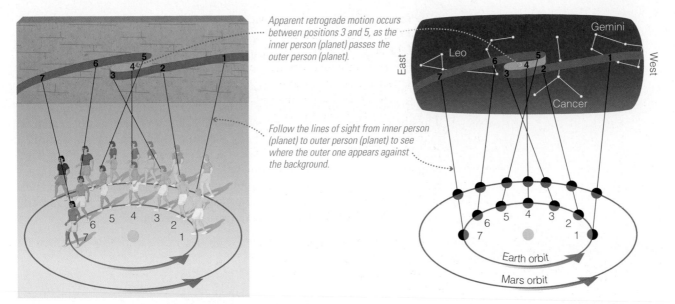

Apparent retrograde motion occurs between positions 3 and 5, as the inner person (planet) passes the outer person (planet).

Follow the lines of sight from inner person (planet) to outer person (planet) to see where the outer one appears against the background.

a This demonstration shows why planets sometimes seem to go backward relative to distant stars. Watch how your friend (in red) usually appears to move forward against the background of the building in the distance but appears to move backward as you (in blue) catch up and pass her in your "orbit."

b This diagram shows how the same idea applies to a planet. Follow the lines of sight from Earth to Mars in numerical order. Notice that Mars appears to move westward relative to the distant stars as Earth passes it in its orbit (from points 3 to 5 in the diagram).

was their inability to detect what we call **stellar parallax.** Extend your arm and hold up one finger. If you keep your finger still and alternately close your left eye and right eye, your finger will appear to jump back and forth against the background. This apparent shifting, called *parallax,* occurs because your two eyes view your finger from opposite sides of your nose. If you move your finger closer to your face, the parallax increases. If you look at a distant tree or flagpole instead of your finger, you may not notice any parallax at all. In other words, parallax depends on distance, with nearer objects exhibiting greater parallax than more distant objects.

If you now imagine that your two eyes represent Earth at opposite sides of its orbit around the Sun and that the tip of your finger represents a relatively nearby star, you have the idea of stellar parallax. That is, because we view the stars from different places in our orbit at different times of year, nearby stars should *appear* to shift back and forth against the background of more distant stars (**Figure 2.18**).

Because the Greeks believed that all stars lay on the same celestial sphere, they expected to observe stellar parallax in a slightly different way. If Earth orbited the Sun, they reasoned, at different times of year we would be closer to different parts of the celestial sphere and would notice changes in the angular separations of stars. However, no matter how hard they searched, they could find no sign of stellar parallax. They concluded that one of the following must be true:

1. Earth orbits the Sun but the stars are so far away that stellar parallax is undetectable to the naked eye.

2. There is no stellar parallax because Earth remains stationary at the center of the universe.

Aside from a few notable exceptions such as Aristarchus, the Greeks rejected the correct answer (the first one) because they could not imagine that the stars could be *that* far away. Today, we can detect stellar parallax with the aid of telescopes, providing direct proof that Earth really does orbit the Sun.

Figure 2.17 | Apparent retrograde motion—the occasional "backward" motion of the planets relative to the stars—has a simple explanation in a Sun-centered solar system.

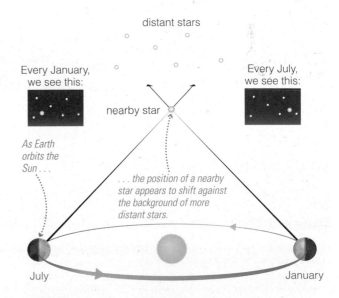

Figure 2.18 | Stellar parallax is an apparent shift in the position of a nearby star as we look at it from different places in Earth's orbit. This figure is greatly exaggerated; in reality, the amount of shift is far too small to detect with the naked eye.

Think about it How far apart are opposite sides of Earth's orbit? How far away are the nearest stars? Using the 1-to-10-billion scale from Chapter 1, describe the challenge of detecting stellar parallax.

The moral of this story is that our ability to do science is inextricably linked with our ability to make careful observations and measurements. The ancient Greeks knew that planetary motion was more easily explained by a Sun-centered system, but they also knew that such a system implied the existence of stellar parallax. Because their measurements were not good enough to detect parallax, they saw no clear reason to abandon their deeply held belief in an Earth-centered universe. As we'll see in the next chapter, evidence sufficient to finally change this belief came only with greatly improved measurements of planetary positions in the sky.

Summary of Key Concepts

 2.1 Understanding the Seasons

What causes the seasons?

The tilt of Earth's axis causes the seasons. The axis points in the same direction throughout the year; therefore, as Earth orbits the Sun, sunlight hits different parts of Earth more directly at different times of year.

Why do the constellations we see depend on the time of year?

The visible constellations vary with the seasons because our night sky lies in different directions in space as we orbit the Sun.

 2.2 Understanding the Moon

Why do we see phases of the Moon?

The phase of the Moon depends on its position relative to the Sun as it orbits Earth. The half of the Moon facing the Sun is always illuminated while the other half is dark, but from Earth we see varying combinations of the illuminated and dark faces.

What causes eclipses?

We see a **lunar eclipse** when Earth's shadow falls on the Moon, and a **solar eclipse** when the Moon blocks our view of the Sun. We do not see an eclipse at every new and full moon because the Moon's orbit is slightly inclined to the ecliptic plane. Eclipses come in different types, depending on where the dark **umbral** and lighter **penumbral** shadows fall.

✳ THE PROCESS OF SCIENCE IN ACTION

2.3 The Puzzle of Planetary Motion

Why did the ancient Greeks reject the real explanation for planetary motion?

Planets generally move eastward relative to the stars from night to night, but sometimes they reverse course for weeks to months in what we call **apparent retrograde motion.** This motion is simply explained in a Sun-centered system, but the Greeks rejected this model in part because they could not detect **stellar parallax**—slight apparent shifts in stellar positions over the course of the year that must occur if Earth orbits the Sun.

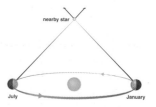

Investigations

Quick Quiz

Choose the best answer to each of the following; answers are in Appendix D. Explain your reasoning with one or more complete sentences.

1. When it is summer in Australia, it is (a) winter in the United States. (b) summer in the United States. (c) spring in the United States.
2. In winter, Earth's axis points toward the star Polaris. In spring, (a) the axis also points toward Polaris. (b) the axis points toward Vega. (c) the axis points toward the Sun.
3. Which of the following is *not* true during Northern Hemisphere summer? (a) Sunlight strikes the ground at a steeper angle in the Northern Hemisphere than it does in the Southern Hemisphere. (b) The Sun follows a longer and higher path through the Northern Hemisphere sky than it does through the Southern Hemisphere sky. (c) Noontime shadows are longer in the Northern Hemisphere than in the Southern Hemisphere.
4. We can be sure that variations in Earth's distance from the Sun are *not* the cause of the seasons, because (a) Earth is always exactly the same distance from the Sun. (b) if distance were responsible for the seasons,

both the Northern and Southern Hemispheres would have summer at the same time. (c) if distance were responsible for the seasons, seasonal temperature changes would be much more extreme.

5. Two stars that are in the same constellation (a) must both be part of the same cluster of stars in space. (b) must both have been discovered at about the same time. (c) may actually be very far away from each other.

6. Suppose you see a crescent moon; how much of the Moon's entire surface (the full globe of the Moon) is in daylight? (a) half (b) one-quarter (c) one-eighth

7. One week after full moon, the Moon's phase is (a) first quarter. (b) third quarter. (c) new.

8. The fact that we always see the same face of the Moon tells us that (a) the Moon does not rotate. (b) the Moon's rotation period is the same as its orbital period. (c) the Moon looks the same on both sides.

9. The reason we do *not* have a solar eclipse at every new moon is that (a) a solar eclipse can occur only at full moon. (b) the Moon's orbit is inclined relative to Earth's orbit around the Sun. (c) the nodes of the Moon's orbit gradually move around the orbit.

10. If there is going to be a total lunar eclipse tonight, then you know that the Moon's phase is (a) full. (b) new. (c) first or third quarter.

11. When we see Saturn going through a period of apparent retrograde motion, it means that (a) Saturn is temporarily moving backward in its orbit around the Sun. (b) Earth is passing Saturn in its orbit, with both planets on the same side of the Sun. (c) Saturn and Earth must be on opposite sides of the Sun.

12. Which of the following best describes why the Greeks rejected the correct explanation for apparent retrograde motion? (a) They did not realize that planets sometimes appear to move backward relative to the constellations. (b) They predicted that if Earth orbited the Sun, then stellar parallax should occur, but they could not detect this parallax. (c) They never even considered the possibility that Earth could orbit the Sun.

Short-Answer/Essay Questions

Explain all answers clearly, using complete sentences and proper essay structure if needed. An asterisk (*) designates a quantitative problem, for which you should show all your work.

13. *Earth-Centered or Sun-Centered?* The phenomena discussed in this chapter are all visible to the naked eye and have been known throughout human history, including during the thousands of years when Earth was assumed to be at the center of the universe. For each of the following, decide whether the phenomenon is consistent or inconsistent with a belief in an Earth-centered system. If consistent, describe how. If inconsistent, explain why, and also explain why the inconsistency did not immediately lead people to abandon the Earth-centered belief.
    a. The daily paths of stars through the sky
    b. Seasons
    c. Phases of the Moon
    d. Eclipses
    e. Apparent retrograde motion of the planets
    f. Stellar parallax

14. *Cause of the Seasons.* Suppose a friend tries to claim that it is summer when Earth is closer to the Sun and winter when Earth is farther away. Refute this claim in at least two different ways: (1) with an argument that invokes seasonal differences between the Northern and Southern Hemispheres and (2) with an argument based on the time of year at which Earth actually is closest and farthest from the Sun. Write out your arguments clearly in one to two paragraphs each.

15. *New Planet.* Suppose we discover a planet in another solar system that has a circular orbit and an axis tilt of 35°. Would you expect this planet to have seasons? If so, would you expect them to be more extreme than the seasons on Earth? If not, why not?

16. *Shadow Phases.* Many people incorrectly guess that the phases of the Moon are caused by Earth's shadow falling on the Moon. How would you go about convincing a friend that the phases of the Moon have nothing to do with Earth's shadow? Describe the observations you would perform to show that Earth's shadow can't be the cause of phases.

17. *View from the Sun.* Suppose you lived on the Sun (and could ignore the heat). Would you still see the Moon go through phases as it orbits Earth? Why or why not?

18. *View from the Moon.* Assume you live on the Moon near the center of the face that looks toward Earth.
    a. If you saw a full Earth in your sky, what phase of the Moon would people on Earth see? Explain.
    b. If people on Earth saw a full moon, what phase would you see for Earth? Explain.
    c. If people on Earth saw a waxing gibbous moon, what phase would you see for Earth? Explain.
    d. If people on Earth were viewing a total lunar eclipse, what would you see from your home on the Moon? Explain.

19. *A Farther Moon.* Suppose the distance to the Moon were twice its actual value. Would it still be possible to have a total solar eclipse? Why or why not?

20. *A Smaller Earth.* Suppose Earth were smaller. Would solar eclipses be any different? If so, how? What about lunar eclipses? Explain.

21. *Observing Planetary Motion.* Find out what planets are currently visible in your evening sky. At least once a week, observe the planets and draw a diagram showing the position of each visible planet relative to stars in a zodiac constellation. From week to week, note how the planets are moving relative to the stars. Can you see any of the wandering features of planetary motion? Explain.

*22. *Seasons Scale.* Figure 2.2 is not shown to scale. Let's examine how it would look if it were.
    a. Use a ruler to measure the diameter of Earth's orbit as shown in the figure (measure in the longest direction, since the perspective distorts the diameter in other directions). Then use the fact that the actual diameter of Earth's orbit around the Sun is about 300,000,000 km to determine the scale on which the orbit is drawn.
    b. How large should Earth itself appear if drawn to the scale you found in part (a)? (*Hint:* Earth's actual diameter is about 12,800 km.)
    c. Suppose someone tries to claim that it is summer whenever your hemisphere (either Northern or Southern) is closer to the Sun than the other hemisphere. Based on your answers to parts (a) and (b), how can you refute their argument?

3 Changes in Our Perspective

Imagine yourself as a satellite, orbiting Earth like the astronaut in the photo and looking down at the thin layer of atmosphere that separates our homes from the blackness of space. You'd see Earth in a way that most of our ancestors could hardly have imagined. Before the advent of science, people assumed Earth to be the center of the universe, not a planet orbiting the Sun. Our perspective began to change as we gathered and sought to interpret increasingly precise observations of the sky. This chapter tells the story of that shift in perspective and discusses how it exemplifies the kind of thinking we now call *science*. We'll examine the key features of scientific thinking and see how science produces reliable knowledge about how the universe works—so reliable, in fact, that we can now launch astronauts into space and return them safely to Earth.

3.1 From Earth-Centered to Sun-Centered

We saw in Chapter 2 that the ancient Greeks considered the possibility that Earth might be a planet going around the Sun, but ultimately rejected it. The Earth-centered view then continued to hold sway for another 1500 years. Why did it take so long to recognize that the ancient belief in an Earth-centered universe was incorrect, and what finally caused our perspective on this issue to change? To understand the answers to these questions, we must first investigate how the ancient Greeks managed to explain observations of planetary motion within their Earth-centered system. We will then be prepared to understand how this system was eventually discarded in an intellectual revolution that culminated about 400 years ago.

How did the Greeks explain planetary motion?

The wandering motions of the planets among the stars were difficult for ancient people to explain, especially the "backward" loops during periods of apparent retrograde motion [Section 2.3]. Nevertheless, the Greeks managed to satisfy themselves that these motions were still compatible with an Earth-centered universe.

Scientific Models   The key to understanding both the Greek and modern explanations of planetary motion lies in an idea first developed by the Greeks that remains central to science today—the idea of creating **models** of nature. Scientific models differ somewhat from the models you may be familiar with in everyday life. In our daily lives, we tend to think of models as miniature physical representations, such as model cars or airplanes. In contrast, a scientific model is a conceptual representation whose purpose is to explain and predict observed phenomena. For example, a model of Earth's climate uses logic and mathematics to represent what we have learned about how the climate works. Its purpose is to explain and predict climate changes, such as the changes that may occur with global warming. Just as a model airplane does not faithfully represent every aspect of a real airplane, a scientific model may not fully explain all our observations of nature. Nevertheless, even the failings of a scientific model can be useful, because they often point the way toward building a better model.

The Greek Geocentric Model   The Greeks constructed conceptual models of the universe in an attempt to explain what they observed in the sky. We do not know precisely when the Greeks first began to think that Earth is round, but this idea was being taught as early as about 500 B.C. by the famous mathematician Pythagoras (c. 560–480 B.C.). More than a century later, Aristotle (384–322 B.C.) cited observations of Earth's curved shadow on the Moon during lunar eclipses as evidence for a spherical Earth. The Greek model was therefore a **geocentric model** of the universe (recall that *geocentric* means "Earth-centered"), with a spherical Earth at the center of a great celestial sphere.

Greek philosophers quickly realized that there had to be more to the heavens than just a single sphere surrounding Earth. Because the Sun and Moon move gradually eastward through the constellations, the Greeks added separate spheres for them, with these spheres turning at different rates from the sphere of the stars. The planets also move relative to the stars, so the Greeks added additional spheres for the planets (**Figure 3.1**).

The difficulty with this geocentric model was that it did not easily account for the apparent retrograde motion of the planets. You might guess that the Greeks would simply have allowed the planetary spheres to sometimes turn forward and sometimes turn backward relative to the sphere of the stars,

Figure 3.1 | This model represents the Greek idea of the heavenly spheres (c. 400 B.C.). Earth is a sphere that rests in the center. The Moon, the Sun, and the planets each have their own spheres. The outermost sphere holds the stars.

A widespread myth gives credit to Columbus for learning that Earth is round, but knowledge of Earth's shape predated Columbus by nearly 2000 years. Not only were scholars of Columbus's time well aware that Earth is round, but they even knew its approximate size: Earth's circumference was first measured in about 240 B.C. by the Greek scientist Eratosthenes. In fact, a likely reason why Columbus had so much difficulty finding a sponsor for his voyages was that he tried to argue a point on which he was dead wrong: He claimed the distance by sea from western Europe to eastern Asia to be much less than the scholars knew it to be. Indeed, when he finally found a patron in Spain and left on his journey, he was so woefully under-prepared that the voyage would almost certainly have ended in disaster if the Americas hadn't stood in his way.

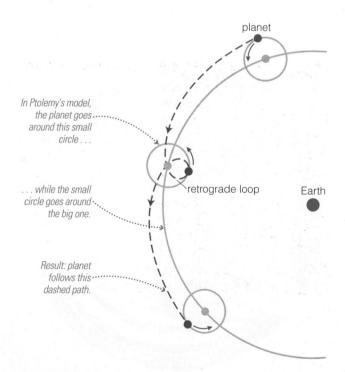

In Ptolemy's model, the planet goes around this small circle . . .

. . . while the small circle goes around the big one.

Result: planet follows this dashed path.

planet

retrograde loop

Earth

Figure 3.2 | This diagram shows how the Ptolemaic model accounted for apparent retrograde motion. Each planet is assumed to move around a small circle that turns upon a larger circle. The resulting path (dashed) includes a loop in which the planet goes backward as seen from Earth.

but they did not because it would have violated their deeply held belief in "heavenly perfection." According to this idea, enunciated most clearly by Plato, heavenly objects could move only in perfect circles.

The Greeks came up with a number of ingenious ideas for explaining planetary motion while preserving Earth's central position and motion in perfect circles. These ideas were refined for centuries and reached their apex in the work of Claudius Ptolemy (c. A.D. 100–170; pronounced "tol-e-mee"). We refer to Ptolemy's model as the **Ptolemaic model,** to distinguish it from earlier geocentric models.

Circles upon Circles   The essence of the Ptolemaic model was that each planet moves on a small circle whose center moves around Earth on a larger circle (**Figure 3.2**). (The small circle is called an *epicycle,* and the larger circle is called a *deferent.*) A planet following this circle-upon-circle motion traces a loop as seen from Earth, with the backward portion of the loop mimicking apparent retrograde motion.

Think about it   Review the real cause for apparent retrograde motion in Section 2.3. Notice that Mars appears biggest and brightest when it is in the middle of its retrograde loop. Why is this the case? Does the Ptolemaic model also account for the fact that planets are brightest in our sky in the middle of their retrograde loops? Explain.

The circle-upon-circle loops successfully explained the general idea of apparent retrograde motion. However, to make the model agree well with observations, Ptolemy had to include a number of other complexities, such as positioning some of the large circles slightly off-center from Earth and adding even smaller circles that moved upon the original set of small circles. As a result, the full Ptolemaic model was mathematically quite complex, and using it to predict planetary positions required long and tedious calculations. Many centuries later, while supervising computations based on the Ptolemaic model, the Spanish monarch Alphonso X (1221–1284) is said to have complained, "If I had been present at the creation, I would have recommended a simpler design for the universe."

Despite its complexity, the Ptolemaic model proved remarkably successful: It could correctly forecast future planetary positions to within a few degrees of arc, which is about the angular extent of your hand held at arm's length against the sky. This was sufficiently accurate to keep the model in use for the next 1500 years. When Ptolemy's book describing the model was translated by Arabic scholars around A.D. 800, they gave it the title *Almagest,* derived from words meaning "the greatest compilation."

How did the Copernican revolution change our view of the universe?

The Ptolemaic model was so successful that few people questioned it during its long period of dominance. Nevertheless, by the late 15th century the model was showing some cracks that ultimately led to its demise. Over a period of about 150 years, the work of numerous scientists ultimately proved that Earth is *not* the center of the universe, thereby revolutionizing our view of the cosmos. Because this revolution began in earnest with the work of the Polish scientist Nicholas Copernicus (1473–1543), we generally refer to it as the **Copernican revolution.** As we will see, the Copernican revolution is in many ways the story of the origin of modern science. It is also the story of several key personalities, beginning with Copernicus himself.

Copernicus   Copernicus was born in Torún, Poland, on February 19, 1473. He began studying astronomy in his late teens, and learned that tables of planetary motion based on the Ptolemaic model had been growing increasingly inaccurate. He began a quest to find a better way to predict planetary positions.

Copernicus was aware of and adopted Aristarchus's ancient Sun-centered idea, first proposed in about 260 B.C. [Section 2.3], probably because it offered such a simple explanation for the apparent retrograde motion of the planets. But he went beyond Aristarchus in working out mathematical details of the model, and discovered simple geometric relationships that allowed him to calculate each planet's orbital period around the Sun and its relative distance from the Sun. The model's success in providing a geometric layout for the solar system convinced him that the Sun-centered idea must be correct.

Despite his own confidence in the model, Copernicus was hesitant to publish his work, fearing that the idea of a moving Earth would be considered absurd. However, he discussed his system with other scholars, including high-ranking officials of the Church, who urged him to publish a book. Copernicus saw the first printed copy of his book, *De Revolutionibus Orbium Coelestium* ("Concerning the Revolutions of the Heavenly Spheres"), on the day he died—May 24, 1543.

Publication of the book spread the Sun-centered idea widely, and many scholars were drawn to its aesthetic advantages. Nevertheless, the Copernican model gained relatively few converts over the next 50 years, for a good reason: It didn't work all that well. The primary problem was that while Copernicus had been willing to overturn Earth's central place in the cosmos, he held fast to the belief that heavenly motion must be in perfect circles. This incorrect assumption forced him to add numerous complexities to his system (including circles on circles much like those used by Ptolemy) in order for it to make decent predictions. As a result, his complete model was no more accurate and no less complex than the Ptolemaic model, and few people were willing to throw out thousands of years of tradition for a new model that worked just as poorly as the old one.

Copernicus (1473–1543)

Tycho Brahe (1546–1601)

Tycho   Part of the difficulty faced by astronomers who sought to improve either the Ptolemaic or the Copernican system was a lack of quality data. The telescope had not yet been invented, and existing naked-eye observations were not very accurate. Better data were needed, and they were provided by the Danish nobleman Tycho Brahe (1546–1601), usually known simply as Tycho (pronounced "tie-koe").

In 1572, Tycho observed a "new star" in the sky and proved that it was much farther away than the Moon, calling into question the ancient Greek belief in unchanging heavens. (Today, we know that Tycho actually saw a *supernova*—the explosion of a distant star [Section 9.2].) In 1577, he observed a comet and proved that it too lay in the realm of the heavens. Others, including Aristotle, had argued that comets were phenomena of Earth's atmosphere. These successes led to royal sponsorship for Tycho's work, which allowed him to build a giant observatory for naked-eye observations. Over a period of three decades, Tycho and his assistants compiled naked-eye observations of planetary positions accurate to within 1 arcminute (1/60 of 1°)—less than the thickness of a fingernail viewed at arm's length.

Despite the quality of his observations, Tycho did not come up with a satisfying explanation for planetary motion. He was convinced that the *planets* must orbit the Sun, but his inability to detect stellar parallax [Section 2.3] led him to conclude, like the ancient Greeks, that Earth must remain stationary. He therefore advocated a model in which the Sun orbits Earth while all other planets orbit the Sun. Few people took this model seriously.

Johannes Kepler
(1571–1630)

Kepler—A Successful Model of Planetary Motion

Tycho failed to explain the motions of the planets satisfactorily, but he succeeded in finding someone who could: In 1600, he hired a young German astronomer, Johannes Kepler (1571–1630).

Like Copernicus, Kepler believed that planetary orbits should be perfect circles. He therefore worked diligently to match circular motions to Tycho's data. After years of effort, he found a set of circular orbits that matched most of Tycho's observations quite well. Even in the worst cases, which were for the planet Mars, Kepler's predicted positions differed from Tycho's observations by only about 8 arcminutes.

Kepler surely was tempted to ignore these discrepancies and attribute them to errors by Tycho. After all, 8 arcminutes is barely one-fourth the angular diameter of the full moon. But Kepler trusted Tycho's careful work. The small discrepancies finally led Kepler to abandon the idea of circular orbits—and to find the correct solution to the ancient riddle of planetary motion. About this event, Kepler wrote,

> If I had believed that we could ignore these eight minutes [of arc], I would have patched up my hypothesis accordingly. But, since it was not permissible to ignore, those eight minutes pointed the road to a complete reformation in astronomy.

Kepler's decision to trust the data over his preconceived beliefs marked an important transition point in the history of science. Once he abandoned perfect circles, he was free to try other ideas, and he ultimately found the correct one: Planetary orbits are not circles but instead take the shapes of the special ovals known as *ellipses*. Much as you can draw a circle by pulling a string around one point (the center), you can draw an ellipse by stretching a string around *two* points; each of these two points is called a *focus* of the ellipse. Kepler then used his knowledge of mathematics to put his new model of planetary motion on a firm footing, expressing the key features of the model with what we now call **Kepler's laws of planetary motion.**

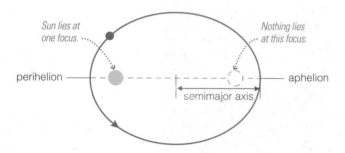

Figure 3.3 | Kepler's first law: The orbit of each planet about the Sun is an ellipse with the Sun at one focus. (The ellipse shown here is more *eccentric,* or stretched out, than any of the actual planetary orbits in our solar system.)

- **Kepler's first law:** *The orbit of each planet about the Sun is an ellipse with the Sun at one focus* (**Figure 3.3**). This law tells us that a planet's distance from the Sun varies during its orbit. Its closest point is called **perihelion** (from the Greek for "near the Sun") and its farthest point is called **aphelion** (from the Greek for "away from the Sun"). The *average* of a planet's perihelion and aphelion distances is the length of its **semimajor axis.**

- **Kepler's second law:** *As a planet moves around its orbit, it sweeps out equal areas in equal times.* As shown in **Figure 3.4**, the "sweeping" refers to an imaginary line connecting the planet to the Sun, and keeping the areas equal means that the planet moves a greater distance when it is near perihelion than it does in the same amount of time near aphelion. This, in turn, means that the planet moves faster when it is nearer to the Sun and slower when it is farther from the Sun.

- **Kepler's third law:** *More distant planets orbit the Sun at slower average speeds, obeying the precise mathematical relationship $p^2 = a^3$; p is the planet's orbital period in years and a is its average distance from the Sun in astronomical units.* The mathematical statement of Kepler's third law allows us to calculate the average orbital speed of each planet (**Figure 3.5**).

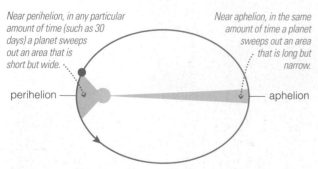

Near perihelion, in any particular amount of time (such as 30 days) a planet sweeps out an area that is short but wide.

Near aphelion, in the same amount of time a planet sweeps out an area that is long but narrow.

perihelion

aphelion

The areas swept out in 30-day periods are all equal.

Figure 3.4 | Kepler's second law: As a planet moves around its orbit, an imaginary line connecting it to the Sun sweeps out equal areas (the shaded regions) in equal times.

 Think about it   Suppose a comet has an orbit that brings it quite close to the Sun at perihelion and beyond Mars at aphelion, but with an average distance (semimajor axis) of 1 AU. How long would the comet take to complete each orbit of the Sun? Would it spend most of its time close to the Sun, far from the Sun, or somewhere in between? Explain.

Kepler published his first two laws in 1609 and his third in 1619. Together, they made a model that could predict planetary positions with far greater accuracy than the models of either Ptolemy or Copernicus. Indeed, Kepler's model works so well that we now see it as more than just an abstract model, and instead as revealing a deep underlying truth about planetary motion.

Galileo—Answering the Remaining Objections The success of Kepler's laws in matching Tycho's data provided strong evidence in favor of Copernicus's placement of the Sun, rather than Earth, at the center of the solar system. Nevertheless, many scientists still voiced reasonable objections to the Copernican view. There were three basic objections, all rooted in the 2000-year-old beliefs of Aristotle and other ancient Greeks.

1. Aristotle had held that Earth could not be moving because, if it were, objects such as birds, falling stones, and clouds would be left behind as Earth moved along.

2. The idea of noncircular orbits contradicted Aristotle's claim that the heavens—the realm of the Sun, Moon, planets, and stars—must be perfect and unchanging.

3. Despite many efforts, no one had detected the stellar parallax that should occur if Earth orbits the Sun.

Galileo Galilei (1564–1642), usually known by his first name, answered all three objections.

Galileo defused the first objection with experiments that almost single-handedly overturned the Aristotelian view of physics. In particular, he used experiments with rolling balls to demonstrate that a moving object remains in motion *unless* a force acts to stop it (an idea now codified in Newton's first law of motion). This insight explained why objects that share Earth's motion through space—such as birds, falling stones, and clouds—should *stay* with Earth rather than falling behind as Aristotle had argued. This same idea explains why passengers stay with a moving airplane even when they leave their seats.

Tycho's supernova and comet observations already had challenged the validity of the second objection by showing that the heavens could change. Galileo shattered the idea of heavenly perfection after he built a telescope in 1609. (The telescope was invented in 1608 by Hans Lippershey, but Galileo's was much more powerful.) Through his telescope, Galileo saw sunspots on the Sun, which were considered "imperfections" at the time. He also used his telescope to prove that the Moon has mountains and valleys like the "imperfect" Earth by studying the shadows cast near the dividing line between the light and dark portions of the lunar face (**Figure 3.6**). If the heavens were in fact not perfect, then the idea of elliptical orbits (as opposed to "perfect" circles) was not so objectionable.

The third objection—the absence of observable stellar parallax—had been of particular concern to Tycho. Based on his estimates of the distances of stars, Tycho believed that his naked-eye observations were sufficiently precise to detect stellar parallax if Earth did in fact orbit the Sun. Refuting Tycho's argument required showing that the stars were more distant than Tycho had thought and therefore too distant for him to have observed stellar parallax. Although Galileo didn't actually prove this fact, he provided strong evidence in its favor. For example, he saw with his telescope that the Milky Way resolved into countless individual stars. This discovery helped him argue that the stars were far more numerous and more distant than Tycho had believed.

In hindsight, the final nails in the coffin of the Earth-centered universe came with two of Galileo's earliest discoveries through the telescope. First, he observed four moons clearly orbiting Jupiter, *not* Earth. Soon thereafter, he

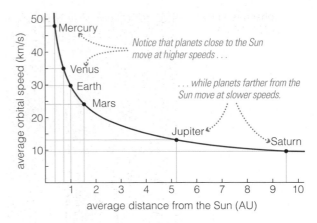

Notice that planets close to the Sun move at higher speeds . . .

. . . while planets farther from the Sun move at slower speeds.

Figure 3.5 | This graph, based on Kepler's third law ($p^2 = a^3$) and modern values of planetary distances, shows that more distant planets orbit the Sun more slowly.

Galileo (1564–1642)

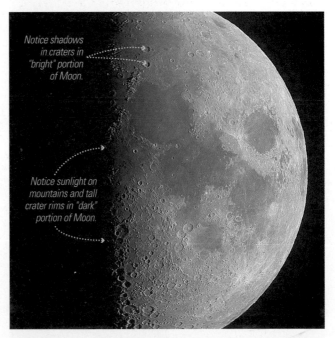

Notice shadows in craters in "bright" portion of Moon.

Notice sunlight on mountains and tall crater rims in "dark" portion of Moon.

Figure 3.6 | Shadows visible near the dividing line between the light and dark portions of the lunar face prove that the Moon's surface is not perfectly smooth.

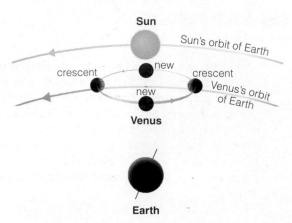

Ptolemaic View of Venus

Copernican View of Venus

a In the Ptolemaic model, Venus orbits Earth, moving around a smaller circle on its larger orbital circle; the center of the smaller circle lies on the Earth–Sun line. If this view were correct, Venus's phases would range only from new to crescent.

b In reality, Venus orbits the Sun, so from Earth we can see it in many different phases. This is just what Galileo observed, allowing him to prove that Venus orbits the Sun.

Figure 3.7 | Galileo's telescopic observations of Venus proved that it orbits the Sun rather than Earth.

observed that Venus goes through phases in a way that proved that it must orbit the Sun and not Earth (**Figure 3.7**).

Although we now recognize that Galileo was correct, the story was more complex in his own time, when Catholic Church doctrine still held Earth to be the center of the universe. On June 22, 1633, Galileo was brought before a Church inquisition in Rome and ordered to recant his claim that Earth orbits the Sun. Nearly 70 years old and fearing for his life, Galileo did as ordered. His life was spared. However, legend has it that as he rose from his knees he whispered under his breath, *Eppur si muove*—Italian for "And yet it moves."

The Church did not formally vindicate Galileo until 1992, but Church officials had given up the argument long before that: Galileo's book, *Dialogue Concerning the Two Chief World Systems,* was removed from the index of banned books in 1824. Today, Catholic scientists are at the forefront of much astronomical research, and official Church teachings are compatible not only with Earth's planetary status but also with the theories of the Big Bang and the subsequent evolution of the cosmos and of life.

Newton—Culmination of the Revolution Kepler's model worked so well and Galileo so successfully countered the remaining objections to the Sun-centered model that, by about the 1630s, scientists were nearly unanimous in accepting the validity of Kepler's laws of planetary motion. However, no one yet knew *why* the planets should move in elliptical orbits with varying speeds. The question became a topic of great debate, and a few scientists even guessed the correct answer—but they could not prove it, largely because the necessary understanding of physics and mathematics didn't exist yet. This understanding finally came through the remarkable work of Sir Isaac Newton (1642–1727), who invented the mathematics of calculus and used it to explain and discover many fundamental principles of physics.

In 1687, Newton published a famous book usually called *Principia,* short for *Philosophiae Naturalis Principia Mathematica* ("Mathematical Principles of Natural Philosophy"). In it, he laid out precise mathematical descriptions of how motion works in general, ideas that we now describe as **Newton's laws of motion.** For reference, **Figure 3.8** (on page 44) illustrates the three laws of motion. Be careful not to confuse *Newton's* three laws, which apply to all motion, with *Kepler's* three laws, which describe only the motion of planets moving about the Sun.

Sir Isaac Newton (1642–1727)

Tools of Science: Telescopes

The story of the Copernican revolution illustrates the importance of precise observations in astronomy. Without Tycho's unprecedented accuracy in naked-eye observations, Kepler could not have come up with his laws of planetary motion. And without Galileo's telescopic observations, we would not have obtained the evidence that sealed the case for the Copernican revolution. The need for ever better observations continues today, and in astronomy this means the development of ever more powerful telescopes.

We characterize the power of a telescope in terms of two fundamental properties: (1) The **light-collecting area** tells us how much total light a telescope can collect at one time. We usually characterize a telescope's "size" as the *diameter* of its light-collecting area; for example, a "10-meter telescope" has a light-collecting area that is 10 meters in diameter. (2) The **angular resolution** of a telescope is the *smallest* angle over which it allows us to tell that two dots—or two stars—are distinct. The human eye has an angular resolution of about 1 arcminute $\left(\frac{1}{60}^{\circ}\right)$ meaning that two stars can appear distinct to our eyes only if they have at least this much angular separation in the sky; if the stars are separated by less than 1 arcminute, they will appear blended together as a single point of light. Today's telescopes can have extremely precise angular resolution. For example, the Hubble Space Telescope has an angular resolution of about 0.05 arcsecond (for visible light), which would allow you to read this book from a distance of almost 1 kilometer.

Telescopes come in two basic designs. A **refracting telescope** uses transparent glass lenses to collect and focus light (**Figure 1**). A **reflecting telescope** uses a precisely curved *primary mirror* to gather light (**Figure 2**); this mirror reflects the gathered light to a *secondary mirror* that lies in front of it, which then reflects the light to a focus at a place where the eye or instruments can observe it. Most research telescopes today are reflecting telescopes. The largest are just over 10 meters in diameter, but astronomers are working on plans for telescopes with diameters as large as 30 to 50 meters.

In addition to having telescopes on the ground, astronomers now put many telescopes into space. One reason for going into space is that Earth's atmosphere can blur light; the familiar twinkling of stars is actually caused by winds and air movements in our atmosphere. In addition, telescopes in space can observe forms of light that do not reach ground, such as ultraviolet and X rays. As our telescope technology continues to improve, we can expect many new discoveries about the universe.

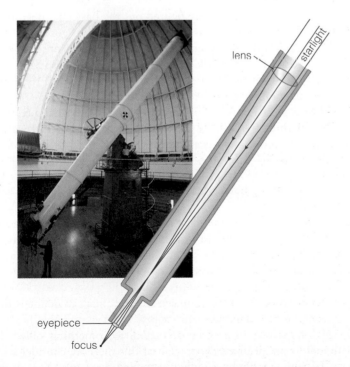

Figure 1 | A refracting telescope collects light with a large transparent lens (see diagram). The photo shows the 1-meter refractor at the University of Chicago's Yerkes Observatory, the world's largest refracting telescope.

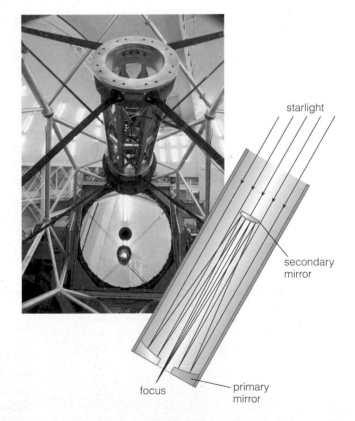

Figure 2 | A reflecting telescope collects light with a precisely curved primary mirror (see diagram). The photo shows the 8-meter Gemini North telescope, located on the summit of Mauna Kea, Hawaii.

Figure 3.8 |
Newton's three laws
of motion.

Newton's first law of motion:
An object moves at constant
velocity unless a net force acts
to change its speed or direction.

Newton's second law of motion:
Force = mass × acceleration

Newton's third law of motion:
For any force, there is always an
equal and opposite reaction force.

Example: A spaceship needs no fuel to
keep moving in space.

Example: A baseball accelerates as the pitcher applies a force by
moving his arm. (Once the ball is released, the force from the pitcher's
arm ceases, and the ball's path changes only because of the forces of
gravity and air resistance.)

Example: A rocket is propelled upward by a
force equal and opposite to the force with which
gas is expelled out its back.

Newton continued on in *Principia* to describe his universal law of
gravitation (see Section 3.3), and then used mathematics to prove that
Kepler's laws are natural consequences of the laws of motion and gravity. In
essence, Newton had created a new model for the inner workings of the
universe in which motion is governed by clear laws and the force of gravity.
Scientists have continued to build on this model ever since, giving us an
ever deeper understanding of nature.

3.2 Hallmarks of Science

The story of how our ancestors gradually figured out the basic architecture of
the cosmos exhibits many features of what we now consider "good science."
We have seen how models were formulated and tested against observations,
and modified or replaced when they failed those tests. The story also illustrates
some classic mistakes, such as the apparent failure of anyone before Kepler to
question the belief that orbits must be circles. The ultimate success of the
Copernican revolution led scientists, philosophers, and theologians to reassess
the various modes of thinking used in the 2000-year process of discovering
Earth's place in the universe. Let's examine how the principles of modern
science emerged from the lessons learned in the Copernican revolution.

How can we distinguish science from nonscience?

It's surprisingly difficult to define the term *science* precisely. The word comes
from the Latin *scientia,* meaning "knowledge," but not all knowledge is
science. For example, you may know what music you like best, but your
musical taste is not a result of scientific study.

Approaches to Science   One reason science is difficult to define is that
not all science works in the same way. For example, you've probably heard it
said that science is supposed to proceed according to something called the
"scientific method." As an idealized illustration of this method, consider what
you would do if your flashlight suddenly stopped working. You might
hypothesize that the flashlight's batteries have died. This type of tentative
explanation, or **hypothesis,** is sometimes called an *educated guess*—in this case,
it is "educated" because you already know that flashlights need batteries. Your
hypothesis allows you to make a simple prediction: If you replace the batteries
with new ones, the flashlight should work. You can test this prediction by

replacing the batteries. If the flashlight now works, you've confirmed your hypothesis. If it doesn't, you must revise or discard your hypothesis, perhaps in favor of another one that you can also test (such as that the bulb is burned out). **Figure 3.9** illustrates the basic flow of this process.

The scientific method can be a useful idealization, but real science rarely progresses in such an orderly way. Scientific progress often begins with someone going out and looking at nature in a general way, rather than by conducting a careful set of experiments. For example, Galileo wasn't looking for anything in particular when he pointed his telescope at the sky and made his startling discoveries. Furthermore, scientists are human beings, and their intuition and personal beliefs inevitably influence their work. Copernicus, for example, adopted the idea that Earth orbits the Sun not because he had carefully tested it but because he believed it made more sense than the prevailing view of an Earth-centered universe. While his intuition guided him to the right general idea, he erred in the specifics because he still held to the ancient belief that heavenly motion must be in perfect circles.

Given that the idealized scientific method is an overly simplistic characterization of science, how can we tell what is science and what is not? To answer this question, we must look a little deeper at the distinguishing characteristics of scientific thinking.

Hallmarks of Science   One way to define scientific thinking is to list the criteria that scientists use when they judge competing models of nature. Historians and philosophers of science have examined (and continue to examine) this issue in great depth, and different experts express somewhat different viewpoints on the details. Nevertheless, everything we now consider to be science shares the following three basic characteristics, which we will refer to as the "hallmarks" of science (**Figure 3.10**):

- Modern science seeks explanations for observed phenomena that rely solely on natural causes.

- Science progresses through the creation and testing of models of nature that explain the observations as simply as possible.

- A scientific model must make testable predictions about natural phenomena that would force us to revise or abandon the model if the predictions did not agree with observations.

Each of these hallmarks is evident in the story of the Copernican revolution. The first shows up in the way Tycho's careful measurements of planetary motion motivated Kepler to come up with a better explanation for those motions. The second is evident in the way several competing models were compared and tested, most notably those of Ptolemy, Copernicus, and Kepler. We see the third in the fact that each model could make precise predictions about the future motions of the Sun, Moon, planets, and stars in our sky. When a model's predictions failed, the model was modified or discarded. Kepler's model gained acceptance in large part because its predictions were so much better than those of the Ptolemaic model in matching Tycho's observations. **Figure 3.11** on pages 46–47 summarizes the Copernican revolution and how it illustrates the hallmarks of science.

Occam's Razor   The criterion of simplicity in the second hallmark deserves additional explanation. Remember that the original model of Copernicus did *not* match the data noticeably better than Ptolemy's model. If scientists had judged Copernicus's model solely on the accuracy of its predictions, they might have rejected it immediately. However, many scientists found elements of the Copernican model appealing, such as the simplicity of its explanation for apparent retrograde motion. They therefore kept the model alive until Kepler found a way to make it work.

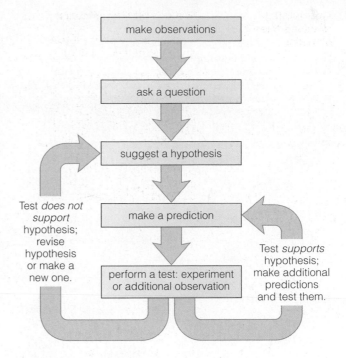

Figure 3.9 | This diagram illustrates what we often call the *scientific method.*

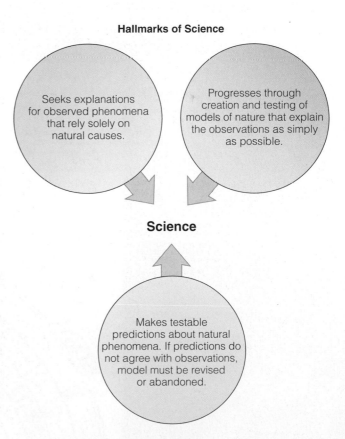

Figure 3.10 | Hallmarks of science.

Ancient Earth-centered models of the universe easily explained the simple motions of the Sun and Moon through our sky, but had difficulty explaining the more complicated motions of the planets. The quest to understand planetary motions ultimately led to a revolution in our thinking about Earth's place in the universe that illustrates the process of science. This figure summarizes the major steps in that process.

1 Night by night, planets usually move from west to east relative to the stars. However, during periods of *apparent retrograde motion,* they reverse direction for a few weeks to months. The ancient Greeks knew that any credible model of the solar system had to explain these observations.

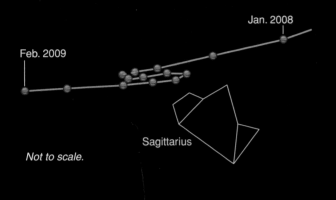

Jan. 2008

Feb. 2009

Sagittarius

Not to scale.

2 Most ancient Greek thinkers assumed that Earth remained fixed at the center of the solar system. To explain retrograde motion, they therefore added a complicated scheme of circles moving upon circles to their Earth-centered model. However, at least some Greeks, such as Aristarchus, preferred a Sun-centered model, which offered a simpler explanation for retrograde motion.

planet

retrograde loop

Earth

The Greek geocentric model explained apparent retrograde motion by having planets move around Earth on small circles that turned on larger circles.

HALLMARK OF SCIENCE **A scientific model must seek explanations for observed phenomena that rely solely on natural causes.** The ancient Greeks used geometry to explain their observations of planetary motion.

(Left page)
A schematic map of the universe from 1539 with Earth at the center and the Sun (Solis) orbiting it between Venus and Mars (Martis).

(Right page)
A page from Copernicus's De Revolutionibus, published in 1543, showing the Sun (Sol) at the center and Earth (Terra) orbiting between Venus and Mars.

(3) By the time of Copernicus (1473–1543), predictions based on the Earth-centered model had become noticeably inaccurate. Hoping for improvement, Copernicus revived the Sun-centered idea. He did not succeed in making substantially better predictions because he retained the ancient belief that planets must move in perfect circles, but he inspired a revolution continued over the next century by Tycho, Kepler, and Galileo.

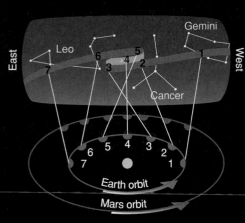

Apparent retrograde motion is simply explained in a Sun-centered system. Notice how Mars appears to change direction as Earth moves past it.

HALLMARK OF SCIENCE **Science progresses through creation and testing of models of nature that explain the observations as simply as possible.** Copernicus developed a simple Sun-centered model in hopes of explaining observations better than the more complicated Earth-centered model.

(4) Tycho exposed flaws in both the ancient Greek and Copernican models by observing planetary motions with unprecedented accuracy. His observations led to Kepler's breakthrough insight that planetary orbits are elliptical, not circular, and enabled Kepler to develop his three laws of planetary motion.

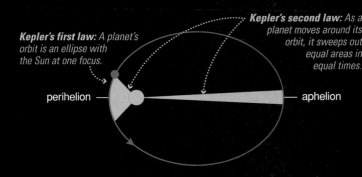

Kepler's first law: *A planet's orbit is an ellipse with the Sun at one focus.*

Kepler's second law: *As a planet moves around its orbit, it sweeps out equal areas in equal times.*

perihelion — aphelion

Kepler's third law: *More distant planets orbit at slower average speeds, obeying $p^2 = a^3$.*

HALLMARK OF SCIENCE **A scientific model makes testable predictions about natural phenomena. If predictions do not agree with observations, the model must be revised or abandoned.** Kepler could not make his model agree with observations until he abandoned the belief that planets move in perfect circles.

(5) Galileo's experiments and telescopic observations overcame remaining scientific objections to the Sun-centered model. Together, Galileo's discoveries and the success of Kepler's laws in predicting planetary motion overthrew the Earth-centered model once and for all.

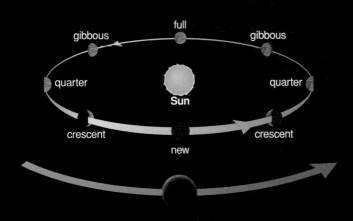

With his telescope, Galileo saw phases of Venus that are consistent only with the idea that Venus orbits the Sun rather than Earth.

In fact, if agreement with data were the sole criterion for judgment, we could imagine a modern-day Ptolemy adding millions or billions of additional circles to the geocentric model in an effort to improve its agreement with observations. A sufficiently complex geocentric model could in principle reproduce the observations with almost perfect accuracy—but it still would not convince us that Earth was the center of the universe. We would still choose the Copernican view over the geocentric view because its predictions would be just as accurate yet would follow from a much simpler model of nature. The idea that scientists should prefer the simpler of two models that agree equally well with observations is called *Occam's razor*, after the medieval scholar William of Occam (1285–1349).

Verifiable Observations   The third hallmark of science forces us to face the question of what counts as an "observation" against which a prediction can be tested. Consider the claim that aliens are visiting Earth in UFOs. Proponents of this claim say that many thousands of eyewitness observations of UFO encounters provide evidence that it is true. But should these personal testimonials count as *scientific* evidence? On the surface, the answer may not be obvious, because all scientific studies involve eyewitness accounts on some level. For example, only a handful of scientists have personally made detailed tests of Einstein's theory of relativity, and it is their personal reports of the results that have convinced other scientists of the theory's validity. However, there's an important difference between personal testimony about a scientific test and an observation of a UFO: The first is verifiable by anyone, at least in principle, while the second is not.

Understanding this difference is crucial to understanding what counts as science and what does not. Even though you may never have conducted a test of Einstein's theory of relativity yourself, there's nothing stopping you from doing so. It might require several years of study to acquire the necessary background to conduct the test, but you could eventually confirm the results reported by other scientists. In other words, while you may currently be trusting the eyewitness testimony of scientists, you always have the option of verifying their testimony for yourself.

In contrast, there is no way for you to verify someone's eyewitness account of a UFO. Moreover, scientific studies of eyewitness testimony show it to be notoriously unreliable; different eyewitnesses often disagree on what they saw even immediately after an event has occurred. As time passes, memories of the event may change further. In some cases in which memory has been checked against reality, people have reported vivid memories of events that never happened at all. This explains something that virtually all of us have experienced: disagreements with a friend about who did what and when. Since both people cannot be right in such cases, at least one person must have a memory that differs from reality.

Because of its demonstrated unreliability, eyewitness testimony alone should never be used as evidence in science, no matter who reports it or how many people offer similar testimony. It can be used in support of a scientific model only when it is backed up by independently verifiable evidence that anyone could, in principle, check.

Objectivity in Science   We generally think of science as being objective, meaning that all people should be able to find the same answers to scientific questions. However, there is a difference between the overall objectivity of science and the objectivity of individual scientists. Science is practiced by human beings, and individual scientists may bring their personal biases and beliefs to their scientific work. For example, most scientists choose their research projects based on personal interests rather than on some objective formula. In extreme cases, scientists have been known to cheat—either deliberately or subconsciously—to obtain a result they desire. In one famous case that occurred a little over a hundred years ago, astronomer Percival Lowell claimed to see a network of artificial canals in blurry telescopic images of Mars, leading him to conclude that there was a great Martian civilization.

Common Misconceptions

EGGS ON THE EQUINOX

One of the hallmarks of science holds that you needn't take scientific claims on faith. In principle, at least, you can always test them for yourself. Consider the claim, repeated in news reports every year, that the spring equinox is the only day on which you can balance an egg on its end. Many people believe this claim, but you'll be immediately skeptical if you think about the nature of the spring equinox. The equinox is merely a point in time at which sunlight strikes both hemispheres equally (see Figure 2.3). It's difficult to see how sunlight could affect an attempt to balance eggs (especially if the eggs are indoors), and neither Earth's gravity nor the Sun's gravity is different on that day than on any other day.

More important, you can test this claim directly. It's not easy to balance an egg on its end, but with practice you'll find that you can do it on any day of the year, not just on the spring equinox. Not all scientific claims are so easy to test for yourself, but the basic lesson should be clear: Before you accept any scientific claim, you should demand a reasonable explanation of the evidence that backs it up.

But no such canals actually exist; Lowell must have allowed his beliefs about extraterrestrial life to influence the way he interpreted what he saw—in essence a form of cheating, though probably not intentional.

Bias can sometimes show up even in the thinking of the scientific community as a whole. Some valid ideas may not be considered by any scientist because they fall too far outside the general patterns of thought, or **paradigm,** of the time. Einstein's theory of relativity is an example. Many scientists in the decades before Einstein had gleaned hints of the theory but did not fully investigate them, in part because they seemed too outlandish.

The beauty of science is that it encourages continued testing by many people. Even if personal biases affect some results, tests by others should eventually uncover any mistakes. Similarly, if a new idea is correct but falls outside the accepted paradigm, sufficient testing and verification of the idea will eventually force a paradigm shift. In that sense, *science ultimately provides a means of bringing people to agreement,* at least on topics that can be subjected to scientific study.

What is a scientific theory?

The most successful scientific models explain a wide variety of observations in terms of just a few general principles. When a powerful yet simple model makes predictions that survive repeated and varied testing, scientists elevate its status and call it a **theory.** Some famous examples are Isaac Newton's theory of gravity, Charles Darwin's theory of evolution, and Albert Einstein's theory of relativity.

Note that the scientific meaning of the word *theory* is quite different from its everyday meaning, in which it is often used to refer to a speculation or hypothesis. In everyday life, someone might get a new idea and say, for example, "I have a new theory about why people enjoy the beach." Without the support of a broad range of evidence that others have tested and confirmed, this "theory" is really only a guess. In contrast, Newton's theory of gravity qualifies as a scientific theory because it uses simple physical principles to explain a great many observations and experiments.

Despite its success in explaining observed phenomena, a scientific theory can never be proved true beyond all doubt, because more sophisticated observations may eventually disagree with its predictions. However, anything that qualifies as a scientific theory must be supported by a large, compelling body of evidence.

In summary, a scientific theory is not at all like a hypothesis or any other type of guess. We are free to change a hypothesis at any time, because it has not yet been carefully tested. In contrast, we can discard or replace a scientific theory only if it makes predictions that fail, and even then we can accept a replacement theory only if it is equally good at explaining the evidence that supported the original theory.

Think about it When people claim that something is "only a theory," what do you think they mean? Does this meaning of *theory* agree with the definition of a theory in science? Do scientists always use the word *theory* in its "scientific" sense? Explain.

THE PROCESS OF SCIENCE IN ACTION

3.3 The Fact and Theory of Gravity

We have discussed the distinction between a hypothesis and a theory in science, but what is the difference between a *fact* and a theory? The basic answer is fairly simple: A fact is something we consider to be demonstrably true, while a theory is a model that explains *why* a body of facts is true. However, even the most successful theories can change as we collect more facts. To further illustrate this distinction between facts and theories, let's look at how the theory of gravity has developed over time.

How does the fact of gravity differ from the theory of gravity?

Gravity is clearly a fact: Things really do fall when you drop them, and planets really do orbit the Sun. However, despite our daily experience with gravity, an adequate *theory* of gravity took a long time to develop. In ancient Greece, Aristotle imagined gravity to be an inherent property of heavy objects and claimed that heavier objects would fall to the ground faster than lighter-weight objects. Galileo later put this idea to the test in a series of experiments that supposedly included dropping weights from the Leaning Tower of Pisa. His results showed that all objects fall to the ground at the same rate, as long as air resistance is unimportant. Aristotle was therefore wrong about gravity, but Galileo's ideas about gravity still fell short of being a useful theory.

See it for yourself Find a piece of paper and a small rock. Hold both at the same height, one in each hand, and let them go at the same instant. The rock, of course, hits the ground first. Next, crumple the paper into a small ball and repeat the experiment. What happens? Explain how this experiment suggests that, without air resistance, gravity causes all falling objects to fall at the same rate.

Newton's Theory of Gravity The breakthrough in our understanding of gravity came from Isaac Newton. By his own account, he experienced a moment of inspiration in 1666 when he saw an apple fall to the ground. He suddenly realized that the gravity making the apple fall was the same force that held the Moon in orbit around Earth. With this insight, Newton eliminated the long-held distinction between the realm of the heavens and the realm of Earth. For the first time, the two realms were brought together as one *universe* governed by a single set of principles.

Newton worked hard to turn his insight into a theory of gravity, which he published in 1687 (in his book *Principia*). Newton expressed the force of gravity mathematically with his **universal law of gravitation.** Three simple statements summarize this law:

- Every mass attracts every other mass through the force called *gravity.*

- The strength of the gravitational force attracting any two objects is *directly proportional* to the product of their masses. For example, doubling the mass of *one* object doubles the force of gravity between the two objects.

- The strength of gravity between two objects decreases with the *square* of the distance between their centers. We therefore say that the gravitational force follows an **inverse square law** with distance. For example, doubling the distance between two objects weakens the force of gravity by a factor of 2^2, or 4.

These three statements tell us everything we need to know about Newton's universal law of gravitation. Mathematically, all three statements can be combined into a single equation, usually written like this:

$$F_g = G \frac{M_1 M_2}{d^2}$$

where F_g is the force of gravitational attraction, M_1 and M_2 are the masses of the two objects, and d is the distance between their centers (**Figure 3.12**). The symbol G is a constant called the **gravitational constant,** and its numerical value has been measured to be $G = 6.67 \times 10^{-11}$ m^3/(kg $\times$ s^2).

The **universal law of gravitation** tells us the strength of the gravitational attraction between the two objects.

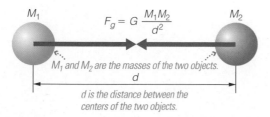

$$F_g = G \frac{M_1 M_2}{d^2}$$

M_1 and M_2 are the masses of the two objects.

d is the distance between the centers of the two objects.

Figure 3.12 | The universal law of gravitation is an *inverse square law,* which means the force of gravity declines with the *square* of the distance d between two objects.

Think about it How does the gravitational force between two objects change if the distance between them triples? If the distance between them drops by half?

Newton's theory of gravity gained rapid acceptance because it explained a great many facts that other scientists had already discovered. For example, it explained Galileo's observations about falling objects and Kepler's laws of planetary motion. Even more impressively, it quickly led to new predictive successes. Shortly after Newton published his theory, Sir Edmund Halley used it to calculate the orbit of a comet that had been seen in 1682, from which he predicted the comet's return in 1758. Halley's Comet returned on schedule, which is why it now bears his name. In 1846, after carefully examining the orbit of Uranus, the French astronomer Urbain Leverrier used Newton's theory to predict that Uranus's orbit was being affected by a previously undiscovered eighth planet. He predicted the location of the planet and sent a letter suggesting a search to Johann Galle of the Berlin Observatory. On the night of September 23, 1846, Galle discovered Neptune within 1° of the position predicted by Leverrier. It was a stunning triumph for Newton's theory.

A Problem Appears Today, we can apply Newton's theory of gravity to objects throughout the universe, including the orbits of extrasolar planets around their stars, of stars around the Milky Way Galaxy, and of galaxies in orbit of each other. There seems no reason to doubt the universality of the law. However, we also now know that Newton's law does not tell the entire story of gravity.

The first hint of a problem with Newton's theory arose not long after Leverrier's success in predicting the existence of Neptune. Astronomers discovered a slight discrepancy between the observed characteristics of the orbit of Mercury and the characteristics predicted by Newton's theory. The discrepancy was very small, and Mercury was the only planet that showed any problem, but there seemed no way to make it go away: Unless the data were wrong, which seemed highly unlikely, Newton's theory was giving a slightly incorrect prediction for the orbit of Mercury.

Einstein's Solution The solution to the problem of Mercury came in 1915, when Albert Einstein (1879–1955) published his *general theory of relativity* (see Tools of Science, p. 169). This theory predicted an orbit for Mercury that matched the observations. Not long after, astronomers put Einstein's new theory to the test during a solar eclipse, finding that it successfully predicted the precise positions of stars visible near the blocked disk of the Sun, while Newton's theory gave a prediction that was slightly off. Scientists have continued to test both Newton's and Einstein's theories ever since. In every case in which the two theories give different answers, Einstein's theory has matched the observations while Newton's theory has not. That is why, today, we consider Einstein's general theory of relativity to have supplanted Newton's theory as our "best" theory of gravity.

Does this mean that Newton's theory of gravity was "wrong"? Remember that Newton's theory successfully explains nearly all observations of gravity in the universe, and it works so well that we can use it to plot the courses of spacecraft to the planets. Moreover, in all cases in which Newton's theory works well, Einstein's theory gives essentially the same answers. The differences in the predictions between the two theories are noticeable only with extremely precise measurements or in cases where gravity is unusually strong. We therefore do not say that Newton's theory was wrong, but rather that it was only an *approximation* to a more exact theory of gravity— Einstein's general theory of relativity. Under most circumstances, the approximation is so good that we can barely tell the difference between the two theories of gravity, but in cases of strong gravity, Einstein's theory works and Newton's fails.

While Einstein's theory of gravity has so far passed every test that it has been subjected to, most scientists suspect that we'll eventually find an even more exact theory of gravity. The reason is that for the most extreme possible case of gravity—that which occurs at the infinitely small and high-density center of a black hole [Section 10.2]—Einstein's theory of relativity gives a different answer than the equally well tested theory of the very small (known as the theory of quantum mechanics). Because these two theories contradict each other in this special case, scientists know that one or both will ultimately have to be modified.

The Bottom Line We have seen that gravity can be considered both a fact and a theory. The *fact* of gravity is obvious in the observations we make of falling objects on Earth and orbiting objects in space. The *theory* of gravity is our best explanation of those observations, and we can use it to predict how gravity will act in many different situations. As we continue to make observations, our theory of gravity can change and improve, but gravity remains a fact regardless of how we revise the theory. Note that gravity is not unique in this way: Scientists make the same type of distinction in many other cases, such as when they talk about the fact of atoms being real and the atomic theory used to explain them, and when they talk about evolution having really occurred and the theory used to explain it.

Summary of Key Concepts

3.1 From Earth-Centered to Sun-Centered

How did the Greeks explain planetary motion?

The Greeks developed **models** in an attempt to explain their observations of nature. The Greek Earth-centered model reached its height with the model of Ptolemy. The **Ptolemaic model** explained apparent retrograde motion by having each planet move on a small circle whose center moves around Earth on a larger circle.

How did the Copernican revolution change our view of the universe?

The Copernican revolution replaced the ancient Earth-centered view of the universe with a new view in which Earth is just one planet going around the Sun. It also made scientists recognize the importance of precise observations, such as those made by Tycho, in testing models of nature. In this case, the model based on **Kepler's laws of planetary motion** proved much better able to explain the observations than the old Earth-centered model or Copernicus's Sun-centered model in which orbits were perfect circles.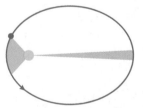

3.2 Hallmarks of Science

How can we distinguish science from nonscience?

Science generally exhibits three hallmarks: (1) Modern science seeks explanations for observed phenomena that rely solely on natural causes. (2) Science progresses through the creation and testing of models of nature that explain the observations as simply as possible. (3) A scientific model must make testable predictions about natural phenomena that would force us to revise or abandon the model if the predictions did not agree with observations.

What is a scientific theory?

A scientific **theory** is a simple yet powerful model that explains a wide variety of observations in terms of just a few general principles and has attained the status of a theory by surviving repeated and varied testing.

 ❋ THE PROCESS OF SCIENCE IN ACTION

3.3 The Fact and Theory of Gravity

How does the fact of gravity differ from the theory of gravity?

Gravity is a *fact* in that objects really do fall to the ground and planets really do orbit the Sun. The *theory* of gravity is used to explain 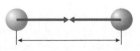 why gravity acts as it does. While the fact of gravity does not change, the theory can be improved with time, as Einstein's general theory of relativity improved on Newton's theory of gravity.

Investigations

Quick Quiz

Choose the best answer to each of the following; answers are in Appendix D. Explain your reasoning with one or more complete sentences.

1. A *geocentric* model of the universe is (a) any model that places Earth at the center of the universe. (b) any model that has Earth as a planet orbiting the Sun. (c) any model that places the Sun at the center of the universe.

2. In Ptolemy's geocentric model, the retrograde motion of a planet occurs when (a) Earth is about to pass the planet in its orbit around the Sun. (b) the planet actually goes backward in its orbit around Earth. (c) the planet is aligned with the Moon in our sky.

3. Which of the following was *not* a major advantage of Copernicus's Sun-centered model over the Ptolemaic model? (a) It made significantly better predictions of planetary positions in our sky. (b) It offered a more natural explanation for the apparent retrograde motion of planets in our sky. (c) It allowed calculation of the orbital periods and distances of the planets.

4. Tycho Brahe's contribution to astronomy included (a) inventing the telescope. (b) proving that Earth orbits the Sun. (c) collecting data that enabled Kepler to discover the laws of planetary motion.

5. Kepler made a major break from ancient beliefs when he (a) decided to look for a system that could explain actual observations. (b) realized that gravity holds the planets in their orbits. (c) abandoned circular orbits in favor of elliptical orbits.

6. Earth is closer to the Sun in January than in July. Therefore, in accord with Kepler's second law, (a) Earth travels faster in its orbit around the Sun in July than in January. (b) Earth travels faster in its orbit around the Sun in January than in July. (c) it is summer in January and winter in July.

7. According to Kepler's *third* law, (a) Mercury travels fastest in the part of its orbit in which it is closest to the Sun. (b) Jupiter orbits the Sun at a faster speed than Saturn. (c) all the planets have elliptical orbits.

8. Galileo's contribution to astronomy included (a) discovering the laws of planetary motion. (b) discovering the law of gravity. (c) making observations and conducting experiments that dispelled scientific objections to the Sun-centered model.

9. Which of the following is *not* true about scientific progress? (a) Science progresses through the creation and testing of models of nature. (b) Science advances only through strict application of the scientific method. (c) Science avoids explanations that invoke the supernatural.

10. Which of the following is *not* true about a scientific theory? (a) A theory must explain a wide range of observations or experiments. (b) Even the strongest theories can never be proved true beyond all doubt. (c) A theory is essentially an educated guess.

11. If Earth were twice as far as it actually is from the Sun, the force of gravity attracting Earth to the Sun would be (a) twice as strong. (b) half as strong. (c) one-quarter as strong.

12. When Einstein's theory of gravity (general relativity) gained acceptance, it demonstrated that Newton's theory had been (a) wrong. (b) incomplete. (c) really only a guess.

Short-Answer/Essay Questions

Explain all answers clearly, with complete sentences and proper essay structure if needed. An asterisk (*) designates a quantitative problem, for which you should show all your work.

13. State each of Kepler's laws of planetary motion and describe in your own words what it means in a way that a friend could understand.

14. What is the difference between a *hypothesis* and a *theory* in science? Explain clearly.

15. Describe each of the three hallmarks of science and give an example of how we can see each one in the unfolding of the Copernican revolution.

16. *Copernican Players.* Using a bulleted list format, make a one-page "executive summary" of the major roles that Copernicus, Tycho, Kepler, Galileo, and Newton played in overturning the ancient belief in an Earth-centered universe.

17. *Influence on History.* Based on what you have learned about the Copernican revolution, write a one- to two-page essay about how you believe it altered the course of human history.

18. *What Makes It Science?* Choose a single idea in the modern view of the cosmos as discussed in Chapter 1, such as "The universe is expanding," "The universe began with a Big Bang," "We are made from elements manufactured by stars," or "The Sun orbits the center of the Milky Way Galaxy once every 230 million years."
    a. Briefly describe how this idea is rooted in each of the three hallmarks of science.
    b. No matter how strongly the evidence may support a scientific idea, we can never be certain beyond all doubt that the idea is true. Describe an observation that might cause us to question the idea you have chosen. Overall, do you think the idea is likely or unlikely to hold up to future observations? Explain.

19. *A Flat Earth.* As an example of scientific thinking, imagine that someone today were to claim that Earth is flat. Describe one or more observations you could make for yourself that would invalidate this model and lend support to the alternative idea that Earth is round.

20. *The Universal Law of Gravitation.*
    a. How does quadrupling the distance between two objects affect the gravitational force between them?
    b. Suppose the Sun were somehow replaced by a star with twice as much mass. What would happen to the gravitational force between Earth and the Sun?
    c. Suppose Earth were moved to one-third of its current distance from the Sun. What would happen to the gravitational force between Earth and the Sun?

*21. *Eris Orbit.* The recently discovered object Eris, which is slightly larger than Pluto, orbits the Sun every 557 years. What is its average distance (semimajor axis) from the Sun? (*Hint*: Use Kepler's third law.)

*22. *Halley Orbit.* Halley's Comet orbits the Sun every 76.0 years.
    a. Find its average distance from the Sun (semimajor axis). (*Hint*: Use Kepler's third law.)
    b. Halley's orbit is a very eccentric (stretched-out) ellipse, so at perihelion Halley's Comet is only about 90 million kilometers from the Sun, compared to more than 5 billion kilometers at aphelion. Does Halley's Comet spend most of its time near its perihelion distance, near its aphelion distance, or halfway in between? Explain.

4 Origin of the Solar System

Learning Goals

4.1 Characteristics of the Solar System

- What does the solar system look like?
- What features of our solar system provide clues to how it formed?

4.2 The Birth of the Solar System

- What theory best explains the orderly patterns of motion in our solar system?
- How does our theory account for the features of planets, moons, and small bodies?

✳ THE PROCESS OF SCIENCE IN ACTION

4.3 The Age of the Solar System

- How do we determine the age of Earth and the solar system?

From space, we see our world as a small blue oasis set against a black void. Such images have inspired great works of art, music, and poetry, and have come to symbolize global awareness of the environment. Images of Earth also inspire us to ask scientific questions. How did our planet come to be, and why is it such an ideal home for life? We will examine these questions in this and the next chapter. We will see that the answers are best understood by not just studying Earth, but also comparing Earth to other worlds. We'll begin in this chapter by exploring general features of our solar system and how they help us understand the modern scientific theory of the birth of our solar system.

Characteristics of the Solar System

Our major goal in this chapter is to understand the scientific theory that best explains the origin of Earth and the solar system. To do so, we must first have a good general picture of what the solar system looks like and a basic understanding of its individual worlds. We'll then use these ideas to come up with a list of major features that any successful theory of the solar system's birth must explain.

What does the solar system look like?

Imagine viewing the solar system from beyond the orbits of the planets. What would we see? Without a telescope, the answer would be "not much." Remember that the Sun and planets are all quite small compared to the distances between them [Section 1.1]—so small that if we viewed them from the outskirts of our solar system, the planets would be only pinpoints of light, and even the Sun would be just a small bright dot in the sky. But if we magnify the sizes of the planets by about a million times compared to their distances from the Sun, and show their orbital paths, we get the central picture in **Figure 4.1** (pp. 56–57). **Table 4.1** then summarizes key data for the planets.

Even on first glance, the figure and table make it clear that our solar system is *not* a random collection of worlds. Instead, our solar system shows several clear patterns. For example, all the planets orbit the Sun in the same direction and in nearly the same plane, and the four inner planets are much smaller and closer together than the next four planets. We'll discuss these and other patterns in more detail shortly. First, let's build our general picture of the solar system by taking a brief tour of its major regions, starting with the Sun and moving outward.

 Think about it Figure 4.1 and Table 4.1 both indicate the axis tilts of the planets. Based on what you learned about the cause of Earth's seasons in Chapter 2, which planets should have seasons similar to those of Earth? Which planets should have no seasons at all? Which planet should have the most extreme seasonal differences?

The Sun   The Sun is by far the largest and brightest object in our solar system. It contains almost 99.9% of the solar system's total mass, making it nearly a thousand times as massive as everything else in the solar system combined. Its mass and brightness also make it the most influential object in our solar system. The Sun's gravity governs the orbits of the planets, and the Sun is the source of virtually all the visible light in our solar system, since the Moon and planets shine only by virtue of the sunlight they reflect. It is also the primary influence on the temperatures of planetary surfaces and atmospheres. Charged particles flowing outward from the Sun (which make up the *solar wind*) help shape planetary magnetic fields and can influence planetary atmospheres. Nevertheless, we can understand the planets without knowing further details about the Sun, so we'll save these details for our study of the Sun as a star in Chapter 8.

The Inner Planets   The four inner planets—Mercury, Venus, Earth, and Mars—are all quite small and close together compared to the outer planets. These four planets also share similar compositions of metal and rock, which is why we often refer to them as the **terrestrial planets.** (*Terrestrial* means "Earth-like.")

Despite their similarities in composition, the four terrestrial planets differ substantially in their details. Mercury is a desolate, cratered world that looks much like our own Moon. Venus, nearly identical in size to Earth, is best known for its extreme temperature and pressure: Its thick, carbon dioxide

Cosmic Context Figure 4.1: The Solar System

The solar system's layout and composition offer four major clues to how it formed. The main illustration below shows the orbits of planets in the solar system from a perspective beyond Neptune, with the planets themselves magnified by about a million times relative to their orbits.

① Large bodies in the solar system have orderly motions. All planets have nearly circular orbits going in the same direction in nearly the same plane. Most large moons orbit their planets in this same direction, which is also the direction of the Sun's rotation.

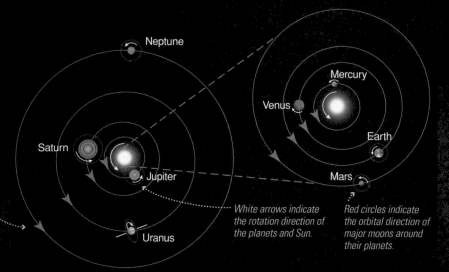

Seen from above, planetary orbits are nearly circular.

White arrows indicate the rotation direction of the planets and Sun.

Red circles indicate the orbital direction of major moons around their planets.

Each planet's axis tilt is shown, with small circling arrows to indicate the direction of the planet's rotation.

Orbits are shown to scale, but planet sizes are exaggerated about 1 million times relative to orbits. The Sun is not shown to scale.

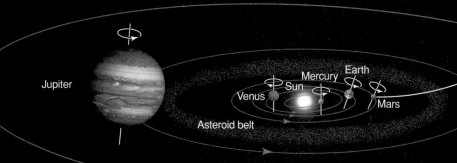

Jupiter

Venus Sun Mercury Earth Mars

Asteroid belt

Neptune

Orange arrows indicate the direction of orbital motion.

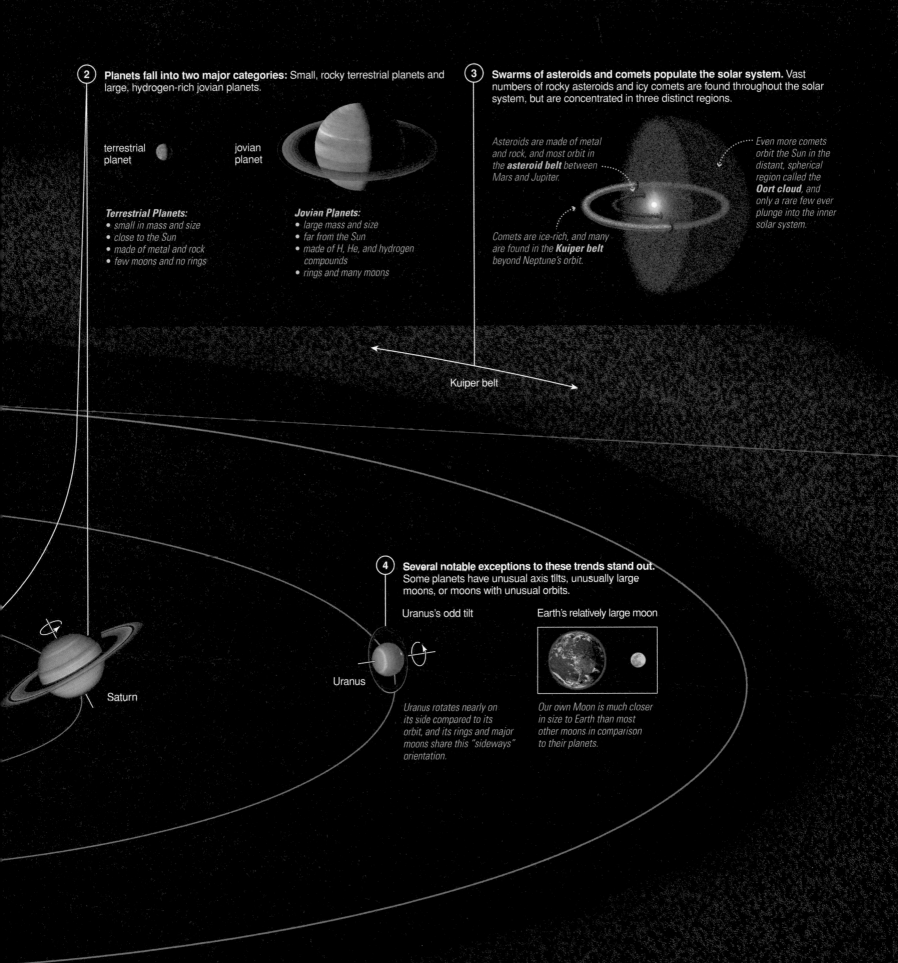

(2) Planets fall into two major categories: Small, rocky terrestrial planets and large, hydrogen-rich jovian planets.

terrestrial planet

jovian planet

Terrestrial Planets:
- small in mass and size
- close to the Sun
- made of metal and rock
- few moons and no rings

Jovian Planets:
- large mass and size
- far from the Sun
- made of H, He, and hydrogen compounds
- rings and many moons

(3) Swarms of asteroids and comets populate the solar system. Vast numbers of rocky asteroids and icy comets are found throughout the solar system, but are concentrated in three distinct regions.

*Asteroids are made of metal and rock, and most orbit in the **asteroid belt** between Mars and Jupiter.*

*Even more comets orbit the Sun in the distant, spherical region called the **Oort cloud**, and only a rare few ever plunge into the inner solar system.*

*Comets are ice-rich, and many are found in the **Kuiper belt** beyond Neptune's orbit.*

Kuiper belt

(4) Several notable exceptions to these trends stand out. Some planets have unusual axis tilts, unusually large moons, or moons with unusual orbits.

Uranus's odd tilt

Earth's relatively large moon

Saturn

Uranus

Uranus rotates nearly on its side compared to its orbit, and its rings and major moons share this "sideways" orientation.

Our own Moon is much closer in size to Earth than most other moons in comparison to their planets.

Table 4.1   Planetary Data*

Photo	Planet	Relative Size	Average Distance from Sun (AU)	Average Equatorial Radius (km)	Mass (Earth = 1)	Average Density (g/m³)	Orbital Period	Rotation Period	Axis Tilt	Average Surface (or Cloud-Top) Temperature[†]	Composition	Known Moons (2008)	Rings?
	Mercury	·	0.387	2440	0.055	5.43	87.9 days	58.6 days	0.0°	700 K (day) 100 K (night)	Rocks, metals	0	No
	Venus	•	0.723	6051	0.82	5.24	225 days	243 days	177.3°	740 K	Rocks, metals	0	No
	Earth	•	1.00	6378	1.00	5.52	1.00 year	23.93 hours	23.5°	290 K	Rocks, metals	1	No
	Mars	·	1.52	3397	0.11	3.93	1.88 years	24.6 hours	25.2°	220 K	Rocks, metals	2	No
	Jupiter	⬤	5.20	71,492	318	1.33	11.9 years	9.93 hours	3.1°	125 K	H, He, hydrogen compounds[§]	63	Yes
	Saturn	⬤	9.54	60,268	95.2	0.70	29.4 years	10.6 hours	26.7°	95 K	H, He, hydrogen compounds[§]	60	Yes
	Uranus	•	19.2	25,559	14.5	1.32	83.8 years	17.2 hours	97.9°	60 K	H, He, hydrogen compounds[§]	27	Yes
	Neptune	•	30.1	24,764	17.1	1.64	165 years	16.1 hours	29.6°	60 K	H, He, hydrogen compounds[§]	13	Yes
	Pluto		39.5	1160	0.0022	2.0	248 years	6.39 days	112.5°	40 K	Ices, rock	3	No
	Eris		67.7	1200	0.0028	2.3	557 years	?	?	?	Ices, rock	1	?

*Including the dwarf planets Pluto and Eris.
[†]Surface temperatures for all objects except Jupiter, Saturn, Uranus, and Neptune, for which cloud-top temperatures are listed.
[§]Includes water (H_2O), methane (CH_4), and ammonia (NH_3).

Venus

Earth

Mercury

Mars

a Heavily cratered Mercury also has smooth volcanic plains and "wrinkles" that are steep cliffs.

b Artist's rendering of Venus's surface as it might appear to our eyes.

c A portion of Earth's surface as it appears without clouds.

d Mars has features that look like dry riverbeds; you can also see impact craters.

Figure 4.2 | The terrestrial planets, shown to scale, along with views highlighting key features.

atmosphere bakes Venus's surface to an incredible 470°C (about 880°F; see Appendix C.6 for a review of temperature scales) while creating a surface pressure equivalent to what exists nearly a kilometer (0.6 mile) beneath the ocean's surface on Earth. Our home planet, Earth, is the only world in our solar system with surface oceans of liquid water. Mars, though it lacks any surface liquid water today, shows clear evidence of having had flowing water in its distant past, which is why scientists are so interested in the search for past or present life on Mars. **Figure 4.2** shows the four terrestrial planets to scale, along with views highlighting some of their features.

The Asteroid Belt   If you look closely at Figure 4.1, you'll see a donut-shaped region of dots between the orbits of Mars and Jupiter. These dots represent some of the more than 400,000 known asteroids that make up the **asteroid belt.** An **asteroid** is essentially a chunk of metal and rock that orbits the Sun much like a planet, but it is much smaller in size (**Figure 4.3**). Despite their large numbers, the total mass of all the asteroids combined is much smaller than the mass of our Moon. Moreover, despite what you may have seen in movies, asteroids are spread out over such a large region of space that there would be little danger of hitting one if you flew a spaceship through the asteroid belt.

The Outer Planets   Beyond the asteroid belt we encounter the realm of the four outer planets: Jupiter, Saturn, Uranus, and Neptune. These planets are much larger and much farther apart than the terrestrial planets. They are also very different in composition from the terrestrial planets, because they contain vast amounts of materials that would be gaseous on Earth, including hydrogen, helium, and *hydrogen compounds* such as water (H_2O), methane (CH_4), and ammonia (NH_3). Their gaseous compositions mean they have no solid surfaces; instead, if you entered any of their atmospheres, you would simply plunge deeper and deeper into the interior until you were crushed by the growing gas pressure. Because Jupiter is the largest of this group of planets, we refer to them jointly as **jovian planets.** (*Jovian* means "Jupiter-like.")

The jovian planets also differ from the terrestrial planets in another important way: While the terrestrial planets have only three moons among

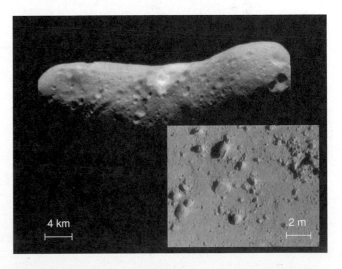

4 km

2 m

Figure 4.3 | The asteroid Eros, photographed from the *NEAR* spacecraft, is probably typical of small asteroids in appearance. The inset shows its surface, on which *NEAR* landed at the end of its mission.

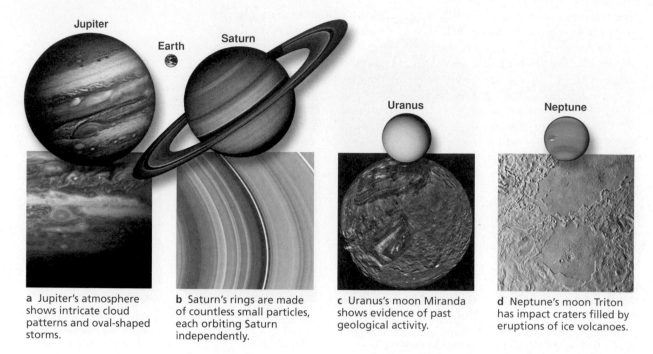

a Jupiter's atmosphere shows intricate cloud patterns and oval-shaped storms.

b Saturn's rings are made of countless small particles, each orbiting Saturn independently.

c Uranus's moon Miranda shows evidence of past geological activity.

d Neptune's moon Triton has impact craters filled by eruptions of ice volcanoes.

Figure 4.4 | The jovian planets shown to scale, with Earth for comparison, along with selected features of the planets, rings, or moons.

them (one for Earth and two very small moons for Mars), the jovian planets each have many moons. Many of these moons are amazing worlds in their own right. For example, Jupiter's moon Io is the most volcanically active world in our solar system, and its moon Europa is thought to have a deep subsurface ocean of liquid water. Saturn's moon Titan is the only moon in the solar system with a thick atmosphere, and its surface contains cold lakes of liquid ethane or methane. In addition to moons, all four jovian planets are orbited by vast numbers of small particles that make up their **rings,** although only Saturn's rings are easily visible from Earth. **Figure 4.4** shows the four jovian planets to scale, along with representative views including rings and moons.

Realm of the Comets People throughout history have been intrigued and inspired by the occasional appearance of a **comet** in our skies (**Figure 4.5**). Comets are made of rock and ice, so when they enter the inner solar system they can grow spectacular *tails* as ice vaporizes and escapes into space.

The comets that we see in the sky must come from somewhere, and study of their orbital paths has led scientists to conclude that they come from two

Figure 4.5 | Comet McNaught and the Milky Way over Patagonia, Argentina, in 2007. (The fuzzy patches above the comet tail are the Small and Large Magellanic Clouds, which are satellite galaxies of the Milky Way.)

vast reservoirs of comets that reside at great distances from the Sun. The first reservoir, known as the **Kuiper belt** (*Kuiper* rhymes with *piper*), is the donut-shaped region represented by the dots beyond the orbit of Neptune in Figure 4.1. The Kuiper belt is thought to contain more than 100,000 comets. Some of these "comets" are large enough for their own gravity to make them round in shape, which qualifies them as *dwarf planets;* as we discussed in Section 1.3, Pluto and Eris are examples of dwarf planets in the Kuiper belt.

The second reservoir of comets is called the **Oort cloud** (*Oort* rhymes with *court*). It is roughly spherical in shape, and its outer regions may extend to nearly one-quarter of the distance to the nearest stars. The Oort cloud is so far from the Sun that our telescopes are not yet capable of seeing the comets that reside within it, but study of comets that come from this region tells us that it may contain as many as a trillion comets.

What features of our solar system provide clues to how it formed?

Having completed our brief tour of the solar system, we now see our solar system as a family of worlds, with a variety of "family traits" that must be explained by any theory of our solar system's formation. More specifically, we can identify four major features of the solar system's family of worlds, each of which is associated with one of the numbered steps in Figure 4.1.

Feature 1: Patterns of Motion   Figure 4.1 shows several clear patterns of motion among the large bodies of our solar system. For example:

- All planetary orbits are nearly circular and lie nearly in the same plane.

- All planets orbit the Sun in the same direction: counterclockwise as viewed from high above Earth's North Pole.

- Most planets rotate in the same direction in which they orbit (counterclockwise as viewed from above the North Pole), with fairly small axis tilts. The Sun also rotates in this same direction.

- Most of the solar system's large moons exhibit similar properties in their orbits around their planets, such as orbiting in their planet's equatorial plane in the same direction that the planet rotates.

Together, these orderly patterns represent the first major feature of our solar system. As we'll see shortly, our theory of solar system formation explains these patterns as consequences of processes that occurred during the early stages in the birth of our solar system.

Feature 2: Two Types of Planets   Our brief planetary tour showed that the planets come in two major types: the relatively small and rocky *terrestrial planets* and the larger and more gaseous *jovian planets.* **Table 4.2** contrasts the general traits of these two types of planets.

Why do the planets come in these two distinct types? As we will see, the answer has to do with temperatures in the solar system at the time during which the planets formed: The rocky terrestrial planets formed in the hotter inner regions of the solar system, while the larger jovian planets—which tend to have icy moons—formed in the colder outer regions.

Feature 3: Asteroids and Comets   We have found that while the Sun and planets are the most massive objects in the solar system, they are far outnumbered by smaller bodies: rocky *asteroids* and icy *comets.* Moreover, rather than being randomly spread throughout the solar system, these small bodies are found mainly in three regions: Most asteroids orbit the Sun within the *asteroid*

Table 4.2   Comparison of Terrestrial and Jovian Planets

Terrestrial Planets	Jovian Planets
Smaller size and mass	Larger size and mass
Higher density	Lower density
Made mostly of rock and metal	Made mostly of hydrogen, helium, and hydrogen compounds
Solid surface	No solid surface
Few (if any) moons and no rings	Rings and many moons
Closer to the Sun (and closer together), with warmer surfaces	Farther from the Sun (and farther apart), with cool temperatures at cloud tops

Tools of Science: Conservation Laws

A candy bar looks quite different if you smash it into little pieces, but the total amount of candy remains the same no matter how it breaks apart. In science, we would say that the total amount of candy is *conserved,* and we could say that the process of smashing the candy bar obeys a "law of conservation of candy." While candy conservation is somewhat obvious, scientists have discovered other conservation laws that reveal deep secrets of nature. Here we'll discuss two that are especially important in astronomy.

The **law of conservation of energy** tells us that energy cannot appear out of nowhere or disappear into nothingness. Objects can gain or lose energy only by exchanging energy with other objects. Energy can change forms, however, so we must know some of the different forms that energy can take (**Figure 1**):

- **Kinetic energy** is energy of motion. Falling rocks, orbiting planets, and cyclists in motion all have kinetic energy in an amount that depends on the object's mass and speed. Note that individual atoms and molecules are always in motion and therefore also have kinetic energy; the collective kinetic energy of many individual particles in a substance like a rock or in the air is called **thermal energy.**

- **Potential energy** is stored energy that might later be converted into some other form. For example, a rock perched on a ledge has **gravitational potential energy** because it will fall if it slips off the edge, and food contains **chemical potential energy** that can be converted into the kinetic energy of the moving cyclist. Einstein discovered that mass itself is a form of potential energy, often

called **mass-energy:** the amount of mass-energy in an object of mass m is given by Einstein's famous equation $E = mc^2$.

- **Radiative energy** is energy carried by light. (The word *radiation* is often used as a synonym for *light*.) All light carries energy, which is why light can affect molecules in our eyes—thereby allowing us to see—or warm the surface of a planet.

A couple of examples should help clarify the idea of energy conservation. Consider a rock that falls off a ledge. Before it falls, it has gravitational potential energy by virtue of its height, but no kinetic energy because it is not moving. As it falls, its gravitational potential energy gradually turns into kinetic energy, so the rock accelerates as it falls. Next consider the Sun, which sends radiative energy into space as it shines. This radiative energy is generated by nuclear fusion in the Sun's core, which releases some of the mass-energy stored in the Sun's hydrogen.

A second important conservation law for astronomy is the **law of conservation of angular momentum.** To understand this law, think about a spinning ice skater (**Figure 2**). We say that she has angular momentum due to the fact that she is spinning, and the amount of angular momentum she has is described by the formula $m \times v \times r$, where m is her mass, v is her velocity as she spins, and r is her "radius" represented (approximately) by how far she extends her arms as she spins. Because there is little friction on the ice to slow her down, the law of conservation of angular momentum tells us that her angular momentum must stay the same no matter what she does with her arms. If she pulls her arms in, thereby decreasing her radius, her velocity of spin must increase to keep the product $m \times v \times r$ constant. For an astronomical example, consider a contracting cloud of interstellar gas: Because its radius decreases as it shrinks in size, its rate of spin must increase in order to keep its angular momentum unchanged.

Energy can be converted from one form to another.

kinetic energy
(including thermal
energy)

radiative energy
(energy of light)

potential energy
(chemical potential energy,
gravitational potential
energy, mass-energy, ...)

Figure 1 | The law of conservation of energy means that can be converted from one form to another, but cannot be created or destroyed.

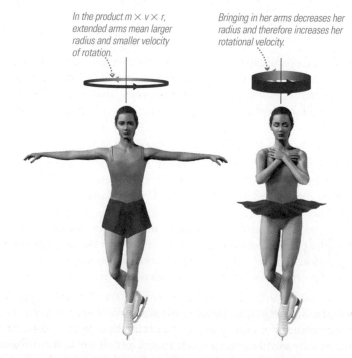

In the product m × v × r, extended arms mean larger radius and smaller velocity of rotation.

Bringing in her arms decreases her radius and therefore increases her rotational velocity.

Figure 2 | A spinning skater conserves angular momentum.

belt between Mars and Jupiter, while comets are found both in the *Kuiper belt* just beyond the orbit of Neptune and in the more distant, spherical *Oort cloud.* Our theory of solar system formation must account both for the vast numbers of small bodies and for their distribution in these three major regions.

Feature 4: Exceptions to the Rules   The fourth key feature of our solar system is that there are a few notable exceptions to the general rules. For example, while most of the planets rotate in the same direction as they orbit, Uranus rotates nearly on its side and Venus rotates "backward" (opposite the direction of planetary orbits). Similarly, while most large moons orbit in their planet's equatorial plane in the same direction as their planet rotates, many small moons have inclined or backward orbits.

One of the most interesting exceptions is our own Moon. While the other terrestrial planets have either no moons (Mercury and Venus) or very tiny moons (Mars has two small moons), Earth has one of the largest moons in the solar system. Our formation theory must be able to account for such exceptions, even while it explains the general rules.

4.2 The Birth of the Solar System

The four major features of our solar system are difficult to attribute to coincidence, and in science we assume that these features have a natural explanation. In this section, we'll discuss how our modern theory of the birth of our solar system successfully explains all the major features.

What theory best explains the orderly patterns of motion in our solar system?

As we discussed briefly in Chapter 1 (see Figure 1.9), scientists now believe that star systems are born through the gravitational collapse of huge clouds of gas in space. This idea was first proposed in 1755 by the German philosopher Immanuel Kant, but it gained acceptance only much more recently, after detailed models based on this idea proved successful at explaining the major features of the solar system. The success of such models elevated the idea to the status of a scientific *theory* [Section 3.2]. We now call it the **nebular theory** of solar system formation, because an interstellar cloud is usually called a *nebula* (Latin for "cloud").

The particular cloud that gave birth to our own solar system about $4\frac{1}{2}$ billion years ago is usually called the **solar nebula.** Recall from Chapter 1 that the universe had already been around for more than 9 billion years by that time, so generations of stars had converted about 2% (by mass) of the original hydrogen and helium from the Big Bang into heavier elements. That is, the solar nebula was made up of about 98% hydrogen and helium, and only 2% everything else combined. The rocky terrestrial planets are therefore made from material that represented only a tiny fraction of the solar nebula.

Think about it   Recall that the universe was born with only two chemical elements in it: hydrogen and helium. Could a solar system like ours have formed with the first generation of stars after the Big Bang? Explain.

The solar nebula probably began as a large and roughly spherical cloud of very cold, low-density gas. Initially, this gas was probably so spread out—perhaps over a region a few light-years in diameter—that gravity alone may not have been able to pull it together to start its collapse. Instead, the collapse may have been triggered by a cataclysmic event, such as the impact of a shock wave from the explosion of a nearby star (a *supernova*).

Once the collapse started, the law of gravity ensured that it would continue. Remember that the strength of gravity follows an inverse square law with distance [Section 3.3]. Because the mass of the cloud remained the same as it shrank, the strength of gravity increased as the diameter of the cloud decreased. For example, when the diameter decreased by half, the force of gravity increased by a factor of four.

Gravity pulls inward in all directions, so you might at first guess that the solar nebula would have remained spherical as it shrank. Indeed, the idea that gravity pulls in all directions explains why the Sun and the planets are spherical. However, gravity is not the only physical law that affects the collapse of a cloud of gas. As the solar nebula shrank in size, three important processes altered its density, temperature, and shape, changing it from a large spread-out cloud to a much smaller spinning disk (**Figure 4.6**):

- *Heating.* The temperature of the solar nebula increased as it collapsed. Such heating represents energy conservation in action (see Tools of Science, p. 62). As the cloud shrank, its gravitational potential energy was converted to the kinetic energy of individual gas particles falling inward. These particles crashed into one another, converting the kinetic energy of their inward fall to the random motions of thermal energy. The Sun formed in the center of the cloud, where temperatures and densities were highest.

- *Spinning.* Like an ice skater pulling in her arms as she spins, the solar nebula rotated faster and faster as it shrank in radius. This increase in rotation rate represents conservation of angular momentum in action (see Tools of Science, p. 62). The rotation of the cloud may have been imperceptibly slow before its collapse began, but over time the cloud's shrinkage made fast rotation inevitable. The rapid rotation also ensured that material in the cloud remained spread out rather than all collapsing into the center.

- *Flattening.* The solar nebula flattened into a disk. This flattening is a natural consequence of collisions between particles in a spinning cloud. A cloud may start with any size or shape, and different clumps of gas within the cloud may be moving in random directions at random speeds. These clumps collide and merge as the cloud collapses, and each new clump has the average velocity of the clumps that formed it. In this way, the random motions of the original cloud become more orderly as the cloud collapses, changing the cloud from its original lumpy shape into a rotating, flattened disk with nearly circular orbits. Note that this process conserves angular momentum, as it must: The rotating disk still has the same total angular momentum as the original cloud.

We can now see that the nebular theory successfully explains our first major feature of the solar system, which is its orderly motions. The planets all orbit the Sun in nearly the same plane because they formed in a flat disk. The direction in which the disk was spinning became the direction of the Sun's

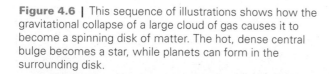

Figure 4.6 | This sequence of illustrations shows how the gravitational collapse of a large cloud of gas causes it to become a spinning disk of matter. The hot, dense central bulge becomes a star, while planets can form in the surrounding disk.

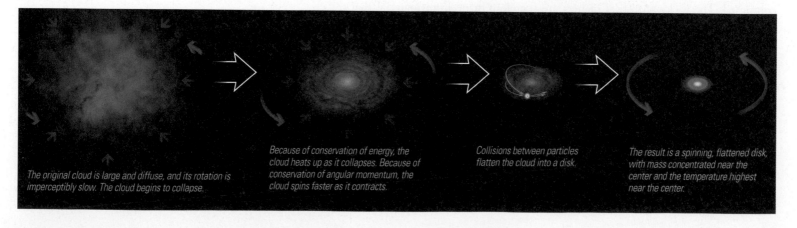

The original cloud is large and diffuse, and its rotation is imperceptibly slow. The cloud begins to collapse.

Because of conservation of energy, the cloud heats up as it collapses. Because of conservation of angular momentum, the cloud spins faster as it contracts.

Collisions between particles flatten the cloud into a disk.

The result is a spinning, flattened disk, with mass concentrated near the center and the temperature highest near the center.

rotation and the orbits of the planets. Computer models show that planets would have tended to rotate in this same direction as they formed—which is why most planets rotate the same way today—though the small sizes of planets compared to the entire disk allowed some exceptions to arise. The fact that collisions in the disk tended to make orbits more circular explains why most planets have nearly circular orbits.

See it for yourself You can demonstrate the development of orderly motion, much as it occurred in the solar system, by sprinkling pepper into a bowl of water and stirring it quickly in random directions. The water molecules constantly collide with one another, so the motions of the pepper grains tend to settle down into a slow rotation representing the average of the original random velocities. Try the experiment several times, stirring the water differently each time. Do the random motions ever cancel out exactly, resulting in no rotation at all? Describe what happens and how this demonstration is similar to what took place in the solar nebula.

How does our theory account for the features of planets, moons, and small bodies?

The nebular theory also accounts for the other three major features of the solar system. To see how, we must investigate what happened after the solar nebula took the shape of a spinning, flattened disk.

In the center of the disk, gravity drew together enough material to form the Sun. In the surrounding disk, however, the gaseous material was too spread out for gravity alone to clump it up. Instead, material had to begin clumping in some other way and to grow in size until gravity could start pulling it together into planets. In essence, planet formation required the presence of "seeds"—solid bits of matter around which gravity could ultimately build planets.

The basic process of seed formation was probably much like that of the formation of snowflakes in clouds on Earth: When the temperature is low enough, some atoms or molecules in a gas may bond together and solidify. The general process in which solid (or liquid) particles form in a gas is called **condensation**—we say that the particles *condense* out of the gas. Different materials condense at different temperatures. As summarized in **Table 4.3**, the ingredients of the solar nebula fell into four major categories:

- *Hydrogen and helium gas (98% of the solar nebula).* These gases never condense under the conditions present in a nebula.

- *Hydrogen compounds (1.4% of the solar nebula).* Materials such as water (H_2O), methane (CH_4), and ammonia (NH_3) can solidify into **ices** at low temperatures (below about 150 K under the low pressure of the solar nebula).

- *Rock (0.4% of the solar nebula).* Rocky material is gaseous at high temperatures, but condenses into solid form at temperatures between about 500 K and 1300 K, depending on the type of rock.

- *Metal (0.2% of the solar nebula).* Metals such as iron, nickel, and aluminum are also gaseous at very high temperatures, but condense into solid form at higher temperatures than rock—typically in the range of 1000 K to 1600 K.

Because hydrogen and helium gas made up 98% of the solar nebula's mass and did not condense, the vast majority of the nebula remained gaseous. However, other materials could condense wherever the temperature allowed (**Figure 4.7**). Close to the forming Sun, where the temperature was above 1600 K, it was too hot for any material to condense. Near what is now Mercury's orbit, the temperature was low enough for metals and some types of rock to condense into tiny solid particles, but it was far too hot for hydrogen compounds to condense into ices. Ices

Table 4.3   Materials in the Solar Nebula

A summary of the four types of materials present in the solar nebula. The squares represent the relative proportion of each type (by mass).

	Examples	Typical condensation temperature	Relative abundance (by mass)
Hydrogen and Helium Gas	hydrogen, helium	do not condense in nebula	98%
Hydrogen Compounds	water (H_2O) methane (CH_4) ammonia (NH_3)	<150 K	1.4%
Rock	various minerals	500–1300 K	0.4%
Metals	iron, nickel, aluminum	1000–1600 K	0.2%

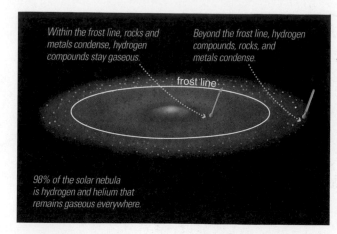

Within the frost line, rocks and metals condense, hydrogen compounds stay gaseous.

Beyond the frost line, hydrogen compounds, rocks, and metals condense.

frost line

98% of the solar nebula is hydrogen and helium that remains gaseous everywhere.

Figure 4.7 | Temperature differences in the solar nebula led to different kinds of condensed materials at different distances from the Sun.

could form only beyond the **frost line,** which lay between the present-day orbits of Mars and Jupiter. The frost line therefore marked the key transition between the warm inner region of the solar system where terrestrial planets formed and the cool outer region where jovian planets formed.

> **Think about it** Consider a region of the solar nebula in which the temperature was about 1300 K. Based on the data in Table 4.3, what fraction of the material in this region was gaseous? What were the solid particles in this region made of? Answer the same questions for a region with a temperature of 100 K. Would the 100 K region be closer to or farther from the Sun than the 1300 K region? Explain.

The first particles to condense were microscopic in size and orbited the Sun with the same orderly, circular paths as the gas from which they condensed. Individual particles therefore moved at nearly the same speed as neighboring particles, so "collisions" were more like gentle touches. Under these circumstances, particles could stick together through electrostatic forces—the same static electricity that makes hair stick to a comb. Small particles thereby began to combine into larger ones. As the particles grew in mass, gravity began to aid the process of their sticking together, accelerating their growth. The general process by which particles stick together and grow larger is called **accretion.** We refer to particles that grew to the size of boulders or larger as **planetesimals,** which means "pieces of planets."

Explaining the Two Types of Planets We can use the processes of condensation and accretion to explain why the solar system ended up with two types of planets. In the inner solar system, where only metal and rock could condense into solid particles, the planetesimals ended up being made of metal and rock. These planetesimals grew rapidly at first, with some probably reaching hundreds of kilometers in size in only a few million years—a long time in human terms, but only about 1/1000 the present age of the solar system.

Further growth became more difficult once the planetesimals reached these relatively large sizes. Gravitational encounters between planetesimals tended to alter their orbits, particularly those of the smaller planetesimals. With different orbits crossing each other, collisions between planetesimals occurred at higher speeds and hence became more destructive. Such collisions produced fragmentation more often than accretion. Only the largest planetesimals avoided being shattered and grew into full-fledged planets. **Figure 4.8** summarizes how the terrestrial planets were built from the solid bits of metal and rock that condensed in the inner solar system.

The planet formation process began similarly in the outer solar system, except the lower temperatures beyond the frost line meant that ices condensed along with metal and rock. Because ices were more abundant than rock and metal (see Table 4.3), the icy planetesimals in the outer solar system grew to larger sizes than the rocky planetesimals of the inner solar system. According to the leading model of jovian planet formation, some of the icy planetesimals grew to masses many times that of Earth. With these large masses, their gravity became strong enough to capture and hold some of the hydrogen and helium gas that made up the vast majority of the surrounding solar nebula. As the growing planets accumulated gas, their gravity grew stronger still, allowing them to capture even more gas. Ultimately, the jovian planets grew so much that they bore little resemblance to the icy seeds from which they started.

This model also explains most of the large moons of the jovian planets. The same processes of heating, spinning, and flattening that made the disk of the solar nebula should also have affected the gas drawn by gravity to the young jovian planets. Each jovian planet came to be

Common Misconceptions

SOLAR GRAVITY AND THE DENSITY OF PLANETS

Some people guess that it was the Sun's gravity that pulled the dense rocky and metallic materials to the inner part of the solar nebula, or that gases escaped from the inner nebula because gravity couldn't hold them. But this is not the case—all the ingredients were orbiting the Sun together under the influence of the Sun's gravity. The orbit of a particle or a planet does not depend on its size or density, so the Sun's gravity cannot be the cause of the different types of planets. Rather, the different temperatures in the solar nebula explain why we have terrestrial and jovian planets.

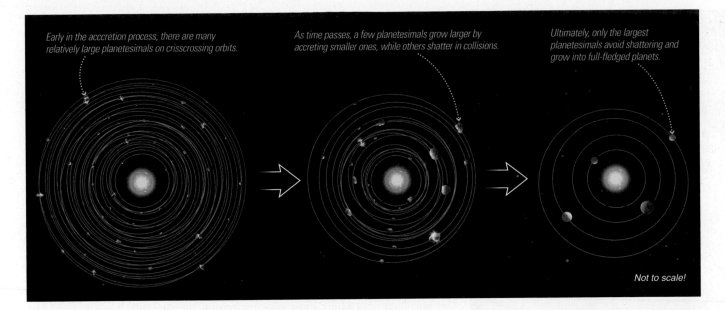

Early in the acccretion process, there are many relatively large planetesimals on crisscrossing orbits.

As time passes, a few planetesimals grow larger by accreting smaller ones, while others shatter in collisions.

Ultimately, only the largest planetesimals avoid shattering and grow into full-fledged planets.

Not to scale!

surrounded by its own disk of gas, spinning in the same direction as the planet rotated (**Figure 4.9**). Moons that accreted from icy planetesimals within these disks therefore ended up with nearly circular orbits going in the same direction as their planet's rotation and lying close to their planet's equatorial plane.

One key question remains for us to answer: Given that the vast majority of the hydrogen and helium gas in the solar nebula never became part of any planet, what happened to it? Models and observations of other star systems suggest that it was cleared away by a combination of energetic light from the young Sun and the *solar wind*—a stream of charged particles continually blown outward in all directions from the Sun [Section 8.1]. The compositions of the planets have remained nearly the same since this gas cleared. If the gas had remained longer, it might have continued to cool until hydrogen compounds condensed into ices even in the inner solar system. In that case, the terrestrial planets might have accreted abundant ice, and perhaps some hydrogen and helium gas as well, changing their basic nature. At the other extreme, if the gas had been blown out much earlier, the raw materials of the planets might have been swept away before the planets could fully form. Although these extreme scenarios did not occur in our solar system, they may sometimes occur around other stars.

Explaining Asteroids and Comets You can probably already see how the nebular theory accounts for the existence of so many asteroids and comets: They are simply "leftover" planetesimals from the era of planet formation. Asteroids are the leftover rocky planetesimals of the inner solar system, while comets are the leftover icy planetesimals of the outer solar system.

The asteroids and comets that exist today probably represent only a small fraction of the leftover planetesimals that roamed the young solar system. Most of the rest must have collided with planets or moons. On worlds with solid surfaces, we see the evidence of past collisions as *impact craters*. Careful study of impact craters on the Moon shows that the vast majority of these collisions occurred in the first few hundred million years of our solar system's history, during the period we call the **heavy bombardment.** Every world in our solar system must have been pelted by impacts during the heavy bombardment (**Figure 4.10**), and most of the craters we see on the Moon and other worlds date from this period.

Figure 4.8 | These diagrams show how planetesimals made of metal and rock gradually accreted to make the terrestrial planets.

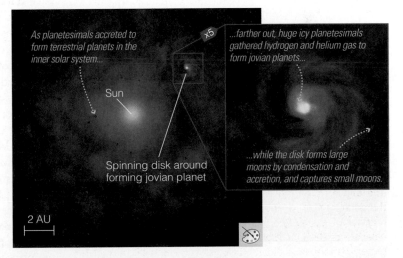

As planetesimals accreted to form terrestrial planets in the inner solar system...

Sun

Spinning disk around forming jovian planet

...farther out, huge icy planetesimals gathered hydrogen and helium gas to form jovian planets...

...while the disk forms large moons by condensation and accretion, and captures small moons.

2 AU

Figure 4.9 | The jovian planets grew as large icy planetesimals captured hydrogen and helium gas from the solar nebula and became surrounded by spinning disks of gas as they formed, much like the disk of the entire solar nebula but smaller in size. This painting shows the gas and planetesimals surrounding one jovian planet in the larger solar nebula.

Figure 4.10 | Around 4 billion years ago, Earth, its Moon, and the other planets were heavily bombarded by leftover planetesimals. This painting shows the young Earth and Moon, with an impact in progress on Earth.

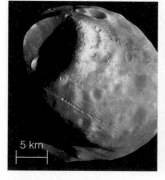
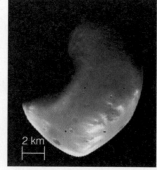

a Phobos   **b** Deimos

Figure 4.11 | The two moons of Mars are probably captured asteroids. Phobos is only about 13 kilometers across and Deimos is only about 8 kilometers across—making each of these two moons small enough to fit within the boundaries of a typical large city.

The impacts of the heavy bombardment did more than just batter the planets. They also brought materials from other regions of the solar system—a fact that is critical to our existence on Earth today. Recall that the terrestrial planets were built from planetesimals made of metal and rock. These planetesimals probably contained little or no water or other hydrogen compounds, because it was too hot for these compounds to condense in our region of the solar nebula. How, then, did Earth come to have the water that makes up our oceans and the gases that first formed our atmosphere? The likely answer is that water, along with other hydrogen compounds, was brought to Earth and other terrestrial planets during their formation by the impacts of water-bearing planetesimals that formed farther from the Sun. Remarkably, the water we drink and the air we breathe probably once were part of planetesimals that accreted beyond the orbit of Mars.

The collisions of the heavy bombardment also explain why asteroids and comets are today found in the distinct regions of the asteroid belt, Kuiper belt, and Oort cloud. Jupiter's gravity stirred up the orbits of leftover planetesimals in the inner solar system, ultimately leaving asteroids concentrated in the asteroid belt. The Kuiper belt represents leftover planetesimals that formed beyond Neptune, where there were no large planets to collide with. The Oort cloud is thought to contain comets that formed in the region between the jovian planets but were "kicked out" to their current great distances when they passed by one of the jovian planets and the planet's gravity accelerated them to high speed in random directions. Once they were far from the Sun, gravitational nudges from other stars further randomized their orbits.

Explaining the Exceptions   We have now explained all the major features of our solar system except the "exceptions to the rules." Today, we think that most of these exceptions arose from collisions and other processes that involved the leftover planetesimals in the young solar system.

Let's start with unusual moons. We have explained the orbits of most large jovian planet moons by their formation in a disk that swirled around the forming planet. But how do we explain moons with unusual orbits, such as those that go in the "wrong" direction (opposite their planet's rotation) or that have large inclinations to their planet's equator? These moons are probably leftover planetesimals that originally orbited the Sun but were then captured into planetary orbit.

It's not easy for a planet to capture a moon, because the law of gravity dictates that a small object passing a large planet will simply fly on by. An object can be captured only if it loses enough orbital energy to settle into a circular or elliptical orbit. For the jovian planets, this type of capture could have occurred when the planets were very young and still surrounded by an extended cloud of relatively dense gas. A passing planetesimal would have been slowed by friction with this gas, and if it was slowed enough, it could have become an orbiting moon. This process is more likely with smaller objects, explaining why most of the jovian moons are only a few kilometers across. Moreover, because of the random nature of the capture process, captured moons would not necessarily orbit in the same direction as their planet or in its equatorial plane—and this is the case for most of the small jovian moons. Mars may also have captured its two small moons, Phobos and Deimos, at a time when it had a much more extended atmosphere than it does today (**Figure 4.11**).

Capture processes cannot explain our own Moon, however, because it is much too large to have been captured by a small planet like Earth. We can also rule out the possibility that our Moon formed simultaneously with Earth: If both had formed together, they would have accreted from planetesimals of the same type and have approximately the same composition and density, but that is not the case. The Moon's density is considerably lower than Earth's, indicating that it has a very different average composition.

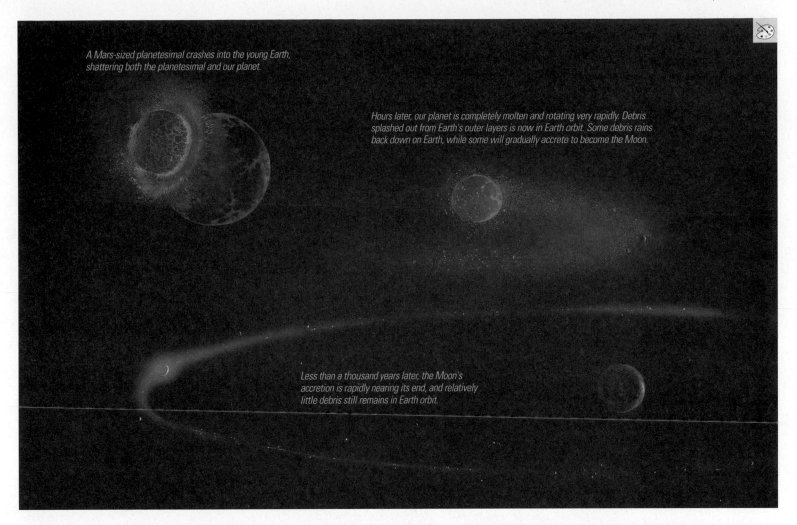

A Mars-sized planetesimal crashes into the young Earth, shattering both the planetesimal and our planet.

Hours later, our planet is completely molten and rotating very rapidly. Debris splashed out from Earth's outer layers is now in Earth orbit. Some debris rains back down on Earth, while some will gradually accrete to become the Moon.

Less than a thousand years later, the Moon's accretion is rapidly nearing its end, and relatively little debris still remains in Earth orbit.

Figure 4.12 | Artist's conception of the giant impact hypothesis for the formation of our Moon. As shown, the Moon formed quite close to a rapidly rotating Earth, but over billions of years tidal forces have slowed Earth's rotation and moved the Moon's orbit outward.

So how did we get our Moon? Today, the leading hypothesis suggests that it formed as the result of a **giant impact** between Earth and a huge planetesimal.

According to models, a few leftover planetesimals may have been as large as Mars. If one of these Mars-size objects struck a young planet, the blow might have tilted the planet's axis, changed the planet's rotation rate, or completely shattered the planet. The giant impact hypothesis holds that a Mars-size object hit Earth at a speed and angle that blasted Earth's outer layers into space. According to computer simulations, this material could have collected into orbit around our planet, and accretion within this ring of debris could have formed the Moon (**Figure 4.12**).

Strong support for the giant impact hypothesis comes from two features of the Moon's composition. First, the Moon's overall composition is quite similar to that of Earth's outer layers—just as we should expect if it were made from material blasted away from those layers. Second, the Moon has a much smaller proportion of easily vaporized ingredients (such as water) than Earth. This fact supports the hypothesis because the heat of the impact would have vaporized these ingredients. As gases, they would not have participated in the process of accretion that formed the Moon, since only solid material could have participated in this accretion.

Giant impacts may also explain other exceptions. Mercury's surprisingly high density may be the result of a giant impact that blasted away its outer, lower density layers. Giant impacts might also have been responsible for tilting the axes of planets (including Earth), for tipping Uranus on its side, and perhaps for Venus's slow and backward rotation.

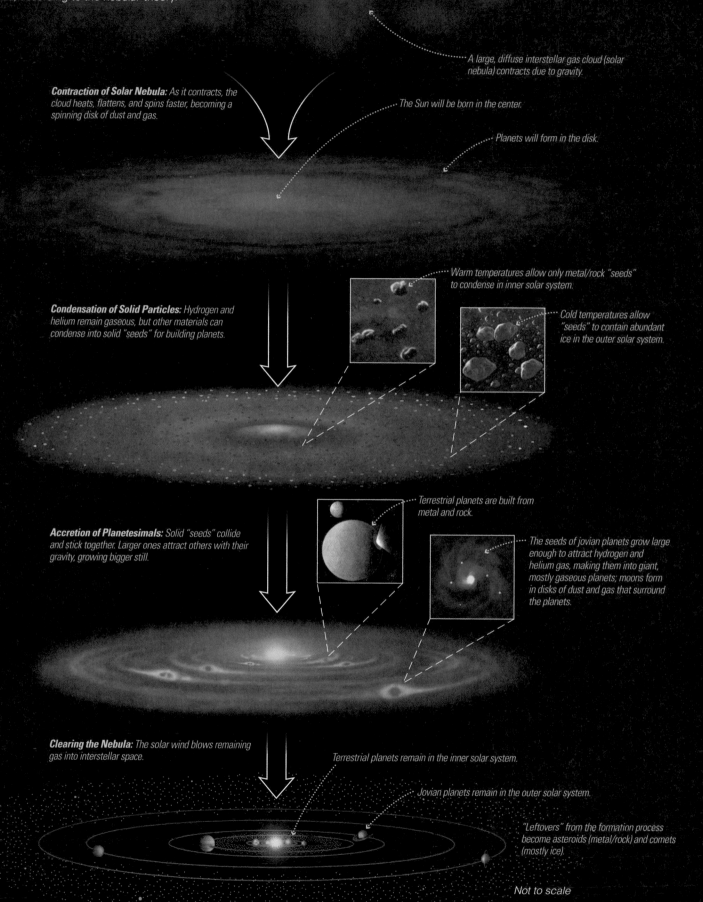

Figure 4.13 A summary of the process by which our solar system formed, according to the nebular theory.

A large, diffuse interstellar gas cloud (solar nebula) contracts due to gravity.

Contraction of Solar Nebula: As it contracts, the cloud heats, flattens, and spins faster, becoming a spinning disk of dust and gas.

The Sun will be born in the center.

Planets will form in the disk.

Warm temperatures allow only metal/rock "seeds" to condense in inner solar system.

Condensation of Solid Particles: Hydrogen and helium remain gaseous, but other materials can condense into solid "seeds" for building planets.

Cold temperatures allow "seeds" to contain abundant ice in the outer solar system.

Terrestrial planets are built from metal and rock.

Accretion of Planetesimals: Solid "seeds" collide and stick together. Larger ones attract others with their gravity, growing bigger still.

The seeds of jovian planets grow large enough to attract hydrogen and helium gas, making them into giant, mostly gaseous planets; moons form in disks of dust and gas that surround the planets.

Clearing the Nebula: The solar wind blows remaining gas into interstellar space.

Terrestrial planets remain in the inner solar system.

Jovian planets remain in the outer solar system.

"Leftovers" from the formation process become asteroids (metal/rock) and comets (mostly ice).

Not to scale

Although we cannot definitively explain these exceptions to the general rules, the main point is clear: The chaotic processes that accompanied planet formation, including the many collisions that surely occurred, are *expected* to have caused at least a few exceptions.

Summary We have found that detailed models based on the nebular theory successfully explain all the major features of our solar system. That is why scientists are confident that the theory captures the essence of what actually happened, even though we may not yet know all the details. **Figure 4.13** summarizes the formation of our solar system according to the nebular theory.

THE PROCESS OF SCIENCE IN ACTION
4.3 The Age of the Solar System

We've said that the solar system formed about $4\frac{1}{2}$ billion years ago. But since no one was around when the solar system was born, how can we possibly know how old it is? The answer is that careful observations and experiments enable scientists to measure the ages of things much older than we are. For example, we can measure ages of trees up to a few thousand years old by counting tree rings, and we can study Earth's climate over a period of a few hundred thousand years by drilling deep into the ice sheets of Greenland and Antarctica, where we see distinct layers laid down each year. To study things that are millions or billions of years old, we must determine the ages of rocks. The technique scientists use to measure these enormous ages is the focus of this chapter's case study in the process of science in action.

How do we determine the age of Earth and the solar system?

The most reliable method for measuring the age of a rock is **radiometric dating,** which relies on careful measurement of the proportions of various atoms in the rock. The method works because some atoms undergo changes with time that allow us to determine how long they have been held in place within the rock's solid structure.

Isotopes and Radioactive Decay You may recall from high school that each chemical element is uniquely characterized by the number of protons in its nucleus. Different **isotopes** of the same element differ only in their number of neutrons. For example, any atom with 6 protons in its nucleus is an atom of carbon, but carbon atoms come in three different isotopes (**Figure 4.14**): carbon-12, which has 6 neutrons in addition to the 6 protons; carbon-13, which has 7 neutrons; and carbon-14, which has 8 neutrons.

Most of the atoms and isotopes we encounter in daily life are stable, meaning that their nuclei stay the same at all times. For example, most of the carbon in our bodies is carbon-12, which is stable. But some isotopes are unstable, meaning that their nuclei are prone to spontaneous change, or *decay*, such as breaking apart or having a proton turn into a neutron. These unstable nuclei are said to be **radioactive.** Carbon-14 is an example of a radioactive isotope, because it undergoes spontaneous change that turns it into nitrogen-14.

Radioactive decay always occurs at the same rate for any particular radioactive isotope, and scientists can measure these rates in the laboratory. We generally characterize the decay rate of an element by stating its **half-life**—the length of time it would take for half its nuclei to decay. Note that it takes only a few months to years of laboratory measurements to pin down the decay rate and determine the half-life, even if the half-life is billions of years. The half-life of carbon-14 is about 5700 years, which makes it useful to archaeologists trying

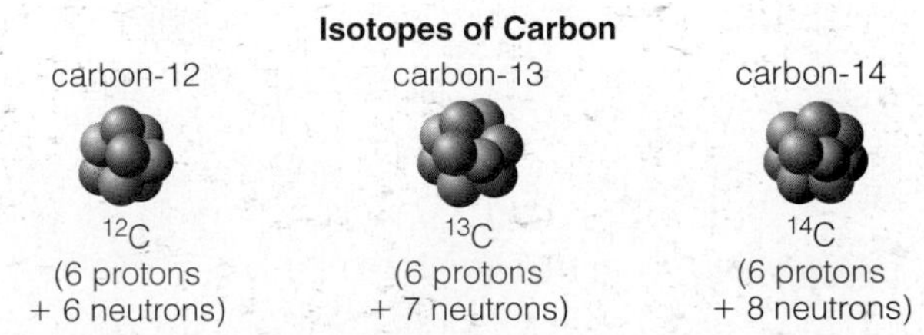

Isotopes of Carbon

carbon-12   carbon-13   carbon-14

^{12}C ^{13}C ^{14}C
(6 protons   (6 protons   (6 protons
+ 6 neutrons)   + 7 neutrons)   + 8 neutrons)

Figure 4.14 | Different isotopes of a chemical element contain the same number of protons, but different numbers of neutrons.

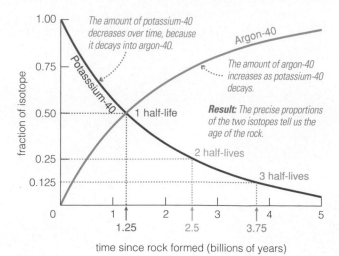

The amount of potassium-40 decreases over time, because it decays into argon-40.

The amount of argon-40 increases as potassium-40 decays.

Result: The precise proportions of the two isotopes tell us the age of the rock.

Figure 4.15 | Potassium-40 is radioactive, decaying into argon-40 with a half-life of 1.25 billion years. The red curve shows the decreasing amount of potassium-40, and the blue curve shows the increasing amount of argon-40. The amount of potassium-40 remaining drops in half with each successive half-life.

to determine the dates of ancient human settlements, but which is far too short for dating rocks that are millions or billions of years old. To see how dating of rocks works, let's consider decay of the radioactive isotope potassium-40 into argon-40, a process with a half-life of 1.25 billion years. (Potassium-40 also decays by other paths, but we focus only on decay into argon-40 to keep the discussion simple.)

Radiometric Dating Consider a small piece of rock that contained 1 microgram of potassium-40 and no argon-40 when it formed (solidified) long ago. The half-life of 1.25 billion years means that half the original potassium-40 would have decayed into argon-40 by the time the rock was 1.25 billion years old, so at that time the rock would contain $\frac{1}{2}$ microgram of potassium-40 and $\frac{1}{2}$ microgram of argon-40. Half of this remaining potassium-40 would then have decayed by the end of the next 1.25 billion years, so after 2.5 billion years the rock would contain $\frac{1}{4}$ microgram of potassium-40 and $\frac{3}{4}$ microgram of argon-40. After three half-lives, or 3.75 billion years, only $\frac{1}{8}$ microgram of potassium-40 would remain, while $\frac{7}{8}$ microgram would have become argon-40. **Figure 4.15** summarizes the gradual decrease in the amount of potassium-40 and corresponding rise in the amount of argon-40.

We can now see the essence of radiometric dating. Suppose you find a rock that contains equal numbers of atoms of potassium-40 and argon-40. If you assume that all the argon came from potassium decay (and if the rock shows no evidence of subsequent heating that could have allowed any argon to escape), then it must have taken precisely one half-life for the rock to end up with equal amounts of the two isotopes. You can therefore conclude that the rock is 1.25 billion years old. The only question is whether you are right in assuming that the rock lacked argon-40 when it formed. In this case, knowing a bit of "rock chemistry" helps. Potassium-40 is a natural ingredient of many minerals in rocks, but argon-40 is a gas that does not combine with other elements and did not condense in the solar nebula. If you find argon-40 gas trapped inside minerals, it must have come from radioactive decay of potassium-40.

Radiometric dating is possible with many other radioactive isotopes as well. In many cases, we can date a rock that contains more than one radioactive isotope; agreement between the ages calculated from the different isotopes gives us confidence that we have dated the rock correctly. We can also sometimes check results from radiometric dating against other methods of measuring ages. For example, some fairly recent archaeological artifacts have dates printed on them, and the dates agree with ages found by radiometric dating. We can validate the $4\frac{1}{2}$-billion-year radiometric age for the solar system as a whole by comparing it to an age based on detailed study of the Sun. Theoretical models of the Sun, along with observations of other stars, show that stars slowly expand and brighten as they age. The model ages are not nearly as precise as radiometric ages, but they confirm that the Sun is between about 4 and 5 billion years old. Overall, the technique of radiometric dating has been checked in so many ways and relies on such basic scientific principles that there is no longer any serious scientific debate about its validity.

Ages of Earth Rocks, Moon Rocks, and Meteorites Radiometric dating allows us to determine the length of time that has passed since the atoms in a rock became locked together in their present arrangement, which in most cases means the time *since the rock last solidified*. The ages of rocks therefore vary greatly. Some Earth rocks are quite young because they formed recently from molten lava, and other Earth rocks have a great variety of ages because they melted and then re-solidified at different times in Earth's history. The oldest Earth rocks are over 4 billion years old, and some small mineral grains date to nearly 4.4 billion years ago. Earth as a whole must be even older than these oldest rocks and minerals.

Moon rocks brought back by the Apollo astronauts date to as far back as 4.4 billion years ago. Although they are older than Earth rocks, these Moon rocks must still be younger than the Moon itself. The ages of these rocks also tell us that the giant impact thought to have created the Moon must have occurred more than 4.4 billion years ago.

To go all the way back to the origin of the solar system, we must find rocks that have not melted or vaporized since they first condensed in the solar nebula. Meteorites that have fallen to Earth are our source of such rocks. Many meteorites appear to have remained unchanged since they condensed and accreted in the early solar system. Careful analysis of radioactive isotopes in meteorites shows that the oldest ones formed about 4.55 billion years ago, so this time must mark the beginning of accretion in the solar nebula. Because the planets accreted within a few tens of millions of years, Earth and the other planets must have finished forming about 4.5 billion years ago.

The Bottom Line   In science, we often want to know the ages of things that are far older than we are. Scientists have therefore developed techniques that in many cases can date extremely old objects with amazing precision. Radiometric dating is one of the most important techniques, and it is also one of the best tested and verified. As a result, we have great confidence in ages established by radiometric dating, including the age of Earth and our solar system.

Summary of Key Concepts

4.1   Characteristics of the Solar System

What does the solar system look like?

Our solar system consists of the Sun, the planets and their moons, and vast numbers of asteroids and comets. The planets are tiny compared to the distances between them. Each world has its own unique character, but there are many clear patterns among the worlds.

What features of our solar system provide clues to how it formed?

Four major features provide clues: (1) The Sun, planets, and large moons generally rotate and orbit in a very organized way. (2) The planets divide clearly into two groups: **terrestrial** and **jovian.** (3) The solar system contains huge numbers of asteroids and comets. (4) There are some notable exceptions to these general patterns.

4.2   The Birth of the Solar System

What theory best explains the orderly patterns of motion in our solar system?

The **nebular theory** holds that the solar system formed from the gravitational collapse of a great cloud of gas and dust. As the **solar nebula** shrank in size, it got hotter, spun faster, and flattened out. The orderly motions we observe today came from the orderly motion of this spinning disk.

How does our theory account for the features of planets, moons, and small bodies?

Planets formed around solid "seeds" of matter that condensed from gas and then grew through **accretion.** In the inner solar system, high temperatures allowed only metal and rock to condense, which explains why terrestrial worlds are made of metal and rock. In the outer solar system, cold temperatures allowed more abundant ices to condense along with metal and rock, making some **planetesimals** that grew large enough for their gravity to draw in hydrogen and helium gas, building massive jovian planets. **Asteroids** are the rocky leftover planetesimals of the inner solar system, and **comets** are the icy leftover planetesimals of the outer solar system. Most of the exceptions to the general trends probably arose from collisions or close encounters with leftover planetesimals. Our Moon probably resulted from a **giant impact** between a Mars-size planetesimal and the young Earth.

✷ THE PROCESS OF SCIENCE IN ACTION

4.3   The Age of the Solar System

How do we determine the age of Earth and the solar system?

The technique of **radiometric dating** allows us to determine the age of a rock by carefully measuring its proportions of a radioactive isotope and the decay product of that isotope. We determine the age of Earth and the solar system by measuring the ages of the oldest meteorites, which are about 4.55 billion years old.

Investigations

Quick Quiz

Choose the best answer to each of the following; answers are in Appendix D. Explain your reasoning with one or more complete sentences.

1. How might our solar system be different if the frost line were much farther out? (a) Earth would be smaller. (b) Jupiter would not exist. (c) It would be no different.
2. Which of the following kinds of object resides closest to the Sun on average? (a) comets (b) asteroids (c) jovian planets
3. Planetary orbits are (a) very eccentric (stretched out) ellipses and in the same plane. (b) fairly circular and in the same plane. (c) fairly circular but oriented in random directions.
4. How many of the planets orbit the Sun in the same direction that Earth does? (a) a few (b) most (c) all
5. Which have more moons on average? (a) jovian planets (b) terrestrial planets (c) Terrestrial and jovian planets both have about the same number of moons.
6. Which ingredients made up about 98% of the solar nebula? (a) rock and metal (b) hydrogen compounds (c) hydrogen and helium
7. In which list are the major steps of solar system formation in the correct order? (a) gravitational collapse of the solar nebula, accretion, condensation (b) gravitational collapse of the solar nebula, condensation, accretion (c) accretion, condensation, gravitational collapse of the solar nebula
8. Which of the following changes did *not* occur during the collapse of the solar nebula? (a) spinning faster (b) heating up (c) concentrating denser materials nearer the Sun
9. Leftover ice-rich planetesimals are called (a) comets. (b) asteroids. (c) meteors.
10. Why didn't a terrestrial planet form at the location of the asteroid belt? (a) There was never enough material in that part of the solar nebula. (b) The solar wind cleared away nebular material there. (c) Jupiter's gravity kept planetesimals from accreting into a planet there.
11. What's the leading theory for the origin of the Moon? (a) It formed at the same time as Earth. (b) It formed from the material ejected by a giant impact on Earth. (c) It split off from a rapidly rotating Earth.
12. About how old is the solar system? (a) 4.5 million years (b) 4.5 billion years (c) 4.5 trillion years

Short-Answer/Essay Questions

Explain all answers clearly, with complete sentences and proper essay structure, if needed. An asterisk (*) designates a quantitative problem, for which you should show all your work.

13. *Planetary Tour.* Based on the brief planetary tour in this chapter, which planet besides Earth do you think is the most interesting, and why? Defend your opinion clearly in two or three paragraphs.
14. *Patterns of Motion.* In one or two paragraphs, summarize the orderly patterns of motion in our solar system and explain why they suggest that the Sun and the planets all formed at one time from one cloud of gas, rather than as individual objects at different times.

15. *Solar System Trends.* Use Table 4.1 to answer each of the following.
    a. Describe the relationship between distance from the Sun and surface temperature. Why does this relationship exist? Explain any notable exceptions to the trend.
    b. Describe in general how the columns for density, composition, and distance from the Sun support the classification of planets into the two categories of terrestrial and jovian.
    c. Describe the relationship between orbital period and distance from the Sun, and explain it in terms of Kepler's third law.
    d. Which column of data would you use to find out which planet has the shortest days? Do you see any notable differences in the length of a day for the different types of planets? Explain.
    e. Which planets would you expect to have seasons? Why?
16. *Two Kinds of Planets.* In words a friend would understand, explain why the jovian planets differ from the terrestrial planets in each of the following aspects: composition, size, density, distance from the Sun, and number of moons.
17. *Pluto.* How does the nebular theory explain the origin of objects like Pluto? How was its formation similar to the formation of jovian and terrestrial planets, and how was it different?
18. *An Early Solar Wind.* Suppose the solar wind had cleared away the solar nebula before the seeds of the jovian planets could gravitationally draw in hydrogen and helium gas. How would the planets of the outer solar system be different? Would they still have many moons? Explain your answer in a few sentences.
19. *History of the Elements.* Our bodies (and most living things) are made mostly of water (H_2O), which contains both hydrogen and oxygen. Summarize the history of a typical hydrogen atom from its creation to the formation of Earth. Do the same for a typical oxygen atom. (*Hint:* Review Chapter 1 to see which elements were created in the Big Bang and where the others were created.)
20. *Understanding Radiometric Dating.* Imagine you had the good fortune to find a rocky meteorite in your backyard. How would you expect its ratio of potassium-40 and argon-40 to be different from that of other rocks in your yard? Explain your answer in a few sentences.
*21. *Mission to Pluto.* The *New Horizons* spacecraft will take about 9 years to travel from Earth to Pluto. About how fast is it traveling on average? Assume that its trajectory is close to a straight line. Give your answer in AU per year and kilometers per hour.
*22. *Size Comparisons.* How many Earths could fit inside Jupiter (assuming you could fill up its total volume)? How many Jupiters could fit inside the Sun? (*Hint:* The equation for the volume of a sphere is $V = \frac{4}{3}\pi r^3$.)
*23. *Radiometric Dating.* For each of the following, assume all the argon-40 comes from radioactive decay of potassium-40, a process with a half-life of 1.25 billion years.
    a. You find a rock that contains equal amounts of potassium-40 and argon-40. How old is it? Explain.
    b. You find a rock that contains three times as much argon-40 as potassium-40. How old is it? Explain.

5 Terrestrial Worlds

Learning Goals

5.1 Terrestrial Surfaces and Atmospheres

- What determines a world's level of geological activity?
- How does an atmosphere affect conditions for life?

5.2 Histories of the Terrestrial Worlds

- Why did the terrestrial worlds turn out so differently?
- What unique features of Earth are important for life?

✺ THE PROCESS OF SCIENCE IN ACTION

5.3 Global Warming

- What is the evidence for global warming?

The photo above shows the surface of Mars viewed from the *Mars Phoenix Lander*. It looks almost Earth-like, but the air is so thin that you could not survive more than a few minutes outside without a pressurized space suit. On top of that, there's no liquid water anywhere on the surface, the temperature is usually well below freezing, and the lack of an ozone layer exposes the ground to dangerous ultraviolet light from the Sun. Venus offers an opposite extreme, with a thick atmosphere that makes the surface hotter than a pizza oven. How did it come to be that Earth is so hospitable to life, while our planetary neighbors are so different? In this chapter, we'll investigate the nature of the terrestrial worlds as we seek to understand what makes our planet so suited to diverse life.

5.1 Terrestrial Surfaces and Atmospheres

We surveyed the four terrestrial planets (Mercury, Venus, Earth, and Mars) briefly in Section 4.1. Because our Moon is nearly three-fourths the size of Mercury in diameter, we often think of it as a fifth terrestrial world. Although the five terrestrial worlds look quite different today, they all formed from the accretion of rocky planetesimals, and all have been shaped by the same basic geological and atmospheric processes. To understand why these differences arise, we will first compare the levels of *geological activity*—the degree to which a planet's surface continues to change—on the terrestrial worlds. Then we will compare their atmospheres to see why the differences between terrestrial worlds make some planets more suitable for life than others. With the insights gained from these comparisons, we will be prepared to appreciate the histories of the terrestrial worlds and the uniqueness of Earth.

What determines a world's level of geological activity?

A quick comparison of the terrestrial worlds reveals that geological activity depends primarily on a world's size. Earth is the largest and most geologically active of the terrestrial worlds, with a surface continually reshaped by volcanic eruptions, earthquakes and other movements of the crust, and erosion. Venus, the next largest, is probably also quite active, though we know relatively little about its geological activity because its thick atmosphere allows us to "see" the surface only with the aid of radar. Mars, middling in size, looks as if it had extensive geological activity in its past, but is far less active today. Mercury and the Moon, the two smallest, undergo so little surface change today that we generally think of them as being geologically dead.

Size, Internal Heat, and Interior Structure Size matters because most geological activity is driven by internal heat. For example, volcanoes erupt when internal heat melts rock and releases gases that drive the molten rock upward. Earthquakes occur when heat-driven internal motions cause rock to slip or shift. All the terrestrial worlds were quite hot when they were young, because accretion of planetesimals converted gravitational potential energy into thermal energy. Decay of radioactive materials inside the terrestrial worlds later provided additional heat. Over time, this internal heat gradually escaped from their surfaces into space. Just as a hot pea cools much more quickly than a hot potato, smaller worlds cool more quickly than larger ones. Mercury and the Moon cooled so quickly that they probably lost the heat needed to power volcanoes before they were 2 billion years old. Mars still retains enough internal heat to drive some geological activity, but much less than it had long ago. Venus and Earth cool so slowly that radioactive decay still keeps their interiors warm, which is why they continue to have strong geological activity.

Common Misconceptions

EARTH IS NOT FULL OF MOLTEN LAVA

Many people guess that Earth is full of molten lava (more technically known as *magma*). This misconception may be related to the fact that molten lava emerges from inside Earth when a volcano erupts. However, Earth's mantle and crust are almost entirely solid. The lava that erupts from volcanoes comes only from a narrow region of partially molten material beneath the lithosphere. The only part of Earth's interior that is fully molten is the outer core, which is so deep within the planet that core material never erupts directly to the surface.

See it for yourself The fact that large objects stay warmer longer than small objects is easy to see with food and drink. The next time you eat something large and hot, cut off a small piece; notice how much more quickly the small piece cools than the rest of it. A similar idea applies to warming up cool objects: Find two ice cubes of the same size; crack one into small pieces with a spoon and then compare how fast the two cubes melt. The reason for these differences in heating and cooling comes from what we call the *surface area to volume ratio*. The volume of an object stores its heat, but heat can enter or leave the object only through its surface area. Explain your observations by considering how the relative amount of surface area differs for objects of different size.

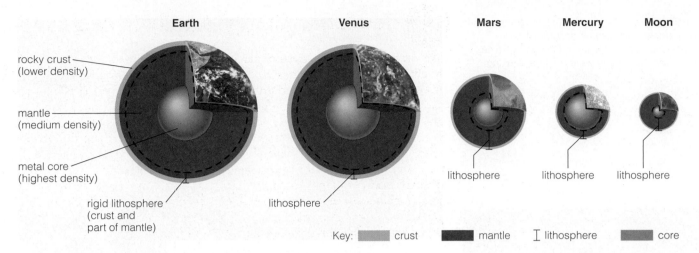

Earth **Venus** **Mars** **Mercury** **Moon**

rocky crust
(lower density)

mantle
(medium density)

metal core
(highest density)

rigid lithosphere
(crust and
part of mantle)

lithosphere

lithosphere lithosphere lithosphere

Key: ▮ crust ▮ mantle ⊺ lithosphere ▮ core

Internal heat also determines what terrestrial worlds are like on the inside. Early in their histories, all these worlds were hot enough to have molten interiors, allowing materials to separate by density in the process called **differentiation** (because it results in layers made of *different* materials). Differentiation led to three major layers within each world:

- **Core.** The highest-density material, consisting primarily of metals such as nickel and iron, sank to the central core.

- **Mantle.** Rocky material of moderate density—mostly minerals that contain silicon, oxygen, and other elements—formed the thick mantle that surrounds the core.

- **Crust.** The lowest-density rock, such as granite and basalt, formed the thin crust that serves as a world's outer skin.

Figure 5.1 shows these layers for the five terrestrial worlds, as determined through a combination of seismic studies on Earth and modeling of the other worlds. Notice that each world also has a region called a **lithosphere** that extends downward from the surface, and it is thicker for the smaller worlds. The lithosphere encompasses the crust and part of the mantle of each world and represents the region in which rock is relatively cool and rigid. Beneath the lithosphere, warmer temperatures make the rock softer, allowing it to flow slowly over millions and billions of years. The rigid rock of the lithosphere essentially "floats" upon the warmer, softer rock below.

Interior heat drives geological activity by supplying the energy needed to melt or move rock and reshape the surface. If the deep interior of a terrestrial world is hot enough, hot rock can gradually rise within its mantle, while cooler rock at the top of the mantle gradually falls (**Figure 5.2**). The process by which hot material expands and rises while cooler material contracts and falls is called **convection.** Keep in mind that mantle convection primarily involves solid rock, not molten rock. Because solid rock flows quite slowly, mantle convection is a very slow process. At the typical rate of mantle convection on Earth—a few centimeters per year—it would take 100 million years for a piece of rock to be carried from the base of the mantle to the top.

Interior heat plays another important role: It can help create a global **magnetic field** because convection in the core can generate electrical currents. Among the terrestrial planets, only Earth has a strong magnetic field, which is important to life because it helps protect our planet from charged particles flowing outward from the Sun. These particles could strip away atmospheric gas and cause genetic damage to living organisms, but Earth's magnetic field deflects most of them (**Figure 5.3**).

Figure 5.1 | Interior structures of the terrestrial worlds, shown to scale and in order of decreasing size. The thicknesses of the crust and lithosphere on Venus and Earth are exaggerated to make them visible in this figure.

Mantle convection: hot rock rises and cooler rock falls.

Figure 5.2 | Earth's hot interior allows the mantle to undergo convection. Arrows indicate the direction of flow in a portion of the mantle.

Figure 5.3 | This diagram shows how Earth's magnetosphere deflects solar wind particles, with some accumulating in charged particle belts encircling our planet. Charged particles can spiral into Earth's atmosphere in polar regions, causing the aurora.

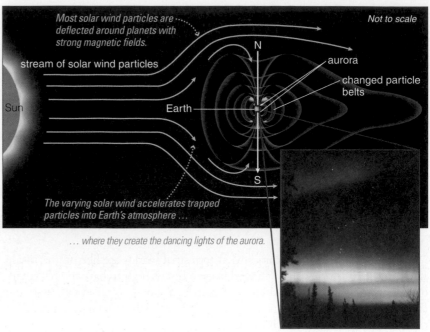

Most solar wind particles are deflected around planets with strong magnetic fields.

Not to scale

stream of solar wind particles

aurora

changed particle belts

Sun

Earth

N

S

The varying solar wind accelerates trapped particles into Earth's atmosphere …

… where they create the dancing lights of the aurora.

200 m

Figure 5.4 | Impact cratering. Meteor Crater in Arizona is more than a kilometer across and almost 200 meters deep. It was created about 50,000 years ago by the impact of a metallic asteroid about 50 meters across.

molten rock in upper mantle

Figure 5.5 | Volcanism. This photo shows the eruption of an active volcano on the flanks of Kilauea on the Big Island in Hawaii. The inset shows the underlying process: Molten rock collects in a *magma chamber* and can erupt upward.

The Four Geological Processes Now that we understand how internal heat drives geological activity, we are ready to examine the specific processes that shape planetary surfaces. The terrestrial worlds show a huge variety of geological surface features, but almost all of them can be explained by just four major geological processes:

- **Impact cratering:** the creation of bowl-shaped *impact craters* by asteroids or comets striking a planet's surface.

- **Volcanism:** the eruption of molten rock from a planet's interior onto its surface.

- **Tectonics:** the disruption of a planet's surface by internal stresses.

- **Erosion:** the wearing down or building up of geological features by wind, water, ice, and other planetary weather phenomena.

Let's examine each of these processes in a little more detail. *Impact cratering* is the only process with an external cause: impacts of objects from space. These objects typically hit the surface at a speed between about 40,000 and 250,000 kilometers per hour, releasing enough energy to vaporize solid rock and blast out a large crater (**Figure 5.4**). As we discussed in Chapter 4, impacts must have battered all the terrestrial worlds during the heavy bombardment, with relatively few impacts occurring after that time. Why, then, do small worlds like the Moon and Mercury have so many impact craters today, while large ones like Venus and Earth have very few? The answer is that geological activity such as volcanic eruptions and erosion gradually erases the impact craters on larger worlds. *We can therefore estimate the age of a surface region from the number of impact craters, with more craters indicating an older surface.* For example, the heavily cratered regions of the Moon must be virtually unchanged since the end of the heavy bombardment some 4 billion years ago, moderately cratered regions of Mars's surface are 2–3 billion years old, and young regions of Earth's surface show only the relative handful of impacts that have occurred during the past few million years.

The second process is *volcanism,* which occurs when underground molten rock finds a path through the lithosphere to the surface (**Figure 5.5**).

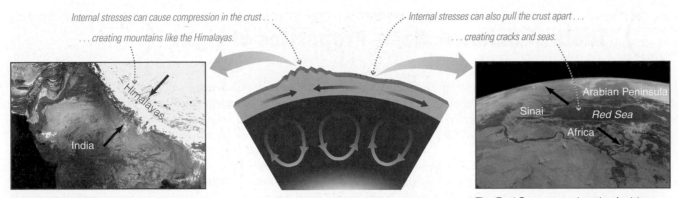

Earth's tallest mountain range, the Himalayas, created as India pushes into the rest of Asia.

The Red Sea, created as the Arabian Peninsula was torn away from Africa.

Figure 5.6 | Tectonics. Tectonic forces can produce a wide variety of features. Mountains created by tectonic compression and valleys or seas created by tectonic stretching are among the most common. Both images are satellite photos.

Volcanism therefore requires substantial internal heat. Eruptions can create tall, steep volcanoes if the molten rock is thick, or flat lava plains if the molten rock is runny. In addition, volcanic eruptions release gases that were trapped beneath the surface when the planet formed. This process, called **outgassing,** is thought to have been responsible for creating the atmospheres of Venus, Earth, and Mars. Measurements of erupting volcanoes on Earth indicate that the most common gases released by outgassing are water vapor (H_2O), carbon dioxide (CO_2), nitrogen (N_2), and sulfur-bearing gases (H_2S or SO_2). Because Venus, Earth, and Mars have similar compositions, all three planets probably released similar proportions of these gases. As we'll discuss later, the three atmospheres differ today because of differences in the ways that various gases have been absorbed by the surfaces or have escaped into space. Outgassing also occurred on Mercury and the Moon, but gravity was too weak to retain the gases on these small worlds, which is why they have no significant atmospheres today.

The third process, *tectonics,* occurs when stretching, compression, or other forces acting on the lithosphere reshape a world's surface. **Figure 5.6** shows two examples of tectonic features on Earth, one created by surface compression (the Himalayas) and one created by surface stretching (the Red Sea). Tectonic activity usually goes hand in hand with volcanism, because both require internal heat, and therefore occurs today only on larger worlds. Much of the tectonic activity on any planet is a direct or indirect result of mantle convection. Tectonics is particularly important on Earth, because the underlying mantle convection fractured Earth's lithosphere into more than a dozen pieces, or *plates.* These plates move over, under, and around each other, leading to a special kind of tectonics that we call **plate tectonics.** While tectonics has affected every terrestrial world, plate tectonics appears to be unique to Earth.

The last of the four major geological processes is *erosion,* which refers to the breakdown or transport of materials through the action of ice, liquid, or gas. The shaping of valleys by glaciers (ice), the carving of canyons by rivers (liquid), and the shifting of sand dunes by wind (gas) are all examples. Erosion plays a far more important role on Earth than on any other terrestrial world, primarily because our planet has both strong winds and plenty of liquid water. Indeed, much of the surface rock on Earth was built by erosion. Over long periods of time, erosion has piled sediments into layers on the floors of oceans and seas, forming **sedimentary rock** (Figure 5.7). Erosion can occur only on worlds with atmospheres, so it does not affect Mercury or the Moon. We see evidence of past water erosion on Mars, but today Mars has only modest wind erosion. Venus has a thick atmosphere but relatively little erosion because it lacks liquid water and has very little wind.

Figure 5.7 | Erosion. The walls of the Grand Canyon consist of layers of sedimentary rock built up by erosion over hundreds of millions of years. These walls are exposed because the Colorado River has carved the deep canyon over the past few million years.

Tools of Science: Basic Properties of Light

As discussed in this chapter, interactions of light and matter explain much about planetary surfaces. In addition, we learn about planets and virtually all other astronomical objects by studying the light we receive from them. It is therefore important to understand basic properties of light.

Light is a form of energy that travels through space at a speed of about 300,000 kilometers per second. Light is characterized by rapidly changing electric and magnetic fields, which is why we often call light an *electromagnetic wave* (**Figure 1**). However, light also comes in distinct pieces, called **photons,** each of which has a **wavelength** (the distance between adjacent peaks of the electric or magnetic field) and a **frequency** (the rate at which the electric and magnetic fields change). A simple formula relates the wavelength and frequency of a photon:

$$\text{wavelength} \times \text{frequency} = \text{speed of light}$$

We therefore find that *longer wavelength means lower frequency* and *shorter wavelength means higher frequency.* The energy of a photon is proportional to its frequency, so higher frequency light has higher energy.

Light can have any wavelength, and we refer to the full range of possibilities as the **electromagnetic spectrum** (**Figure 2**). Note that visible light has only a small range of wavelengths (from about 400 to 700 nanometers). Light with slightly longer wavelength than visible light is called **infrared** (because it is beyond the red end of the visible light spectrum); **radio waves** represent light with even longer wavelength. Going in the other direction, **ultraviolet light** (beyond the violet end of the visible light spectrum) has slightly shorter wavelength than visible light, followed by the even shorter wavelengths of **X rays** and **gamma rays.**

Light can be produced in many ways, but for the moment we'll focus on light emitted by objects such as planets and stars. This light has a spectrum that depends only on the object's surface temperature, so it is called **thermal radiation. Figure 3** shows thermal radiation spectra for objects of different temperature. These spectra obey two laws:

1. Hotter objects emit more light per unit surface area at all wavelengths. For example, a 15,000 K star emits a lot more light per unit area at every wavelength than a 3000 K star.

2. Hotter objects emit photons with a higher average energy, which means a shorter average wavelength. For example, the peak wavelength of emission for the 15,000 K star is much shorter than that of the 3000 K star.

Notice that a star like the Sun emits more strongly in visible light than at any other wavelength, while a typical planet emits infrared light but no visible light at all. This fact is crucial to understanding the greenhouse effect and planetary atmospheres, as we'll discuss in this chapter.

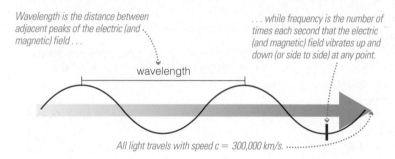

Wavelength is the distance between adjacent peaks of the electric (and magnetic) field . . .

. . . while frequency is the number of times each second that the electric (and magnetic) field vibrates up and down (or side to side) at any point.

wavelength

All light travels with speed c = 300,000 km/s.

Figure 1 | Light is an electromagnetic wave, but also comes in individual pieces called *photons,* each characterized by a wavelength and frequency.

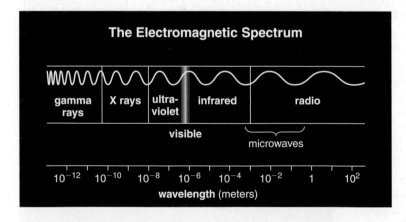

The Electromagnetic Spectrum

| gamma rays | X rays | ultra-violet | infrared | radio |

visible

microwaves

10^{-12} 10^{-10} 10^{-8} 10^{-6} 10^{-4} 10^{-2} 1 10^{2}

wavelength (meters)

Figure 2 | The electromagnetic spectrum.

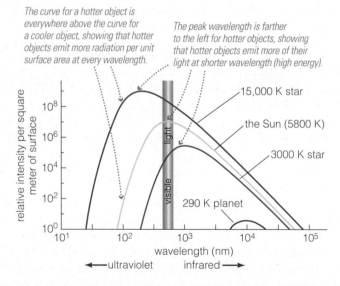

The curve for a hotter object is everywhere above the curve for a cooler object, showing that hotter objects emit more radiation per unit surface area at every wavelength.

The peak wavelength is farther to the left for hotter objects, showing that hotter objects emit more of their light at shorter wavelength (high energy).

15,000 K star

the Sun (5800 K)

3000 K star

290 K planet

relative intensity per square meter of surface

10^{8}

10^{6}

10^{4}

10^{2}

10^{0}

10^{1} 10^{2} 10^{3} 10^{4} 10^{5}

visible light

wavelength (nm)

←ultraviolet   infrared→

Figure 3 | Idealized thermal radiation spectra for objects of different temperature.

How does an atmosphere affect conditions for life?

Venus, Earth, and Mars all have atmospheres that profoundly affect their surfaces. On Earth, the atmosphere also creates the conditions that make our existence possible. The atmosphere provides the air we breathe, and its pressure and temperature allow liquid water to rain down and flow on Earth's surface. The atmosphere is also crucial to life in less obvious ways. Without the atmosphere, dangerous solar radiation would render Earth's surface lifeless, and it would be so cold that most of the water would be perpetually frozen.

Remarkably, our atmosphere plays all these roles despite being very thin. About two-thirds of the air in Earth's atmosphere lies within 10 kilometers of the surface. You could represent this air on a standard globe with a layer only as thick as a sheet of paper. Let's examine how interactions between light and the atmosphere protect and warm Earth's surface. We will then be prepared to understand how atmospheres affect the other terrestrial worlds.

Surface Protection   The Sun emits visible light that allows us to see, but it also emits dangerous ultraviolet and X-ray radiation. In space, astronauts need thick space suits to protect them from the hazards of this radiation. On Earth, the atmosphere protects us (**Figure 5.8**).

X rays carry enough energy to *ionize*, or knock electrons free from, almost any atom or molecule. That is why they can damage living tissue. Fortunately, atoms and molecules in Earth's atmosphere absorb X rays so easily that none reach the ground. Ultraviolet light is not so easily absorbed. Most gases are transparent to ultraviolet light, allowing it to pass through unhindered. We owe our protection from ultraviolet light to a relatively rare gas called **ozone** (O_3). Ozone resides primarily in a middle layer of Earth's atmosphere (the *stratosphere*), where it absorbs most of the dangerous ultraviolet radiation from the Sun.

Visible light from the Sun passes easily through Earth's atmosphere and thereby provides light and heat to Earth's surface. However, not all visible-light photons pass straight through Earth's atmosphere. A few are scattered randomly around the sky, which is why the daytime sky is bright; without scattering, the Sun would be just a very bright disk set against a black, starry sky. Scattering also explains why the daytime sky is blue (**Figure 5.9**). Sunlight consists of all the colors of the rainbow, but the colors are not all scattered equally. Gas molecules scatter blue light much more effectively than red light. When the Sun is overhead, this scattered blue light reaches our eyes from all directions and the sky appears blue. At sunset or sunrise, the sunlight must pass through a greater amount of atmosphere on its way to us. Most of the blue light is scattered away, leaving only red light to color the sky.

The atmospheres of Venus and Mars similarly absorb X rays from the Sun, but they lack ozone or other gases to absorb ultraviolet light. Both visible and ultraviolet light can therefore penetrate their atmospheres, though Venus's clouds are so thick that they scatter most of this light away before it reaches the ground.

The Greenhouse Effect   Visible light warms Earth's surface, but not as much as you might guess. Calculations based on Earth's distance from the Sun and the percentages of visible light absorbed and reflected show that, by itself, visible light would heat Earth to an average surface temperature of only −16°C (+3°F)—well below the freezing point of water. Earth's actual global average temperature is about 15°C (59°F), plenty warm enough for liquid water to flow and life to thrive. Why is Earth so much warmer than it would be from visible light warming alone? The answer is that our atmosphere traps additional heat through what we call the **greenhouse effect.**

Figure 5.10 shows the basic idea behind the greenhouse effect. Some of the visible light that reaches the ground is reflected and some is absorbed. The

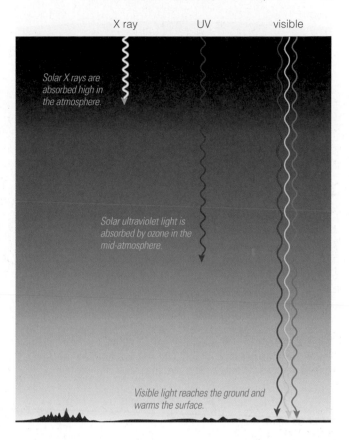

Figure 5.8 | This diagram summarizes how different forms of light from the Sun are affected by Earth's atmosphere.

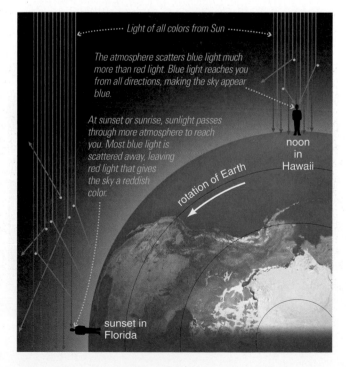

Figure 5.9 | This diagram summarizes why the sky is blue and sunsets (and sunrises) are red.

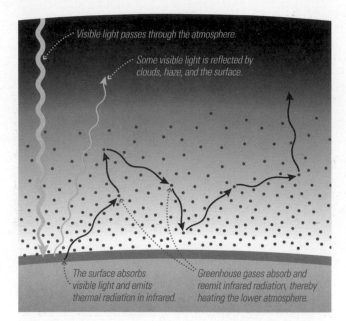

Visible light passes through the atmosphere.

Some visible light is reflected by clouds, haze, and the surface.

The surface absorbs visible light and emits thermal radiation in infrared.

Greenhouse gases absorb and reemit infrared radiation, thereby heating the lower atmosphere.

Figure 5.10 | The greenhouse effect. The lower atmosphere becomes warmer than it would be if it had no greenhouse gases such as water vapor, carbon dioxide, and methane.

absorbed energy must be returned to space—otherwise the ground would rapidly heat up—but planetary surfaces are too cool to emit visible light. Instead, planets emit energy primarily in the form of infrared light.

The greenhouse effect occurs when the atmosphere contains gases that absorb infrared light. Such gases are called **greenhouse gases,** and they include water vapor (H_2O), carbon dioxide (CO_2), and methane (CH_4). These gases absorb infrared light effectively because their molecular structures make them prone to begin rotating or vibrating when struck by an infrared photon. After absorbing the photon, a greenhouse gas molecule reemits a similar photon in some random direction. This photon can then be absorbed by another greenhouse gas molecule, which does the same thing. The net result is that greenhouse gases tend to slow the escape of infrared radiation from the lower atmosphere, while their molecular motions heat the surrounding air. In this way, the greenhouse effect makes the surface and the lower atmosphere warmer than they would be from sunlight alone. The more greenhouse gases present, the more the surface will be warmed.

Think about it Molecules that consist of two atoms of the same type—such as the N_2 and O_2 that make up 98% of Earth's atmosphere—are poor infrared absorbers. But imagine if this were not the case: How would Earth be different if nitrogen and oxygen absorbed infrared light effectively?

The greenhouse effect also warms both Venus and Mars, because both have atmospheres made primarily of carbon dioxide. Venus's thick atmosphere creates an extreme greenhouse effect that raises its surface from a temperature that would otherwise be well below freezing to an astounding 470°C. In contrast, Mars's thin atmosphere creates a weak greenhouse effect that currently warms Mars by only about 6°C.

5.2 Histories of the Terrestrial Worlds

Given that the same basic geological and atmospheric processes can occur on any terrestrial world, how did the terrestrial worlds end up so different from one another? We can answer this question by briefly studying the histories of the terrestrial worlds.

Why did the terrestrial worlds turn out so differently?

We can ultimately trace most of the differences among the terrestrial worlds to differences either in size or in distance from the Sun. Size has the greatest influence on geology, so we'll work our way up from the two smallest worlds.

The Moon and Mercury: Active No More We know that our Moon is geologically dead because much of its surface still bears many impact craters from the period of heavy bombardment, which ended nearly 4 billion years ago. However, parts of the lunar surface are not as heavily cratered, indicating that the Moon experienced some geological activity after the heavy bombardment ended. You can see these smoother, less cratered regions with the naked eye (**Figure 5.11**). They are called **lunar maria,** and they got their name because they look much like oceans seen from afar (*maria* is Latin for "seas").

A combination of models and studies of Moon rocks have allowed scientists to trace the origin of the maria to a period several hundred million years after the end of the heavy bombardment, when the Moon was roughly

200 km

Figure 5.11 | The familiar face of the full Moon shows numerous dark, smooth maria.

a billion years old. By that time, the decay of radioactive elements had built up enough interior heat to melt much of the Moon's mantle. Molten rock welled up through surface fractures made by the largest impacts of the heavy bombardment, and lava flooded the huge craters made by those impacts. The relatively few craters within the maria today were made by impacts that occurred after the maria formed, when the heat from radioactive decay was no longer sufficient to produce lava flows.

 Think about it Did the Moon have more or fewer craters before the maria formed than it does today? Explain.

Mercury is not much bigger than the Moon, and therefore has suffered much the same fate. In fact, Mercury looks so much like the Moon that it can be difficult to tell which world you are looking at in surface photos (**Figure 5.12**). Impact craters are visible almost everywhere, though they are less crowded together than in the most ancient regions of the Moon. That fact, along with smooth regions within and between craters, suggests that molten lava covered up some of the craters that formed during the heavy bombardment. As on the Moon, these lava flows probably occurred when heat from radioactive decay accumulated enough to melt part of the mantle.

Mercury also differs from the Moon in having tremendous cliffs that run hundreds of kilometers in length, with vertical faces up to 3 or more kilometers high. These cliffs probably formed when early tectonic forces compressed the crust, causing the surface to crumple. Scientists suspect that these forces arose from shrinkage of the entire planet, because there are no "stretch marks" suggesting that other parts of the surface have expanded. Why did Mercury shrink more than the Moon? The answer seems to be a difference in composition. Mercury is much denser than the Moon, indicating that it has a surprisingly large, metallic core (see Figure 5.1), perhaps because a giant impact blasted away its outer layers early in its history [Section 4.2]. Metal expands and contracts more with temperature than rock, and models suggest that as this large core cooled, it would have contracted by as much as 20 kilometers in radius. The mantle and lithosphere then had to contract along with the core, generating tectonic stresses that created the great cliffs.

Mars: Once Warm and Wet?

Mars has about half the diameter of Earth, which means it is much larger than Mercury and the Moon. As we therefore expect, its surface shows far more geological activity than the two smaller worlds. Numerous large impact craters scar much of the southern hemisphere, but a relative lack of impact craters in other regions indicates that parts of the Martian surface were reshaped after the heavy bombardment. Several towering volcanoes dot the surface; the largest, Olympus Mons, rises three times as high as Mount Everest on Earth, and its base covers an area the size of Arizona (**Figure 5.13**). Mars also shows evidence of past tectonics, such as the prominent system of valleys called *Valles Marineris*, which is as long as the United States is wide and almost four times as deep as Earth's Grand Canyon (**Figure 5.14**). However, the most intriguing surface features on Mars are the ones produced by erosion, which suggest that Mars was once relatively warm and wet—conditions that may have been conducive to life.

Erosion features on Mars include numerous channels that appear to have been carved by running water (**Figure 5.15**), crater rims that appear to have been eroded by rain, and crater floors with sculpted sediments suggesting they once held lakes. Surface studies provide additional evidence for

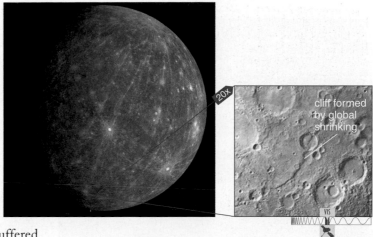

Figure 5.12 | Mercury's surface shows many craters along with smooth lava plains. Bright streaks are caused by material ejected from large craters. Mercury also has many tremendous cliffs (inset), thought to have been created when the entire planet cooled and shrank. (Global image from *MESSENGER*, inset from *Mariner 10*.)

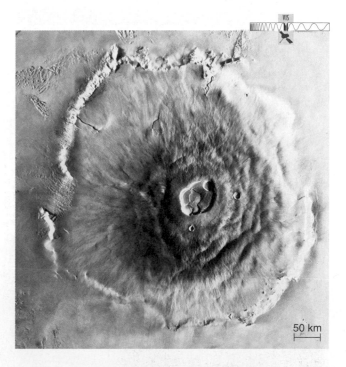

Figure 5.13 | Olympus Mons, the tallest volcano in the solar system, covers an area the size of Arizona and rises higher than Mount Everest on Earth. Note the tall cliff around its rim and the central volcanic crater from which lava erupted.

Figure 5.14 | Valles Marineris is a huge system of valleys on Mars, created in part by tectonic stresses. This image shows a perspective view looking north across the center of the canyon, obtained by *Mars Express*.

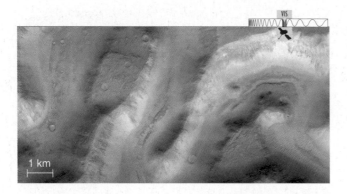

Figure 5.15 | This photo, taken by the *Mars Reconnaissance Orbiter*, shows what appears to be a dried-up meandering riverbed, now filled with dunes of windblown dust. The numerous small craters indicate that the riverbed dried out billions of years ago.

Figure 5.16 | Mineral evidence of past water on Mars. This image, taken by the *Opportunity* rover, shows tiny mineral spheres that formed within sedimentary rock layers like those in the background, then eroded out and rolled downhill; rover tracks and rock grinding locations are visible uphill.

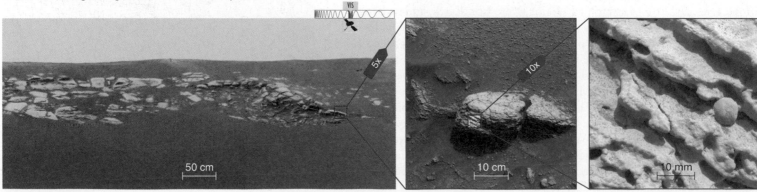

past water. In 2004, the robotic rovers *Spirit* and *Opportunity* landed on opposite sides of Mars. The twin rovers carried cameras, instruments to identify rock composition, and a grinder to expose fresh rock for analysis. The rovers, designed for just 3 months of operation, were still going strong as this book was being completed, nearly 6 years after their arrival. Both rovers found strong evidence that liquid water was once plentiful on Mars. For example, rocks at the *Opportunity* landing site contain tiny spheres with specific mineral compositions suggesting that they formed in standing water (**Figure 5.16**).

No liquid water exists on the surface of Mars today. We know this not only because we've studied most of the surface in detail, but also because the surface conditions do not allow it. In most places and at most times, Mars is so cold that any liquid water would immediately freeze into ice. Even when the temperature rises above freezing, as it often does at midday near the equator, the air pressure is so low that liquid water would quickly evaporate. If you put on a space suit and took a cup of water outside your pressurized spaceship, the water would rapidly freeze or boil away (or a combination of both). The only water we find on Mars today is frozen. Much of this water ice is locked up in the polar caps (which also contain frozen carbon dioxide), but orbital radar measurements and studies from the *Phoenix* lander show that substantial amounts of ice also lie underground at high latitudes.

We conclude that most erosion features on Mars must date from one or more periods in the past when Mars had a thicker atmosphere and a stronger greenhouse effect, so the greater air pressure and warmer temperatures could have allowed liquid water to flow. The timing and extent of these warm and wet periods remain subject to considerable scientific debate, but there's little doubt that Mars eventually underwent a major and permanent climate change, turning a world that was once wet and warm into a frozen wasteland.

The idea that Mars once had a thicker and warmer atmosphere makes sense, because Mars is large enough for gravity to have retained the water vapor and carbon dioxide outgassed by volcanoes in its early history. If Martian volcanoes outgassed carbon dioxide and water in the same proportions as volcanoes on Earth, Mars would have had enough water to fill oceans tens or even hundreds of meters deep. The big question, then, is what happened to all that atmospheric gas.

The leading hypothesis ties most of the carbon dioxide loss to a change in Mars's magnetic field (**Figure 5.17**). Early in its history, Mars had enough internal heat to have molten convecting metals in its core, much like Earth today. These convecting metals should have produced a magnetic field that protected the Martian atmosphere from the solar wind. Mars could therefore have had a much thicker atmosphere while this protection held. However, because Mars is much smaller than Earth, its interior cooled until core convection ceased, which greatly weakened the magnetic field. The solar wind was then able to strip carbon dioxide gas from the top of the Martian atmosphere and into space, drastically weakening the greenhouse effect and turning the surface of Mars into a frozen wasteland.

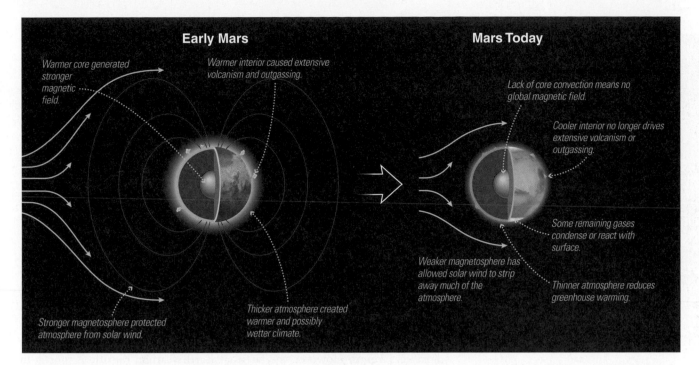

Early Mars

Warmer core generated stronger magnetic field.

Warmer interior caused extensive volcanism and outgassing.

Stronger magnetosphere protected atmosphere from solar wind.

Thicker atmosphere created warmer and possibly wetter climate.

Mars Today

Lack of core convection means no global magnetic field.

Cooler interior no longer drives extensive volcanism or outgassing.

Some remaining gases condense or react with surface.

Weaker magnetosphere has allowed solar wind to strip away much of the atmosphere.

Thinner atmosphere reduces greenhouse warming.

The plunge in temperature that came with the weakening greenhouse effect explains why Mars lacks liquid water today, but another water-related mystery remains: While much of the water is now frozen in the polar caps and underground, it appears that Mars may also have lost a great deal of its original water to space. Because Mars lacks ultraviolet-absorbing gases, ultraviolet light from the Sun can break apart atmospheric water molecules. Hydrogen atoms that broke away, being lightweight, would have rapidly escaped to space, leaving oxygen from the water molecules in the atmosphere. Eventually, this oxygen either was stripped away by the solar wind or was drawn out of the atmosphere by chemical reactions with surface rock. The oxygen absorbed by these rocks literally rusted the surface of Mars, giving the planet its distinctive red tint.

We have much more to learn about Martian history, but a general lesson seems clear: Mars's fate was shaped primarily by its relatively small size. It was big enough for volcanism and outgassing to release water and atmospheric gas early in its history, but too small to maintain the internal heat needed to keep this water and gas. As its interior cooled, its volcanoes quieted and it lost its magnetic field, allowing gas to be stripped away to space. If Mars had been as large as Earth, it might still have a moderate climate today.

 Think about it Could Mars still have flowing water today if it had formed closer to the Sun? Use your answer to explain how Mars's fate has been shaped not just by its size, but also by its distance from the Sun.

Venus: Runaway Greenhouse Effect On the basis of size alone, we would expect Venus and Earth to be quite similar: Venus is only about 5% smaller than Earth in radius, and its overall composition is about the same as that of Earth. But Venus ended up profoundly different from Earth because it is about 30% closer to the Sun.

Figure 5.18 shows a typical surface region of Venus, as revealed by cloud-penetrating radar observations from spacecraft. As expected from its similarity to Earth in size, Venus has many of the same types of geological features as Earth, including occasional impact craters, volcanoes, and a lithosphere that has been contorted by tectonic forces. However, Venus lacks two major features of Earth's geology. First, as we discussed earlier, Venus has little erosion because of its lack of liquid water and its slow surface

Figure 5.17 | This figure shows the leading hypothesis explaining the lack of liquid water on Mars today: Mars underwent dramatic climate change as its interior cooled, ensuring that rain could never fall again.

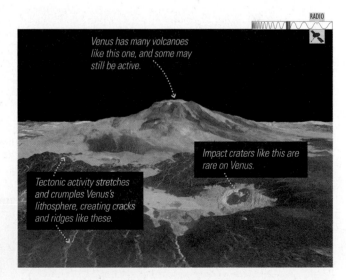

Venus has many volcanoes like this one, and some may still be active.

Impact craters like this are rare on Venus.

Tectonic activity stretches and crumples Venus's lithosphere, creating cracks and ridges like these.

RADIO

Figure 5.18 | This computer-generated image shows typical volcanic and tectonic features on Venus, along with a rare impact crater. The image is based on *Magellan* radar data and exaggerates vertical structure by ten times to reveal features. Bright regions represent rough terrain.

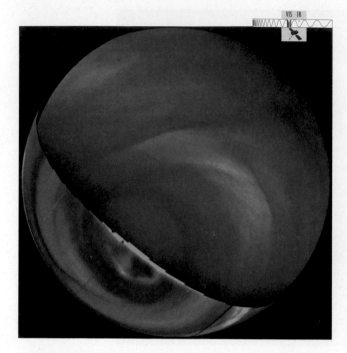

Figure 5.19 | Venus has the thickest atmosphere of any terrestrial planet and is continually shrouded by clouds. This composite from *Venus Express* combines a visible-wavelength image of the day side (upper right, shaded blue) and an infra-red image of the night side (lower left, shaded red). Venus's south pole lies midway along the boundary between the two images.

winds, which are a consequence of its extremely slow rotation (one Venus rotation takes 243 Earth days). Second, we find no evidence of Earth-like plate tectonics on Venus, suggesting that Venus has a thicker and stronger lithosphere than Earth.

These geological differences are minor compared to the differences between the two planets' atmospheres. Given the similar sizes and compositions of Venus and Earth, volcanoes should have outgassed similarly large amounts of water and carbon dioxide on both planets. Venus's thick atmosphere (**Figure 5.19**) does indeed contain a huge amount of carbon dioxide—nearly 200,000 times as much carbon dioxide as Earth's atmosphere—but it has virtually no water. Earth's atmosphere has comparatively little of either gas. So what happened to all the outgassed water on Venus, and to the outgassed water and carbon dioxide on Earth?

We can easily account for both missing gases on Earth. The huge amounts of water vapor released into our atmosphere condensed into liquid form, which rained down and now resides primarily in the oceans. The huge amount of carbon dioxide released into our atmosphere is also still here, but in solid form: Carbon dioxide dissolves in water, where it can undergo chemical reactions to make *carbonate rocks* (rocks rich in carbon and oxygen) such as limestone. Earth has about 170,000 times as much carbon dioxide locked up in rocks as in its atmosphere—which means that Earth has almost as much total carbon dioxide as Venus. However, the fact that Earth's carbon dioxide is mostly in rocks rather than in the atmosphere makes all the difference in the world: If this carbon dioxide were in our atmosphere, our planet would be nearly as hot as Venus and uninhabitable.

We are left with the question of what happened to Venus's water. We know that Venus has essentially no water today: The surface is too hot for liquid water and the atmosphere contains very little water vapor. This absence of water explains why Venus retains so much carbon dioxide in its atmosphere. Without oceans, carbon dioxide cannot dissolve or become locked away in carbonate rocks. Assuming that a huge amount of water really was outgassed on Venus, this water somehow disappeared. The leading hypothesis for the disappearance of this water is that, much like on Mars, ultraviolet light from the Sun broke apart molecules of water vapor and the released hydrogen atoms escaped to space. Acting over billions of years, this process can easily explain the loss of an ocean's worth of water from Venus—as long as the water was in the atmosphere as a gas rather than in liquid oceans.

But why didn't Venus develop oceans like Earth? To understand the answer, think about what would happen if we could magically move Earth to the orbit of Venus (**Figure 5.20**). The greater intensity of sunlight would almost immediately raise Earth's global average temperature by about 30°C, from its current 15°C to about 45°C (113°F). Although this is still well below the boiling point of water, the higher temperature would lead to increased evaporation

Figure 5.20 | This diagram shows how, if Earth were placed at Venus's distance from the Sun, the runaway greenhouse effect would cause the oceans to evaporate completely.

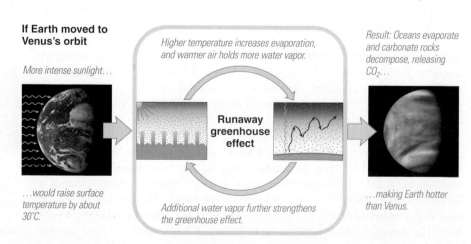

If Earth moved to Venus's orbit

More intense sunlight…

Higher temperature increases evaporation, and warmer air holds more water vapor.

Runaway greenhouse effect

Result: Oceans evaporate and carbonate rocks decompose, releasing CO_2…

…would raise surface temperature by about 30°C.

Additional water vapor further strengthens the greenhouse effect.

…making Earth hotter than Venus.

of water from the oceans and would allow the atmosphere to hold more water vapor before the vapor condensed to make rain. Because water vapor is also a greenhouse gas, the added water vapor would strengthen the greenhouse effect and drive temperatures even higher. The higher temperatures, in turn, would lead to even more ocean evaporation and more water vapor in the atmosphere, strengthening the greenhouse effect even further. In other words, we'd have a positive feedback loop in which each little bit of additional water vapor in the atmosphere would lead to higher temperature and even more water vapor. The process would career rapidly out of control as a **runaway greenhouse effect,** resulting in the complete vaporization of Earth's oceans.

The runaway greenhouse effect would cause Earth to heat up until the oceans were completely evaporated and the carbonate rocks had released all their carbon dioxide back into the atmosphere. By the time the process was complete, temperatures on the relocated Earth would be even higher than they are on Venus today, thanks to the combined greenhouse effects of carbon dioxide and water vapor in the atmosphere. The water vapor would then gradually disappear, as ultraviolet light broke water molecules apart and the hydrogen escaped to space. In short, moving Earth to Venus's orbit would essentially turn our planet into another Venus.

We have arrived at a simple explanation of why Venus is so much hotter than Earth. Even though Venus is only about 30% closer to the Sun than Earth is, this difference was enough to be critical. On Earth, it was cool enough for water to rain down and make oceans. The oceans then dissolved carbon dioxide and chemical reactions locked it away in carbonate rocks, leaving our atmosphere with only enough greenhouse gases to make our planet pleasantly warm. On Venus, the greater intensity of sunlight made it just enough warmer that oceans either never formed or soon evaporated, leaving Venus with a thick atmosphere full of greenhouse gases.

 Think about it We've seen that moving Earth to Venus's orbit would cause our planet to become Venus-like. If we could somehow move Venus to Earth's orbit, would it become Earth-like? Why or why not?

What unique features of Earth are important for life?

If you think about the other terrestrial worlds, you can probably identify a number of features that are unique to Earth. Four of them turn out to be particularly important to our existence: surface liquid water, atmospheric oxygen, plate tectonics, and a climate that has remained relatively stable throughout Earth's history. Let's explore each of these features and how it is important to life.

All life on Earth requires liquid water, and the abundance of liquid water on Earth's surface makes possible a great abundance of life. As we have discussed, this water originally outgassed from Earth's volcanoes, then rained down on the surface to make the oceans, which neither froze nor evaporated thanks to our moderate greenhouse effect and distance from the Sun.

Oxygen (O_2), which is crucial to animal life, makes up about 20% of Earth's atmosphere. But oxygen is not a product of volcanic outgassing. It is such a highly reactive gas that it would be absorbed by rocks and disappear from the atmosphere in just a few million years if it were not continuously resupplied. The ongoing source of oxygen for our atmosphere is life itself: Plants and many microorganisms release oxygen through photosynthesis. Today, photosynthetic organisms return oxygen to the atmosphere in approximate balance with the rate at which animals and chemical reactions consume it, which is why the oxygen concentration remains relatively steady. Note that oxygen is also what makes possible Earth's protective ozone layer, since ozone (O_3) is produced from ordinary oxygen (O_2).

Common Misconceptions

WHY IS THE SKY BLUE?

Some people guess that the sky is blue because of light reflecting from the oceans, but that would not explain blue skies over inland areas. Others claim that "air is blue," a vague statement that is also wrong: If air molecules emitted blue light, then air would glow blue even in the dark; if they were blue because they reflected blue light and absorbed red light, then no red light could reach us at sunset. The real explanation for our blue sky is light scattering, as shown in Figure 5.9, and this scattering also explains our red sunsets.

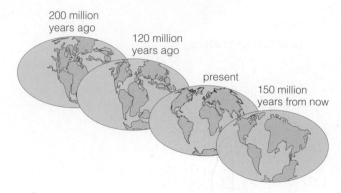

200 million years ago

120 million years ago

present

150 million years from now

Figure 5.21 | Plate tectonics rearranges the continents over millions of years, as shown in these maps of Earth at different times.

Think about it Suppose that, somehow, all photosynthetic life died out. What would happen to the oxygen in our atmosphere? Could animals, including us, still survive?

The third and fourth items on our list of Earth's unique features—plate tectonics and climate stability—turn out to be closely linked. Plate tectonics is a slow process; on average, Earth's plates move only a few centimeters per year, which is about the speed at which your fingernails grow. Nevertheless, over millions of years, these motions rearrange the continents, open ocean basins between continents, build mountain ranges, and much more (**Figure 5.21**). Even more important, plate tectonics turns out to be crucial to climate stability and therefore to the ongoing evolution of life on Earth.

Earth's climate is not perfectly stable—our planet has endured numerous ice ages and warm periods in the past. Nevertheless, even in the deepest ice ages and warmest warm periods, Earth's surface temperature has remained in a range in which some liquid water could still exist and harbor life. On the surface, the key to this climate stability is simple: Earth has always kept just enough carbon dioxide in its atmosphere to keep the temperature in a range suitable for liquid water and life. This long-term climate stability is even more remarkable when we consider that models suggest that the Sun has brightened substantially (about 30%) over the past 4 billion years, yet Earth's temperature has managed to stay in the same range throughout this time. Apparently, the strength of the greenhouse effect somehow self-adjusted to keep the climate stable.

The mechanism by which Earth self-regulates its temperature is called the **carbon dioxide cycle,** or the **CO$_2$ cycle** for short. **Figure 5.22** shows how the cycle works. Atmospheric carbon dioxide dissolves in rainwater, which erodes rocks and leads to minerals being carried to the oceans. In the oceans, these minerals combine with dissolved carbon dioxide and fall to the ocean floor, making carbonate rocks. Over millions of years, motions associated with plate tectonics push the carbonate rocks down into the mantle, where they melt and release their carbon dioxide. This carbon dioxide then returns to the atmosphere by being outgassed from volcanoes.

The CO$_2$ cycle acts as a long-term thermostat for Earth, because the overall rate at which carbon dioxide is pulled from the atmosphere is very sensitive to temperature: the higher the temperature, the higher the rate at which carbon dioxide is removed. As a result, a small change in Earth's temperature will be offset by a change in the CO$_2$ cycle. Consider first what happens if Earth warms up a bit. Warmer temperature means more evaporation and rainfall, pulling more CO$_2$ out of the atmosphere. The reduced atmospheric CO$_2$ concentration leads to a weakened greenhouse effect, which counteracts the initial warming and cools the planet back down. Similarly, if Earth cools a bit, rainfall decreases and less CO$_2$ is dissolved in rainwater, allowing the CO$_2$ released by volcanism to build back up in the atmosphere. The increased CO$_2$ concentration strengthens the greenhouse effect and warms the planet back up.

We can now see why plate tectonics is so intimately connected to our existence. Plate tectonics is a crucial part of the CO$_2$ cycle; without plate tectonics, CO$_2$ would have remained locked up in seafloor rocks rather than being recycled through outgassing. Earth's climate might then have undergone changes as dramatic as those that occurred on Venus and Mars. We know that Mars lacks plate tectonics because of its small size. But why doesn't Venus have plate tectonics, given its similar size to Earth? The leading hypothesis is that the lack of water on Venus—which we've traced to the high temperatures due to the runaway greenhouse effect—has given Venus a thicker and stronger lithosphere that has been able to resist fracturing into plates. If this hypothesis is correct, then two primary characteristics

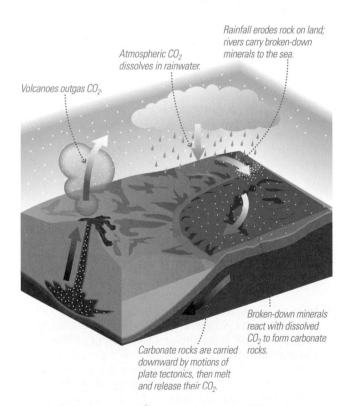

Volcanoes outgas CO$_2$.

Atmospheric CO$_2$ dissolves in rainwater.

Rainfall erodes rock on land; rivers carry broken-down minerals to the sea.

Broken-down minerals react with dissolved CO$_2$ to form carbonate rocks.

Carbonate rocks are carried downward by motions of plate tectonics, then melt and release their CO$_2$.

Figure 5.22 | This diagram shows how the CO$_2$ cycle continually moves carbon dioxide from the atmosphere to the ocean to rock and back to the atmosphere. Note that plate tectonics plays a crucial role in the cycle.

The Role of Planetary Size

Small Terrestrial Planets

Large Terrestrial Planets

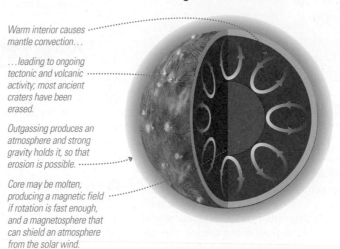

Interior cools rapidly…

…so that tectonic and volcanic activity cease after a billion years or so. Many ancient craters therefore remain.

Lack of volcanism means little outgassing, and low gravity allows gas to escape more easily; no atmosphere means no erosion.

Warm interior causes mantle convection…

…leading to ongoing tectonic and volcanic activity; most ancient craters have been erased.

Outgassing produces an atmosphere and strong gravity holds it, so that erosion is possible.

Core may be molten, producing a magnetic field if rotation is fast enough, and a magnetosphere that can shield an atmosphere from the solar wind.

The Role of Distance from the Sun

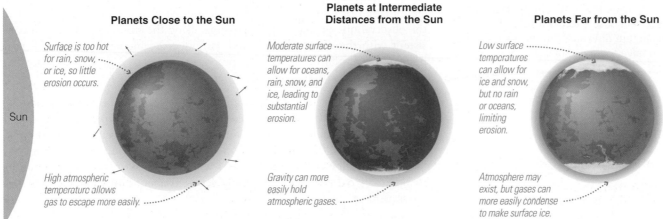

Planets Close to the Sun

Planets at Intermediate Distances from the Sun

Planets Far from the Sun

Sun

Surface is too hot for rain, snow, or ice, so little erosion occurs.

High atmospheric temperature allows gas to escape more easily.

Moderate surface temperatures can allow for oceans, rain, snow, and ice, leading to substantial erosion.

Gravity can more easily hold atmospheric gases.

Low surface temperatures can allow for ice and snow, but no rain or oceans, limiting erosion.

Atmosphere may exist, but gases can more easily condense to make surface ice.

Figure 5.23 | This illustration shows how a terrestrial world's size and distance from the Sun determine its geological history. Earth has oceans and life because it is large enough and at a suitably moderate distance from the Sun.

of our planet are responsible for allowing life: (1) a size that was large enough to retain internal heat and drive plate tectonics and (2) a distance from the Sun that allowed outgassed water vapor to rain down to form oceans. In that case, life could be common in the universe if planets with similar size and orbit to Earth are common around other stars. **Figure 5.23** summarizes how a planet's size and distance from the Sun determine its fate.

✳ THE PROCESS OF SCIENCE IN ACTION

5.3 Global Warming

Our planet seems to regulate its own climate quite effectively over long time scales, but fossil and geological evidence tells us that substantial and rapid swings in global climate can occur on shorter ones. Past climate changes have been due to natural causes. Today, Earth is undergoing climate change for a new reason: human activity that is rapidly increasing the atmospheric concentration of carbon dioxide and other greenhouse gases, leading to

Figure 5.24 | Average global temperatures from 1880 through 2008. Notice the clear global warming trend of the past few decades. (Data from the National Climate Data Center.)

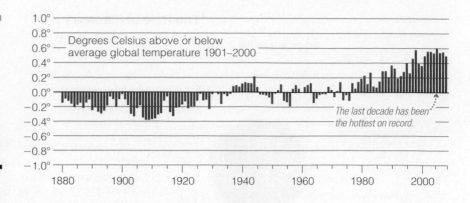

Common Misconceptions

THE GREENHOUSE EFFECT IS BAD

Discussions about environmental problems sometimes make the greenhouse effect sound hazardous, but in itself the greenhouse effect is not a bad thing. In fact, we could not exist without it, since it is responsible for keeping our planet warm enough for liquid water to flow in the oceans and on the surface. Why, then, is the greenhouse effect discussed as an environmental problem? The reason is that human activity is adding more greenhouse gases to the atmosphere—and scientists agree that the additional gases are warming Earth's climate. When you combine that fact with the fact that a much more extreme greenhouse effect is responsible for the searing 470°C temperature of Venus, it becomes clear that it's possible to have too much of a good thing.

global warming (**Figure 5.24**). Global warming is one of the most important issues of our time, so we will devote this chapter's case study of the process of science in action to its study and to the role of models in predicting its implications for the future.

What is the evidence for global warming?

As with most topics in science, our understanding of global warming rests on a combination of observations and modeling. Models are designed to explain the observations as accurately as possible, using the simplest explanation consistent with the observed facts, and we judge the models by how well they predict the results of additional observations [Section 3.2]. The case linking global warming with human activity rests on three basic facts.

1. The greenhouse effect is a simple and well-understood scientific model (see Figure 5.10). We can be confident in our understanding of it because it so successfully explains the observed surface temperatures of other planets. Given this basic model, there is no doubt that a rising concentration of greenhouse gases would make our planet warm up more than it would otherwise; the only debate is about how soon and how much.

2. The burning of fossil fuels and other human activity is clearly increasing the amounts of greenhouse gases in the atmosphere (**Figure 5.25**). Ob-

Figure 5.25 | This diagram shows the atmospheric concentration of carbon dioxide over the past 400,000 years. The data for the past half century come from direct measurements; most of the earlier data come from studies of air bubbles trapped in Antarctic ice. The concentration is measured in parts per million (ppm), which is the number of molecules among every 1 million air molecules.

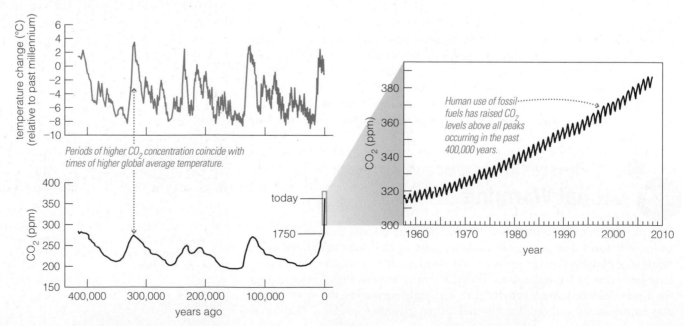

servations show that the atmospheric concentration of carbon dioxide is currently significantly higher (by about 30%) than it has been at any time during the past million years, and it is rising rapidly.

3. Climate models that ignore human activity fail to match the observed rise in global temperatures. In contrast, climate models that include the enhanced greenhouse effect from human production of greenhouse gases match the observed temperature trend quite well (**Figure 5.26**). Comparisons between observations and models therefore clearly indicate that global warming results from human activity.

With scientific evidence in hand that links global warming to human activity, how do we proceed? Again, the comparison between observations and models is crucial. When making predictions for the future, we need to consider the whole range of models that are currently consistent with the data. There are still uncertainties in our models of climate change, but they generally predict that if current trends in the greenhouse gas concentration continue—that is, if we do nothing to slow our emissions of carbon dioxide and other greenhouse gases—the warming trend will accelerate. By the end of this century, the global average temperature will be 3°C–5°C (6°F–10°F) higher than it is now, giving our children and grandchildren the warmest climate that any generation of *Homo sapiens* has ever experienced.

Models can also be used to predict the consequences of a continued rise in greenhouse gas concentrations. Although a temperature increase of a few degrees might not sound so bad, small changes in *average* temperature can lead to much more dramatic changes in regional climate patterns. We already see such dramatic changes occurring in polar regions, and some scientists suspect that recent upswings in the severity of hurricanes and other storms may also be tied to global warming. Another serious threat comes from increase in sea level: Water expands very slightly as it warms—so slightly that we don't notice the change in a glass of water, but enough that sea level has risen some 20 centimeters in the past hundred years. This effect alone could cause sea level to rise as much as another meter during this century, with potentially devastating effect on coastal communities and low-lying countries such as Bangladesh. The added effect of melting ice could increase sea level much more. Recent data suggest that the Greenland ice sheet is melting much more rapidly than models have predicted. If this trend continues, sea level could rise as much as *several meters* by the end of this century—enough to flood most of Florida (**Figure 5.27**). Looking further ahead, complete melting of the polar ice caps would increase sea level by some 70 meters (more than 200 feet). Although such melting would probably take centuries or millennia, it suggests the disconcerting possibility that future generations will have to send deep-sea divers to explore the underwater ruins of our major cities.

Fortunately, climate models suggest that we still have time to avert the most serious consequences of global warming, provided we dramatically and rapidly curtail our greenhouse gas emissions. That can happen only if humans broadly understand the consequences of their actions. A century ago, as cars were first mass produced and electricity became common, our use of fossil fuels seemed to provide only benefit to humanity; no one at the time foresaw the problems that would later be created by the release of toxic pollutants and carbon dioxide. If we hope to avoid other problems in the future, we must continue to study the potential consequences of our actions, through a combination of scientific modeling and observations, and make our choices accordingly.

Think about it   What do *you* think we should do to alleviate the threat of global warming?

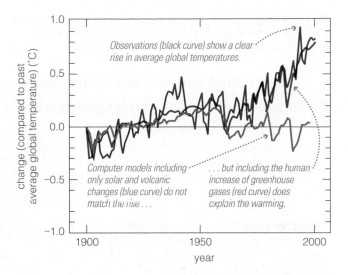

Figure 5.26 | This graph compares observed temperature changes (black curve) with the predictions of climate models. The blue curve represents models that include only natural factors, such as changes in the brightness of the Sun and effects of volcanoes. The red curve represents models that also include the human contribution due to increasing greenhouse gas concentrations. Only the red curve matches the observations well, providing very strong evidence that global warming is a result of human activity.

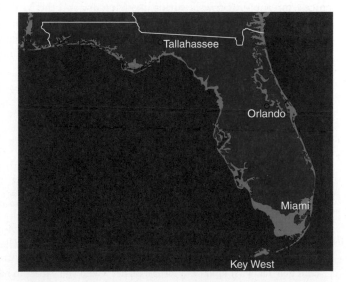

Figure 5.27 | This diagram shows the change in Florida's coastline that would occur if sea levels rose by 1 meter. Some models predict that this rise could occur within a century. The light blue regions show portions of the existing coastline that would be flooded.

Summary of Key Concepts

5.1 Terrestrial Surfaces and Atmospheres

What determines a world's level of geological activity?

A world's size determines its level of geological activity. Larger worlds retain their internal heat longer and therefore experience more **volcanism** and **tectonics.** They also experience more **erosion** because they retain thick atmospheres. **Impact cratering** occurred extensively on all terrestrial worlds, but its effects are visible mainly on worlds where other geological activity has been minimal.

How does an atmosphere affect conditions for life?

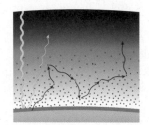

Two crucial effects are (1) protecting the surface from dangerous solar radiation, such as X rays and ultraviolet light, and (2) the **greenhouse effect,** which traps heat in a planet's atmosphere and makes it warmer.

5.2 Histories of the Terrestrial Worlds

Why did the terrestrial worlds turn out so differently?

A terrestrial world's current state depends on its size and its distance from the Sun. Small worlds like Mercury and the Moon are now geologically dead but experienced volcanism and tectonics when they had warmer interiors. Moderately sized Mars is large

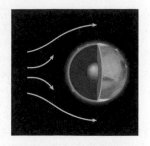

enough to have had a thick atmosphere that made it wet and warm in the past, but it became a frozen wasteland after it lost its atmosphere. Venus and Earth are almost the same size, but because Venus is slightly closer to the Sun, a **runaway greenhouse effect** caused it to become much hotter than Earth.

What unique features of Earth are important for life?

Unique features of Earth on which we depend for survival are (1) surface liquid water, made possible by Earth's moderate temperature; (2) atmospheric oxygen, a product of photosynthetic life; (3) plate tectonics, driven by internal heat; and (4) climate stability, a result of the **carbon dioxide cycle,** which relies on plate tectonics.

✳ THE PROCESS OF SCIENCE IN ACTION

5.3 Global Warming

What is the evidence for global warming?

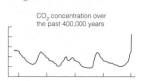

CO_2 concentration over the past 400,000 years

Scientists are convinced that human activity causes **global warming** because recent observations of rising global temperatures agree with models that include human production of greenhouse gases and disagree with models that do not account for human activity.

Investigations

Quick Quiz

Choose the best answer to each of the following; answers are in Appendix D. Explain your reasoning with one or more complete sentences.

1. Which heat source continues to contribute to Earth's *internal* heat? (a) accretion (b) radioactive decay (c) sunlight

2. In general, what kind of terrestrial planet would you expect to have the thickest lithosphere? (a) a large planet (b) a small planet (c) a planet located far from the Sun

3. Which of a planet's fundamental properties has the greatest effect on its level of volcanic and tectonic activity? (a) size (b) distance from the Sun (c) rotation rate

4. Which describes our understanding of flowing water on Mars? (a) It was never important. (b) It was important once, but no longer. (c) It is a major process on the Martian surface today.

5. What do we conclude if a planet has few impact craters of any size? (a) The planet was never bombarded by asteroids or comets. (b) Its atmosphere stopped impactors of all sizes. (c) Other geological processes have erased many craters.

6. How many of the five terrestrial worlds are considered to be geologically dead? (a) none (b) two (c) four

7. Which terrestrial world has the most atmospheric gas? (a) Venus (b) Earth (c) Mars

8. Which of the following is a strong greenhouse gas? (a) nitrogen (b) water vapor (c) oxygen

9. The oxygen in Earth's atmosphere was released by (a) volcanic outgassing. (b) the CO_2 cycle. (c) life.

10. Most of the CO_2 that was outgassed from Earth's volcanoes (a) is still in the atmosphere. (b) escaped to space. (c) is locked up in rocks.

11. What is the leading theory for Venus's lack of water? (a) Venus formed with very little water. (b) Its water is underground. (c) Its water molecules were split by sunlight, allowing the hydrogen to escape to space.

12. What gas is primarily responsible for recent global warming? (a) ozone (b) water vapor (c) CO_2

Short-Answer/Essay Questions

Explain all answers clearly, with complete sentences and proper essay structure if needed. An asterisk (*) designates a quantitative problem, for which you should show all your work.

13. *Miniature Mars.* Suppose Mars had turned out to be significantly smaller than its current size—say, the size of our Moon. How would this have affected its number of geological features due to each of the four major geological processes? Would this smaller Mars still be a good candidate for harboring extraterrestrial life?

14. *Two Paths Diverged.* Use the fundamental properties of size and distance from the Sun to explain why Earth has oceans and very little atmospheric carbon dioxide, while Venus has a thick carbon dioxide atmosphere.

15. *Change in Formation Properties.* Choose either size or distance from the Sun and suppose it had been different for Earth. (For example, suppose Earth had been smaller in size or a greater distance from the Sun.) Describe how this change might have affected Earth's subsequent history and the possibility of life on Earth.

16. *Predictive Geology.* Suppose another star system has a rocky terrestrial planet that is twice as large as Earth but at the same distance from its star (which is just like our Sun). Describe the type of geology you would expect the planet to have.

17. *Experiment: Geological Properties of Silly Putty.* Roll room-temperature Silly Putty into a ball and measure its diameter. Place the ball on a table and gently place one end of a heavy book on it. After 5 seconds, measure the height of the squashed ball. Repeat the experiment two more times, the first time warming the Silly Putty in hot water before you start and the second time cooling it in ice water before you start. How do the different temperatures affect the rate of squashing? How does the experiment relate to planetary geology? Explain.

18. *Experiment: Planetary Cooling in a Freezer.* To simulate the cooling of planetary bodies of different sizes, use a freezer and two small plastic containers of similar shape but different size. Fill each container with cold water and put both in the freezer at the same time. Checking every hour or so, record the time and your estimate of the thickness of the "lithosphere" (the frozen layer) in the two tubs. How long does it take the water in each tub to freeze completely? Describe in a few sentences the relevance of your experiment to planetary geology. Extra credit: Plot your results on a graph, with time on the x-axis and lithospheric thickness on the y-axis. What is the ratio of the two freezing times?

19. *Amateur Astronomy: Observing the Moon.* Any amateur telescope has adequate resolution to identify geological features on the Moon. The light highlands and dark maria should be evident, and shadowing is visible near the line between night and day. Try to observe the Moon near the first- or third-quarter phase. Sketch or photograph the Moon at low magnification, and then zoom in on a region of interest. Again sketch or photograph your field of view, label the features, and identify the geological process that created them. Look for craters, volcanic plains, and tectonic features. Estimate the size of each feature by comparing it to the size of the whole Moon (radius = 1738 km).

20. *Global Warming.* What, if anything, should we be doing that we are not doing already to alleviate the threat of global warming? Write a one-page editorial summarizing and defending your opinion.

*21. *Internal vs. External Heating.* Earth's internal heat leaks out through the surface with a power of about 3 trillion watts. Divide this amount by Earth's total surface area to find the rate of heat loss per square meter. Compare your result to the amount of power contributed by sunlight, which provides an average of about 400 watts per square meter in daylight.

*22. *Plate Tectonics.* Plates typically move relative to one another at a speed of 1 centimeter per year. How long would it take for two continents 3000 kilometers apart to collide?

6

The Outer Solar System

Learning Goals

6.1 Jovian Planets, Rings, and Moons
- What are jovian planets like?
- Why are jovian moons so geologically active?

6.2 Asteroids, Comets, and the Impact Threat
- Why are asteroids and comets grouped into three distinct regions?
- Do small bodies pose an impact threat to Earth?

✱ THE PROCESS OF SCIENCE IN ACTION
6.3 Extinction of the Dinosaurs
- Did an impact kill the dinosaurs?

The small blue dot of light just inside Saturn's rings at the left (about the 10:00 position) is Earth, far in the distance.

Beyond the realm of the terrestrial worlds, we enter the outer solar system, dominated by the four large jovian planets. These planets are not alone out there. Saturn, shown above, is orbited by dozens of moons—including one, Titan, that has a thick atmosphere—and the countless small particles that make up its beautiful rings. The other jovian planets also have rings and numerous moons, some with amazing features such as active volcanoes or subsurface oceans. The rest of the outer solar system contains vast numbers of small bodies—asteroids and comets. Although these objects orbit the Sun independently, their orbits are shaped by the gravity of the jovian planets. In this chapter, we'll explore the fascinating objects of the outer solar system and see how, on occasion, they can influence life on Earth.

6.1 Jovian Planets, Rings, and Moons

The mythological namesakes of the jovian planets were rulers among gods: Jupiter was the king of the gods, Saturn was Jupiter's father, Uranus was the lord of the sky, and Neptune ruled the sea. The true majesty of the jovian planets may be even greater. The smallest of these four worlds, Neptune, has a volume more than 50 times that of Earth. The largest, Jupiter, could swallow up more than 1000 Earths. These worlds also differ from the terrestrial planets in composition and nature: They are essentially giant balls of gas, with no solid surface on which to stand. In this section, we'll investigate these worlds and their wondrous rings and moons.

What are jovian planets like?

Most of what we know about the jovian planets comes from the observations made by spacecraft. We have sent spacecraft to all four of the jovian planets, and one—the *Cassini* spacecraft—has been orbiting and studying Saturn since 2004.

Size and Composition Figure 6.1 shows the four jovian planets to scale, along with some basic data. Notice that Jupiter and Saturn are quite similar to each other, but fairly different from Uranus and Neptune. The most important differences are in composition. Jupiter and Saturn are both made almost entirely of hydrogen and helium, with just a few percent of their masses in the form of hydrogen compounds and even smaller amounts of rock and metal. In fact, their overall compositions are much more similar to the composition of the Sun than to the compositions of the terrestrial planets. Some people even call Jupiter a "failed star" because it has a starlike composition but lacks the nuclear fusion needed to make it shine. Its lack of fusion is due to its size: Although Jupiter is large for a planet, it is much less massive than any star. As a result, its gravity is too weak to compress its interior to the extreme temperatures and densities needed for nuclear fusion. (Jupiter would have needed to grow to about 80 times its current mass to have become a star.)

Uranus and Neptune are much smaller than Jupiter and Saturn, and while they also contain substantial amounts of hydrogen and helium, they are made primarily of hydrogen compounds such as water (H_2O), methane (CH_4), and ammonia (NH_3), along with smaller amounts of metal and rock.

We can understand the differences in composition and size among the jovian planets by looking at their formation according to the nebular theory

Figure 6.1 | Jupiter, Saturn, Uranus, and Neptune, shown to scale with Earth for comparison.

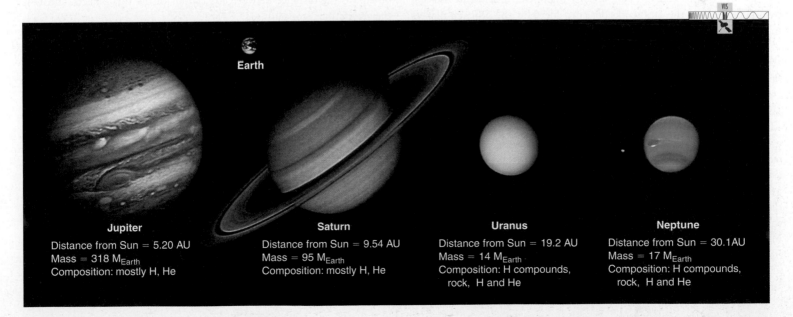

Earth

Jupiter
Distance from Sun = 5.20 AU
Mass = 318 M_{Earth}
Composition: mostly H, He

Saturn
Distance from Sun = 9.54 AU
Mass = 95 M_{Earth}
Composition: mostly H, He

Uranus
Distance from Sun = 19.2 AU
Mass = 14 M_{Earth}
Composition: H compounds, rock, H and He

Neptune
Distance from Sun = 30.1 AU
Mass = 17 M_{Earth}
Composition: H compounds, rock, H and He

[Section 4.2]. Recall that the jovian planets formed beyond the frost line (see Figure 4.7), where it was cold enough for hydrogen compounds to condense into ices. Because hydrogen compounds were so much more abundant than metal and rock, some of the ice-rich planetesimals of the outer solar system grew large enough for their gravity to draw in the hydrogen and helium gas that surrounded them. All four jovian planets are thought to have grown from ice-rich planetesimals of about the same mass—roughly 10 times the mass of Earth—so their differences in composition must stem from the amounts of hydrogen and helium gas that they captured. Jupiter and Saturn captured so much hydrogen and helium gas that these gases now make up the vast majority of their masses. Uranus and Neptune pulled in much less hydrogen and helium gas, leaving their bulk compositions similar to the compositions of the ice-rich planetesimals around which they grew. But if they all started from planetesimals of about the same size, why did Uranus and Neptune pull in so much less gas? The answer likely lies in their distance from the Sun: Because the density of the solar nebula was lower at greater distances, it took longer for planetesimals to accrete at the distances of Uranus and Neptune than at the distances of Jupiter and Saturn. Because all the planets stopped accreting gas at the same time—when the solar wind blew the remaining gas into interstellar space—the more distant planets had less time to capture gas and ended up smaller in size, with less hydrogen and helium.

 Think about it Suppose that solar wind had taken longer to clear the solar nebula. How would you expect the jovian planets to be different in that case? Explain.

Interior Structure   The jovian planets are often called "gas giants," making it sound as if they are entirely gaseous, like air on Earth. The reality is more

---

Tools of Science: Newton's Version of Kepler's Third Law

Have you ever wondered how we measure the masses of the jovian planets, the Sun, or other distant objects in the universe? Newton's universal law of gravitation gives us the key tool, as long as we can observe an orbiting object. To see how the idea works, let's consider how the orbits of planets tell us the mass of the Sun.

The Sun's gravity holds the planets in orbit. The stronger the gravity, the stronger the pull on a planet and therefore the faster it travels in its orbit around the Sun. That is why Kepler's third law holds true: Planets closer to the Sun orbit at higher speed than planets farther away, because gravity is stronger closer to the Sun. Now consider what would happen if we replaced the Sun with a more massive star. In that case, the pull of gravity would be even stronger at any particular distance, so all the planets would have to orbit faster if they stayed at their current orbital distances.

Newton made these ideas quantitative by using his universal law of gravitation to determine the relationship between the period and orbital distance of an orbiting object. In doing so, he discovered that Kepler's third law in its original form ($p^2 = a^3$) is actually just a special case of a more general formula that follows directly from the universal law of gravitation. This formula is called **Newton's version of Kepler's third law**:

$$p^2 = \frac{4\pi^2}{G(M_1 + M_2)}\, a^3$$

where M_1 and M_2 are the masses of two objects that are gravitationally attracting each other, p is the orbital period of one object around the other, a is the distance between the centers of the two objects, and G is the gravitational constant.

Let's see how this formula allows us to calculate the mass of the Sun. Because the Sun is so much more massive than any planet, the sum of the masses, $M_{Sun} + M_{planet}$, is pretty much just M_{Sun}. Therefore, we can use the orbital period p and distance of any planet a to calculate the mass of the Sun; note that we'll get the same answer for the Sun's mass no matter which planet we use.

The same idea works for other distant objects. We can calculate the mass of Jupiter from the orbital period and distance of any of its moons. We can calculate the mass of Earth from the orbital period and distance of our Moon. We can even calculate the sum of the masses of two stars in a distant binary star system from their orbits around each other. When Newton extended Kepler's law, he achieved one of a scientist's main goals: to take what's been learned in one situation and apply it more universally. Today, Newton's version of Kepler's third law is the primary means by which we determine masses throughout the universe.

complex, because the strong gravity of these planets compresses most of the "gas" into forms of matter quite unlike anything we are familiar with in everyday life on Earth. To understand the idea, imagine what would happen if you could plunge into Jupiter.

The visible clouds of Jupiter extend downward only a few tens of kilometers into Jupiter's atmosphere. Below that, you'd never encounter a solid surface, but you'd find increasingly higher temperatures and pressures as you descended. These temperatures and pressures would soon destroy any real spacecraft entering Jupiter. When the *Galileo* spacecraft dropped a scientific probe into Jupiter in 1995, the probe provided valuable data about Jupiter's atmosphere, but it survived to only a depth of about 200 kilometers below the cloud tops.

We use computer models to determine the structure of Jupiter's interior at greater depths, and **Figure 6.2** shows the layers you'd encounter as you continued downward into Jupiter. The layers do not differ much in composition—all except the core are mostly hydrogen and helium—but they differ in the phase (such as liquid or gas) of their hydrogen. The uppermost layer has conditions in which hydrogen remains in its familiar gaseous form; this layer extends about 10% of the way downward. In most of the rest of Jupiter, the temperatures and pressures are so extreme that hydrogen is forced into a compact form that has the electrical properties of metal but still flows like a liquid. Jupiter's strong magnetic field is generated in this layer of metallic hydrogen. Below that, if you could find some way to survive the extreme pressures and temperatures, you'd finally encounter Jupiter's core, which contains a mix of hydrogen compounds, rock, and metal. However, the high temperature and pressure ensure that this mix bears little resemblance to familiar solids or liquids. The core contains about 10 times as much mass as the entire Earth, but it is only about the same size as Earth because it is compressed to such high density.

All four jovian planets have cores of about the same mass, so their interiors differ mainly in the layers around their cores. Saturn's interior layers are very similar to those of Jupiter, except Saturn's weaker gravity means that the pressures necessary for liquid and metallic hydrogen occur at greater depths. Pressures within Uranus and Neptune are not high enough to form liquid or metallic hydrogen at all. Each of these two planets has only a thick layer of gaseous hydrogen surrounding its core of hydrogen compounds, rock, and metal. In fact, the core material in Uranus and Neptune may be liquid, making for very odd "oceans" buried deep inside these planets. Some scientists have speculated that life might exist in these cores, but they are so deeply buried that it would be difficult to explore them.

Jovian Weather Jovian atmospheres have dynamic winds and weather, with colorful clouds and enormous storms. Weather on these planets is driven not only by energy from the Sun (as on the terrestrial planets) but also by heat generated within the planets themselves. While the amount of heat escaping from the interiors of the terrestrial planets is very small compared to the amount of heat provided by sunlight, the heat escaping from within the jovian planets can rival solar heating. All but Uranus generate a great deal of internal heat. No one knows the precise source of this internal heat, but it probably comes from conversion of gravitational potential energy into thermal energy, which is occurring either because these planets continue to shrink slowly but imperceptibility in size or because they have ongoing differentiation as heavier materials continue to sink toward their cores.

The jovian planets all have very cloudy atmospheres, but their clouds are different from those on Earth. Clouds form when a gas condenses to make tiny liquid droplets or solid flakes. Earth's atmosphere contains only one ingredient that can condense into clouds: water vapor. The jovian planets have several gases that can condense to form clouds. Because different gases condense at different temperatures, these planets have distinctive cloud layers at different altitudes. For example, Jupiter has three primary cloud layers (**Figure 6.3**). The

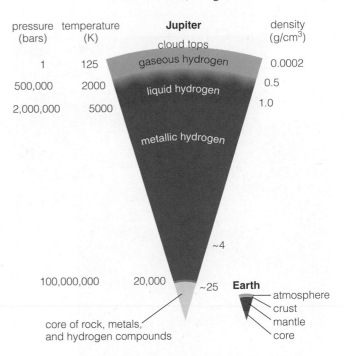

Figure 6.2 | Jupiter's interior structure, labeled with the pressure, temperature, and density at various depths. Earth's interior structure is shown to scale for comparison. (Notes on units: 1 *bar* is approximately the atmospheric pressure at sea level on Earth; the density of liquid water is 1 g/cm³.)

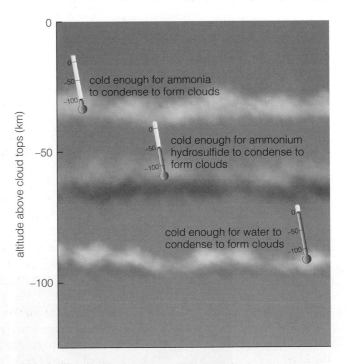

Figure 6.3 | This diagram shows how different cloud layers form at different altitudes in Jupiter's atmosphere. The different layers form because temperature is lower at higher altitudes; different gases condense at different temperatures. (The tops of the ammonia clouds are usually considered the zero altitude for Jupiter, which is why lower altitudes are negative.)

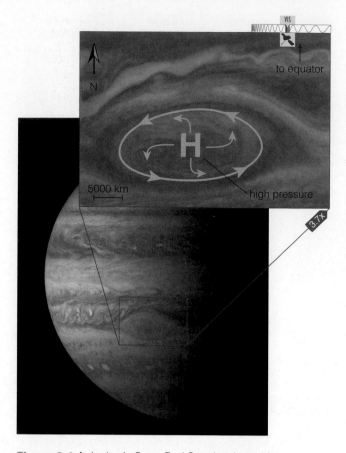

Figure 6.4 | Jupiter's Great Red Spot is a huge high-pressure storm that is large enough to swallow Earth. The overlaid diagram shows a weather map of the region. The global image was taken by the *Cassini* spacecraft and the Red Spot close-up by *Voyager*.

lowest layer occurs at a depth of about 100 kilometers, where temperatures are Earth-like and water can condense to form clouds. The temperature is lower at higher altitudes; about 50 kilometers above the water clouds, it is cold enough for the gas ammonium hydrosulfide (NH_4SH) to condense into clouds. These ammonium hydrosulfide clouds give Jupiter its distinctive tan and red colors. Higher still, the temperature is so cold that ammonia (NH_3) condenses to make an upper layer of white clouds. Jupiter's striped appearance occurs because atmospheric flow patterns create cloudy and clear zones at different latitudes, much as Earth's equatorial rainforests are cloudier than the desert regions on either side. Saturn has the same set of three cloud layers as Jupiter, but these layers occur deeper in Saturn's cooler atmosphere, making the colors more subdued. Uranus and Neptune contain more methane gas, and they're so cold that some of it condenses into methane clouds. Methane gas absorbs red light from the Sun and therefore gives Uranus and Neptune their distinctive blue colors.

All the jovian planets also have strong winds and powerful storms. The most famous storm is Jupiter's *Great Red Spot*, which is more than twice as wide as Earth. It is somewhat like a giant hurricane, except that its winds circulate around a high-pressure region rather than a low-pressure region (**Figure 6.4**). It is also extremely long-lived compared to storms on Earth: Astronomers have seen it throughout the three centuries during which telescopes have been powerful enough to detect it. No one knows why the Great Red Spot has lasted so long. However, storms on Earth tend to lose their strength when they pass over land—perhaps Jupiter's biggest storms last for centuries simply because there's no solid surface there to sap their energy.

Rings All of the jovian planets have rings, but only Saturn's rings are easily visible from Earth. They are also the best-studied rings, so we'll use them to discuss general properties of ring systems.

Earth-based views make Saturn's rings appear to be continuous, concentric sheets of material separated by a large gap called the *Cassini division* (**Figure 6.5a**). Spacecraft images reveal these sheets to be made of many individual rings, each separated from the next by a narrow gap (**Figure 6.5b**). But even these images are somewhat deceiving. If we could wander into Saturn's rings, we'd find that they are made of countless icy particles, ranging in size from dust grains to large boulders (**Figure 6.5c**). Each individual ring particle orbits Saturn independently in accord with Kepler's laws, so the rings are much like a collection of vast numbers of tiny moons.

Close-up photographs show an astonishing number of rings, as well as gaps, ripples, and other features in the rings. Scientists are still struggling to explain all these features, but some general ideas are now clear. Rings and gaps are caused by particles bunching up at some orbital distances and being forced out at others. This bunching happens when gravity nudges the orbits of ring particles in some particular way. One source of nudging comes from small moons located within the gaps in the rings themselves, sometimes called *gap moons* (**Figure 6.6**).

Figure 6.5 | Zooming in on Saturn's rings.

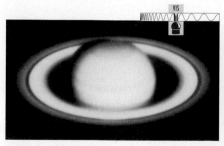

a This Earth-based telescopic view of Saturn makes the rings look like large concentric sheets.

b This image of Saturn's rings from the *Cassini* spacecraft reveals many individual rings separated by narrow gaps.

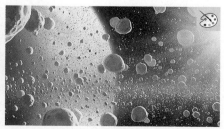

c This artist's conception shows what we would see if we could enter the rings, which are made of countless small particles.

Ring particles also may be nudged by the gravity of larger, more distant moons. For example, a ring particle orbiting about 120,000 kilometers from Saturn's center will circle the planet in exactly half the time it takes Saturn's moon Mimas to orbit. Every time Mimas returns to a certain location, the ring particle will also be at its original location and therefore will experience the same gravitational nudge from Mimas. The periodic nudges reinforce one another and clear a gap in the rings—in this case, the large gap visible from Earth (the Cassini division). This type of reinforcement due to repeated gravitational nudges is called an **orbital resonance;** the name comes from the idea that the repeated nudges *resonate* with one another, amplifying their effects. Other orbital resonances, caused by moons both within the rings and farther out from Saturn, probably explain most of the intricate structures visible in ring photos.

Where do rings come from? Scientists once guessed that ring particles might be chunks of rock and ice left over from the time at which the planets formed, but we now know that particles of the size we find in the rings today could not have survived for billions of years. Ring particles are continually being ground down in size, primarily by the impacts of the countless sand-size particles that orbit the Sun—the same types of particles that become meteors in Earth's atmosphere and cause micrometeorite impacts on the Moon [Section 5.2]. Millions of years of such tiny impacts would have ground Saturn's existing ring particles to dust long ago. In the case of Uranus, ring particles are also lost as a result of drag in the planet's thin upper atmosphere.

We are left with only one reasonable possibility: New ring particles must be continually replacing those that are destroyed. The most likely source is numerous small *moonlets*—moons the size of gap moons (see Figure 6.6)—that formed in the disks of material orbiting the young jovian planets. Tiny impacts are gradually grinding away these small moonlets, like the ring particles themselves, but they are large enough to still exist despite 4.5 billion years of such sandblasting. This allows them to contribute ring particles in two ways. First, each tiny impact releases particles from a small moonlet's surface, and these released particles become new dust-size ring particles. Second, occasional larger impacts can shatter a small moonlet completely, creating a supply of boulder-size ring particles. **Figure 6.7** summarizes the way small moonlets are thought to supply the particles that make up the rings around the jovian planets.

Why are jovian moons so geologically active?

In addition to their rings, the jovian planets are orbited by more than 150 known moons. Jupiter has the most, with more than 60 moons known to date. It's helpful to organize these moons into three groups by size: small moons less than about 300 kilometers in diameter, medium moons ranging from about 300 to 1500 kilometers in diameter, and large moons more than 1500 kilometers in diameter. The vast majority of the jovian moons are small; many are no more than a few kilometers in diameter. Most of these small moons are probably asteroids or comets that were captured into orbit when the jovian planets were still surrounded by swirling clouds of gas [Section 4.2].

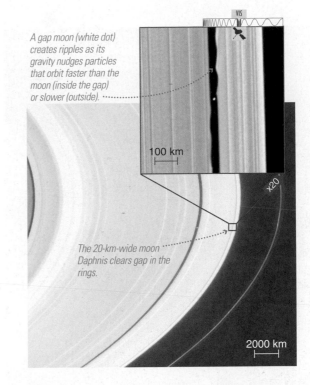

A gap moon (white dot) creates ripples as its gravity nudges particles that orbit faster than the moon (inside the gap) or slower (outside).

100 km

The 20-km-wide moon Daphnis clears gap in the rings.

2000 km

Figure 6.6 | Small moons within the rings have important effects on ring structure (*Cassini* photos).

Figure 6.7 | This illustration summarizes the origin of rings around the jovian planets.

jovian planet

Tidal forces near the planet prevent small moonlets from accreting into larger moons.

Moonlets are occasionally disrupted by impacts.

Ongoing small impacts blast off dust and debris to form the rings.

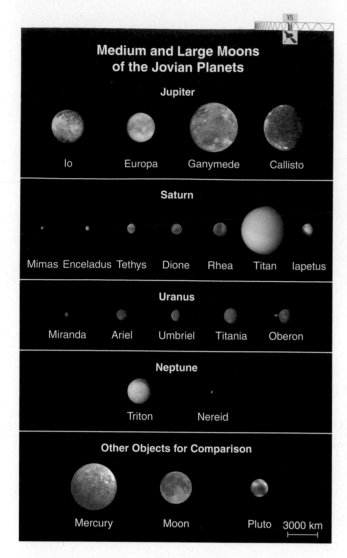

Medium and Large Moons of the Jovian Planets

Jupiter

Io   Europa   Ganymede   Callisto

Saturn

Mimas Enceladus Tethys   Dione   Rhea   Titan   Iapetus

Uranus

Miranda   Ariel   Umbriel   Titania   Oberon

Neptune

Triton   Nereid

Other Objects for Comparison

Mercury   Moon   Pluto   3000 km

Figure 6.8 | The medium and large moons of the jovian planets, with sizes (but not distances) shown to scale. Mercury, the Moon, and Pluto are included for comparison.

Figure 6.9 | Io is the most volcanically active body in the solar system.

Figure 6.8 shows all of the medium and large moons to scale. These worlds are all large enough for gravity to have shaped them into spheres, so they would qualify as dwarf planets or planets if they orbited the Sun independently. Because they have solid surfaces, these moons are shaped by the same four geological processes as the terrestrial worlds—impact cratering, volcanism, tectonics, and erosion. The most surprising aspect of these moons is their level of geological activity. Because even the largest of them are only slightly larger than the Moon and Mercury, we might expect all of these moons to be geologically dead, like those two terrestrial worlds. Instead, we find evidence of tremendous past geological activity on the jovian moons, and some of them remain geologically active today.

How can such small worlds have so much geological activity? The best way to understand the answer is to look at a few of the most interesting cases. As we will see, their geological activity can be traced to two major factors. First, because they formed in the cold outer solar system, the jovian moons have compositions that include substantial amounts of ice in addition to metal and rock, and much less heat is required for "ice geology" than for rock geology. Second, interactions between orbiting moons helps to create a source of heating, called *tidal heating*, that is not present in the terrestrial worlds.

The Galilean Moons of Jupiter   We begin our tour of selected jovian moons with the *Galilean moons* of Jupiter—the four moons discovered by Galileo [Section 3.1]. For anyone who thinks of moons solely as barren, cratered worlds like our own Moon, these moons shatter the stereotype. Io is by far the most volcanically active world in our solar system. Europa looks like a ball of cracked ice and probably has a deep ocean of liquid water beneath its icy crust. Ganymede and Callisto have more impact craters but may also contain subsurface oceans.

Let's start by considering Io (**Figure 6.9**). Its many active volcanoes tell us that Io must be quite hot inside. However, Io is only about the size of our Moon, so it should have long ago lost any heat from its birth and is too small for radioactivity to provide much ongoing heat. It must therefore have some other source of heat for its interior. Scientists have identified this source as **tidal heating,** which arises from tidal forces exerted by Jupiter.

You probably know that the Moon's gravity creates tides on Earth. Tides arise because the strength of gravity declines with distance, so the gravitational attraction between the Moon and the side of Earth facing the Moon is stronger than that between the Moon and the side of Earth facing away from the Moon. This difference in attraction creates a "stretching force," or **tidal force,** that stretches the entire Earth to create two tidal bulges: one facing the

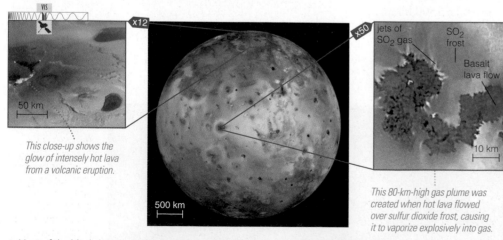

50 km

x12

This close-up shows the glow of intensely hot lava from a volcanic eruption.

500 km

x50

jets of SO_2 gas SO_2 frost   Basalt lava flow

10 km

This 80-km-high gas plume was created when hot lava flowed over sulfur dioxide frost, causing it to vaporize explosively into gas.

a Most of the black, brown, and red spots on Io's surface are recently active volcanic features. White and yellow areas are sulfur dioxide (SO_2) and sulfur deposits, respectively, from volcanic gases. (Photographs from the *Galileo* spacecraft; some colors slightly enhanced or altered.)

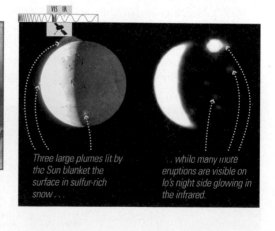

Three large plumes lit by the Sun blanket the surface in sulfur-rich snow . . .

. . . while many more eruptions are visible on Io's night side glowing in the infrared.

b Two views of Io's volcanoes taken by *New Horizons* on its way to Pluto.

Moon and one opposite the Moon. Earth's rotation carries every point on the surface through the two bulges each day, which is why there are two high tides each day. Moreover, although the tidal rise and fall of the oceans is far more noticeable, tides also cause land to rise and fall twice each day (by about a centimeter). This daily motion generates friction inside Earth, and over millions of years this friction is gradually slowing Earth's rotation.

Just as the Moon's gravity causes it to exert a tidal force on Earth, Earth's gravity exerts a tidal force on the Moon. In fact, because Earth is much more massive than the Moon, Earth's tidal force on the Moon has a much greater effect than the Moon's tidal force on Earth. This strong tidal force explains the Moon's synchronous rotation (see Figure 2.11): The Moon probably once rotated faster, but Earth's tidal force generated friction inside the Moon that slowed its rotation until it eventually kept the same face toward Earth at all times.

Other planets and moons also exert tidal forces on each other. Jupiter's tidal force on the Galilean moons is so strong that it has caused all four moons to keep the same face toward Jupiter at all times. Tides cause heating on Io, Europa, and Ganymede because these three moons have elliptical orbits around Jupiter, which means the strength of Jupiter's gravity and tidal force acting on them varies as these moons orbit. This varying tidal force flexes the interiors of these moons continuously, generating friction and heat. But why do these moons have elliptical orbits, when almost all other large moons have nearly circular orbits? The answer lies in an orbital resonance much like those that affect Saturn's rings: During the time Ganymede takes to complete one orbit of Jupiter, Europa completes exactly two orbits and Io completes exactly four orbits (**Figure 6.10**). The three moons therefore line up periodically, and over time the repeated gravitational tugs with these alignments have stretched the orbits from circles into ellipses. Tidal heating is strongest for Io because it is closest to Jupiter. Precise calculations show that tidal heating can indeed generate enough heat to explain Io's incredible volcanic activity.

Tidal heating is weaker for Europa but still strong enough to make Europa one of the most interesting places in the solar system. Visually, Europa's surface of water ice shows cracks and other evidence suggesting that ice sometimes melts and refreezes (**Figure 6.11**). We can model what lies under the surface ice with a combination of data from the *Galileo* spacecraft and calculations based on tidal heating. These models indicate that Europa has a metallic core and rocky mantle, surrounded by a layer of water (**Figure 6.12**). The water is frozen solid near the surface, but tidal heating should keep the water liquid a few kilometers beneath the surface, creating a deep subsurface ocean. Several pieces of evidence support the hypothesis that this ocean exists, including close-up photos of the surface and careful studies of Europa's magnetic field.

If Europa's ocean really exists, it probably contains more than twice as much liquid water as all of Earth's oceans combined. Moreover, just as there

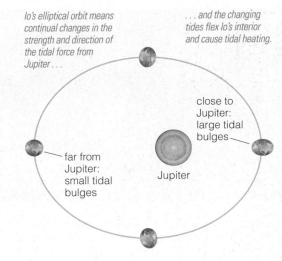

Io's elliptical orbit means continual changes in the strength and direction of the tidal force from Jupiter . . .

. . . and the changing tides flex Io's interior and cause tidal heating.

close to Jupiter: large tidal bulges

far from Jupiter: small tidal bulges

Jupiter

a Tidal heating arises because Io's elliptical orbit (exaggerated in this diagram) causes varying tides.

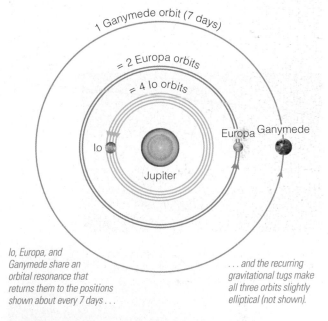

1 Ganymede orbit (7 days)
= 2 Europa orbits
= 4 Io orbits

Io   Europa Ganymede

Jupiter

Io, Europa, and Ganymede share an orbital resonance that returns them to the positions shown about every 7 days . . .

. . . and the recurring gravitational tugs make all three orbits slightly elliptical (not shown).

b Io's orbit is elliptical because of the orbital resonance it shares with Europa and Ganymede.

Figure 6.10 | These diagrams explain the cause of tidal heating on Io. Tidal heating has a weaker effect on Europa and Ganymede, because they are farther from Jupiter and tidal forces weaken with distance.

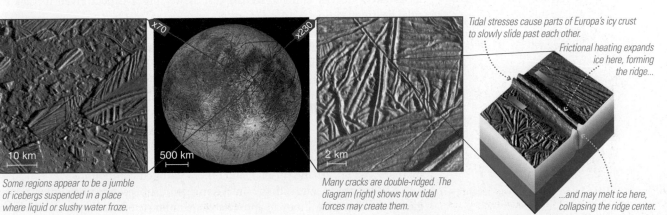

Some regions appear to be a jumble of icebergs suspended in a place where liquid or slushy water froze.

Many cracks are double-ridged. The diagram (right) shows how tidal forces may create them.

Tidal stresses cause parts of Europa's icy crust to slowly slide past each other.

Frictional heating expands ice here, forming the ridge...

...and may melt ice here, collapsing the ridge center.

Figure 6.11 | Europa's icy crust may hide a deep, liquid water ocean beneath its surface. These photos are from the *Galileo* spacecraft; colors are enhanced in the global view.

10 km   500 km   2 km   x70   x230

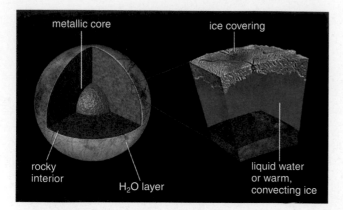

Figure 6.12 | This diagram shows one model of Europa's interior structure. There is little doubt that the H₂O layer is real, but questions remain about whether the material beneath the icy crust is actually liquid water, relatively warm convecting ice, or some of each.

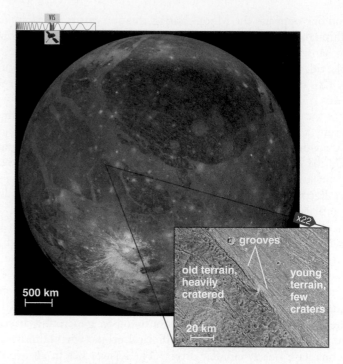

Figure 6.13 | Ganymede, the largest moon in the solar system, has both old and young regions on its surface of water ice. The dark regions are heavily cratered and must be billions of years old, while the light regions are younger landscapes where eruptions of water have presumably erased ancient craters; the long grooves in the light regions were probably formed by water erupting along surface cracks. Notice that the boundary between the two types of terrain can be quite sharp.

Callisto is heavily cratered, indicating an old surface that nonetheless may hide a deeply buried ocean.

Close-up photo of craters.

Figure 6.14 | Callisto, the outermost of the four Galilean moons, has a heavily cratered icy surface.

are undersea volcanoes on Earth's seafloor, Europa may also have volcanic vents deep in its ocean. Knowing that primitive life thrives near seafloor vents on Earth has made scientists wonder if Europa might also support life in its ocean, a possibility we will explore further in Chapter 15.

Ganymede and Callisto also have surfaces of water ice, and some data suggest that these moons have subsurface oceans. However, tidal heating alone is not strong enough to explain how Ganymede could have an ocean, and tidal heating plays no role at all on Callisto, which does not participate in the orbital resonance that affects the other three Galilean moons. If these moons have subsurface oceans, the heat may be supplied by ongoing radioactive decay. With or without oceans, each of these worlds shows fascinating surface geology. Ganymede has some regions that are dark and densely cratered, suggesting that they look much the same today as they did billions of years ago, while other regions are light-colored with very few craters, suggesting that liquid water has recently flowed and refrozen (**Figure 6.13**). Callisto, by contrast, looks like a heavily cratered iceball, just as we might expect given its lack of tidal heating (**Figure 6.14**).

Saturn's Active Moons: Titan and Enceladus   We next turn to two of Saturn's moons, Titan and Enceladus. Titan is the second largest moon in the solar system (after Ganymede). It is also unique among the moons of our solar system in having a thick atmosphere—so thick that it hides the surface from view, except at a few specific wavelengths of light (**Figure 6.15**). The atmosphere is about 90% nitrogen, not that different from the 77% nitrogen content of Earth's atmosphere. However, on Earth the rest of the atmosphere is mostly oxygen, while the rest of Titan's atmosphere consists of argon, methane (CH_4), ethane (C_2H_6), and other hydrogen compounds.

The methane and ethane in Titan's atmosphere are both greenhouse gases, and they give Titan an appreciable greenhouse effect [Section 5.1], which

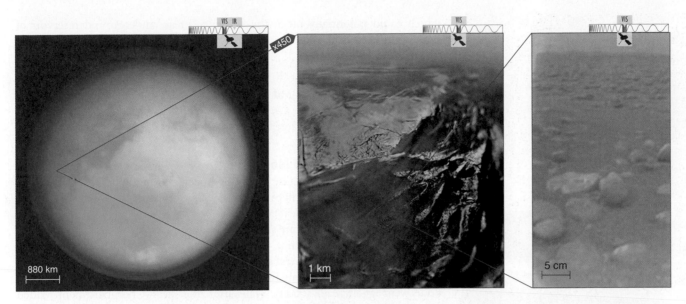

Figure 6.15 | This sequence zooms in on the *Huygens* landing site on Titan. Left: A global view taken by the *Cassini* orbiter using filters designed to peer through the atmosphere at the specific near-infrared wavelengths least affected by the atmosphere. Center: An aerial mosaic of images taken by the probe during descent. Right: A surface view taken by the probe after landing; the "rocks," which are 10 to 20 centimeters across, are presumably made of ices.

makes it warmer than it would be otherwise. Still, because of its great distance from the Sun, Titan's surface temperature is a frigid 93 K (−180°C). The surface pressure on Titan is about 1.5 times the sea level pressure on Earth, which would be fairly comfortable if not for the lack of oxygen and the cold temperatures.

A moon with a thick atmosphere is intriguing enough, but we have at least two other reasons for special interest in Titan. First, its complex atmospheric chemistry probably produces numerous organic chemicals—the chemicals that are the basis of life. Second, although it is far too cold for liquid water to exist on Titan, conditions are right for liquid methane or ethane to exist. NASA and the European Space Agency (ESA) therefore joined forces to explore Titan with the *Cassini* spacecraft that orbits Saturn and a European-built probe, called *Huygens* (pronounced "Hoy-guns"), that parachuted to a soft landing on Titan in January 2005. During its descent, the probe photographed river valleys merging together, flowing down to what looks like a shoreline. On the ground, instruments discovered that the surface has a hard crust but is a bit squishy below, like sand with liquid mixed in. The view from the surface shows "ice boulders" rounded by erosion. All these results support the idea of a wet climate—but wet with liquid methane or ethane rather than liquid water. More recent observations from *Cassini* have confirmed the presence of liquid methane or ethane lakes on Titan (**Figure 6.16**).

We can attribute Titan's geological activity to the combination of its relatively large size and icy composition. Because all of Saturn's other moons are much smaller, scientists had expected them to show very little evidence of active geology. The reality has proven quite different. Each of Saturn's six medium-size moons shows evidence of substantial geological activity in its past. Most amazingly, one moon—Enceladus—shows clear evidence of *ongoing* geological activity, despite being barely 500 kilometers across, small enough to fit inside the borders of Colorado. Enceladus features strange grooves near its south pole that vent huge clouds of water vapor and ice crystals (**Figure 6.17**). These fountains are driven by internal heat, presumably due to tidal heating created by orbital resonances between Enceladus and other moons of Saturn. Moreover, the fountains must have some subsurface source,

Figure 6.16 | Radar image of Titan's north polar region. The blue and black regions may be lakes of liquid methane or ethane at a temperature of −180°C.

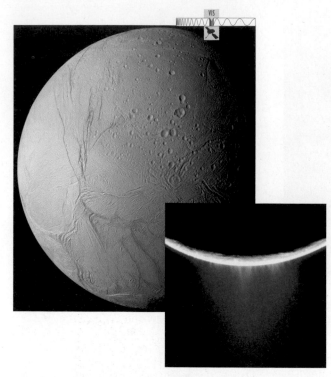

Figure 6.17 | *Cassini* photo of Enceladus. The blue "tiger stripes" near the bottom of the main photo are regions of fresh ice that must have recently emerged from below. The colors are exaggerated; the image is a composite made at near-ultraviolet, visible, and near-infrared wavelengths. The inset shows Enceladus backlit by the Sun, with fountains of ice particles and water vapor spraying downward.

which could potentially mean the existence of an underground reservoir of liquid water. In that case, there would be a slim possibility that Enceladus could harbor life.

The *Cassini* spacecraft continues to study Titan and Enceladus as it orbits Saturn. Do a Web search to find recent results. What new things have we learned about these worlds?

Moons of Uranus and Neptune   We know far less about the moons of Uranus and Neptune, because they have been photographed close up only once, during the *Voyager 2* flybys of the 1980s. Nevertheless, we again see evidence of surprising geological activity.

Uranus has five medium-size moons (and no large moons), and at least three of them show evidence of past volcanism or tectonics. Miranda, the smallest of the five, is the most surprising (**Figure 6.18**). Despite its small size, it shows tremendous tectonic features and relatively few craters. Apparently, it underwent geological activity well after the heavy bombardment ended [Section 4.2], erasing its early craters.

The surprises continue with Neptune's moon Triton (**Figure 6.19**). Triton is a strange moon to begin with: It is a large moon, but it does not follow the orbital patterns of all other large moons in the outer solar system. Instead, it orbits Neptune backward (opposite to Neptune's rotation) and at a high inclination to Neptune's equator. These are telltale signs of a moon that was captured rather than formed in the disk of gas around its planet. No one knows how a moon as large as Triton could have been captured, but it still seems almost certain that Triton once orbited the Sun rather than Neptune. Recent research suggests that Triton might have been one member of a binary Kuiper belt object that passed too close to Neptune, causing Triton to be captured while its companion was flung off into deep space.

Triton's geology is just as surprising as its origin. Triton is smaller than our own Moon, yet its surface shows evidence of relatively recent geological activity. Some regions show evidence of past volcanism, while others show wrinkly ridges (nicknamed "cantaloupe terrain") that appear tectonic in nature. Triton even has a very thin atmosphere that has left some wind streaks on its surface. The likely source of Triton's geological activity is tidal heating, which would have been important if it had originally been captured into an elliptical orbit.

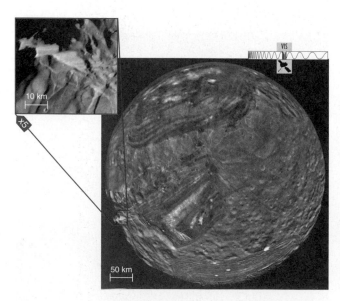

Figure 6.18 | The surface of Miranda shows astonishing tectonic activity despite the moon's small size. The cliff walls shown in the inset are higher than those of the Grand Canyon on Earth.

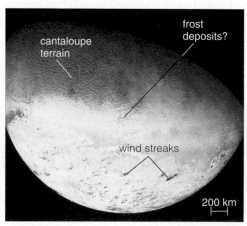

Triton's southern hemisphere as seen by *Voyager 2.*

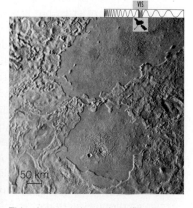

This close-up shows lava-filled impact basins similar to the lunar maria, but the lava was water or slush rather than molten rock.

Figure 6.19 | Neptune's moon Triton shows evidence of a surprising amount of past geological activity.

Ice Geology The jovian moons have taught us an important lesson about geology. If these moons were similar to the terrestrial worlds, their small sizes would have left them geologically dead long ago. However, there is a crucial difference between the jovian moons and the terrestrial worlds: composition.

Most of the jovian moons contain ices that can melt or deform at far lower temperatures than rock. As a result, they can experience geological activity even when their interiors have cooled to temperatures far below what they were at their births. Indeed, except on Io, most of the volcanism that has occurred in the outer solar system probably did not produce any hot lava at all; it produced icy lava that was essentially liquid water, perhaps mixed with methane and ammonia. The key point is that ice geology is possible at far lower temperatures than rock geology. This fact, combined with the occasional extra heating supplied by tidal heating, explains how the jovian moons have had such interesting geological histories despite their small sizes.

6.2 Asteroids, Comets, and the Impact Threat

The small bodies of the solar system include asteroids, comets, and dwarf planets. The small sizes of these objects might at first make them seem insignificant, but there is strength in numbers. Moreover, their small sizes allow them to be more easily perturbed by gravitational tugs from other objects, and as a result they sometimes head inward to the inner solar system, where they can appear as spectacular comets or, sometimes, come crashing down to Earth.

Why are asteroids and comets grouped into three distinct regions?

As we discussed in Chapter 4, the small bodies of the solar system are found in three major regions: the *asteroid belt* between Mars and Jupiter, the *Kuiper belt* just beyond the orbit of Neptune, and the spherical *Oort cloud* that extends to great distances from the Sun. As we investigate the nature and origin of the swarms of small bodies in each of these regions, we'll see that they are influenced by more than just the Sun.

The Asteroid Belt Asteroids come in a wide range of sizes. The largest, Ceres, is just under 1000 kilometers in diameter and qualifies as a dwarf planet because its gravity has made it round. About a dozen other asteroids are large enough that we would call them medium moons if they orbited a planet. Smaller asteroids are far more numerous (**Figure 6.20**); there are probably more than a million asteroids with diameters greater than 1 kilometer and many more even smaller in size.

Why are these many asteroids concentrated in the asteroid belt rather than spread throughout the inner solar system? The simple answer is that the asteroid belt was the only place where rocky planetesimals could survive. Nearly all of the rocky planetesimals that formed within Mars's orbit ultimately accreted into one of the four inner planets. In contrast, the asteroids in the asteroid belt stay clear of any planet and can therefore survive on their current orbits for billions of years. But this leaves us with a deeper question: Why didn't another planet form in this region and sweep up asteroids, just as the terrestrial worlds did in the inner solar system?

Asteroid orbits provide a key clue: Rather than being distributed randomly, most asteroids share a few particular orbital periods and distances, leaving gaps in which there are very few asteroids. Careful study shows that the gaps occur at orbital periods that are simple fractions of Jupiter's 12-year orbital period, such

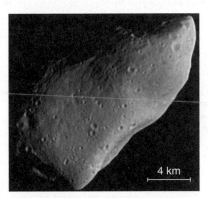

a Gaspra

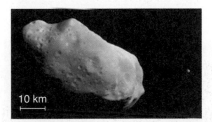

b Ida and its tiny moon Dactyl (the dot to right)

c Mathilde

Figure 6.20 | Spacecraft photos of three asteroids.

Common Misconceptions

DODGE THOSE ASTEROIDS!

Science fiction movies often show brave spacecraft pilots navigating through crowded fields of asteroids, dodging this way and that as they heroically pass through with only a few bumps and bruises. It's great drama, but not very realistic. The asteroid belt looks crowded when we draw it on paper as in Figure 4.1, but in reality it is an enormous region of space. Despite their large numbers, asteroids are thousands to millions of kilometers apart on average—so far apart that it would take incredibly bad luck to crash into one by accident. Indeed, spacecraft must be carefully guided to fly close enough to an asteroid to take a decent photograph. Future space travelers will have plenty of dangers to worry about, but dodging asteroids is not likely to be one of them.

Figure 6.21 | Anatomy of a comet. The inset photo is the nucleus of Halley's Comet, photographed by the *Giotto* spacecraft; the coma and tails shown are those of Comet Hale–Bopp in a ground-based photo. A comet grows tails only if it happens to come close to the Sun. Most comets never do this, remaining perpetually frozen in the outer solar system.

as one-half of Jupiter's period or one-third of Jupiter's period. These simple fractions are telltale signs of orbital resonances, much like those in Saturn's rings and among the Galilean moons of Jupiter. We conclude that Jupiter's gravity is the primary influence on the orbits of asteroids in the asteroid belt.

Jupiter's gravity also explains why a planet never formed in the wide space between Mars and Jupiter. When the solar system was forming, this region probably contained enough rocky material to form another terrestrial planet. However, resonances with the young planet Jupiter disrupted the orbits of this region's planetesimals, preventing them from accreting into a full-fledged planet. Over the next $4\frac{1}{2}$ billion years, ongoing orbital disruptions gradually kicked pieces of this "unformed planet" out of the asteroid belt altogether. Once booted from the asteroid belt, these objects either crashed into a planet or moon or were flung out of the solar system. The asteroid belt thereby lost most of its original mass, which explains why the total mass of all its asteroids is now much less than that of any terrestrial planet.

The Kuiper Belt The next major region of small bodies is the Kuiper belt, which is similar to the asteroid belt in many ways, except its leftover planetesimals are ice-rich in composition. As a result, when a small Kuiper belt object is gravitationally perturbed and flung into the inner solar system, its ice begins to vaporize and we may see it as a long-tailed comet in our sky (**Figure 6.21**).

As in the asteroid belt, the orbits of most Kuiper belt objects are shaped by the gravity of a jovian planet—in this case, Neptune. However, the Kuiper belt probably lacks a large planet not because objects were flung off but because the low densities at such great distance from the Sun slowed the rate at which planetesimals grew. As a result, none of them grew large enough to draw in surrounding hydrogen and helium gas before sunlight and the solar wind cleared the solar nebula.

Although no jovian planet formed in the Kuiper belt, some icy planetesimals grew large enough for gravity to make them round, qualifying them as dwarf planets. As we discussed in Section 1.3, Pluto and Eris are the two largest of these dwarf planets known to be in the Kuiper belt, though larger ones may yet be discovered.

The Oort Cloud The Oort cloud probably extends to a distance nearly 1000 times as far from the Sun as Pluto, which puts its most distant objects nearly one-quarter of the way to the nearest stars. We have never seen a comet in the Oort cloud, because our telescopes are not yet capable of detecting such small objects at such great distances from the Sun. However, based on studies of the numbers of comets that enter the inner solar system, the Oort cloud must contain close to a *trillion* (10^{12}) comets. How did so many small bodies end up orbiting the Sun at such great distances? The only answer that makes scientific sense comes from thinking about what happened to the icy leftover planetesimals that roamed the region in which the jovian planets formed.

The leftover planetesimals that cruised the spaces between Jupiter, Saturn, Uranus, and Neptune were doomed to suffer either a collision or a close gravitational encounter with one of the young jovian planets. The planetesimals that escaped being swallowed up tended to be flung off in all directions. Some may have been cast away at such high speeds that they completely escaped the solar system and now drift through interstellar space. The rest ended up on orbits with very large average distances from the Sun. These became the comets of the Oort cloud. The random directions in which they were flung, along with small gravitational nudges from other stars, explain the roughly spherical shape of the Oort cloud.

Think about it Consider the explanations given for the origins of the Kuiper belt and the Oort cloud. Assuming these explanations are correct, which region contains objects that formed *closer* to the Sun? Explain.

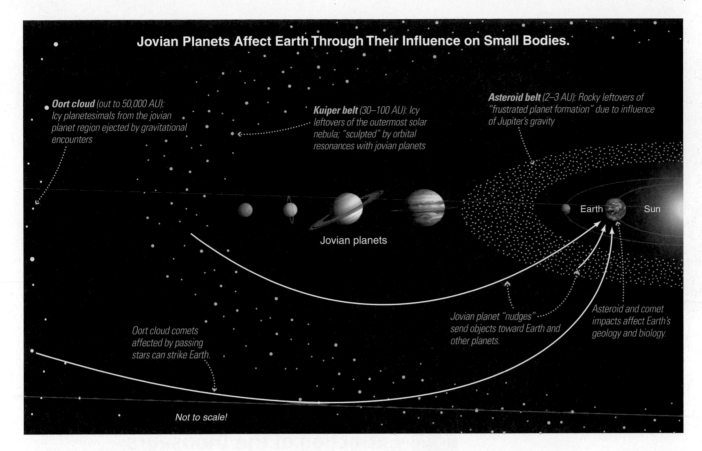

Jovian Planets Affect Earth Through Their Influence on Small Bodies.

Oort cloud (out to 50,000 AU): Icy planetesimals from the jovian planet region ejected by gravitational encounters

Kuiper belt (30–100 AU): Icy leftovers of the outermost solar nebula; "sculpted" by orbital resonances with jovian planets

Asteroid belt (2–3 AU): Rocky leftovers of "frustrated planet formation" due to influence of Jupiter's gravity

Jovian planets

Earth   Sun

Oort cloud comets affected by passing stars can strike Earth.

Jovian planet "nudges" send objects toward Earth and other planets.

Asteroid and comet impacts affect Earth's geology and biology.

Not to scale!

Figure 6.22 | The connections between the jovian planets, small bodies, and Earth. The gravity of the jovian planets helped shape both the asteroid belt and the Kuiper belt, and the Oort cloud consists of comets ejected from the jovian planet region by gravitational encounters with these large planets. Ongoing gravitational influences sometimes send asteroids or comets heading toward Earth.

Altering Orbits   The existence of impact craters provides proof that small bodies are not always confined to the asteroid belt, Kuiper belt, and Oort cloud. The objects that have crashed into the planets and moons must have had orbits that crossed planetary orbits, and there are still many such objects today. Some of these objects may simply be leftovers from formation that have been lucky enough to avoid a collision to date. More are probably objects that once orbited in the asteroid belt, Kuiper belt, or Oort cloud but experienced a close encounter or an orbital resonance that altered their orbit. This leads us to an important conclusion: Because the jovian planets have been the primary influence on the distribution of small bodies in the solar system, they are also ultimately responsible for most of the impacts that still occur on Earth and other worlds today. **Figure 6.22** summarizes the ways in which the jovian planets have been able to exert influence on the geology and biology of Earth.

Do small bodies pose an impact threat to Earth?

Space is filled with plenty of objects that could hit our planet. As of 2009, astronomers had identified more than 5000 asteroids that pass near Earth's orbit, making them potential threats to our future. The threat from comets is lower, but because they come from the great distance of the Kuiper belt or Oort cloud, we may not see them until a collision is imminent. How serious is the threat posed by these small bodies?

There's no doubt that impacts can cause great damage. Some have even been implicated in mass extinctions of life on Earth, an issue we'll discuss in the next section. The impact of a 10-kilometer-wide asteroid or comet could wipe out human civilization. Fortunately, the chance of such an impact in our lifetime is quite small. Geological data show that impacts of this size happen many tens of millions of years apart, on average. We're far more likely to do ourselves in than to be done in by a large asteroid or comet.

Figure 6.23 | This photo shows forests burned and flattened by the 1908 impact over Tunguska, Siberia. Atmospheric friction caused the small asteroid to explode completely before it hit the ground, so it left no impact crater.

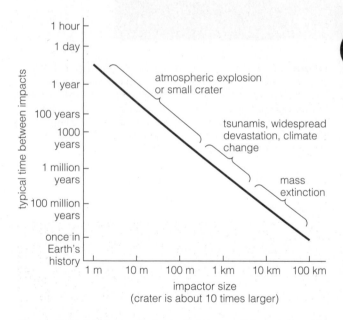

Figure 6.24 | This graph shows that larger objects (asteroids or comets) hit Earth less frequently than do smaller ones. The labels describe the effects of impacts of different sizes.

Smaller impacts can be expected more frequently. While such impacts would not wipe out our civilization, they could kill thousands or millions of people. We know of one close call in modern times. In 1908, a tremendous explosion occurred over Tunguska, Siberia (**Figure 6.23**). The explosion, estimated to have released energy equivalent to that of several atomic bombs, is thought to have been caused by a small asteroid about 40 meters across. If the asteroid had exploded over a major city instead of Siberia, it would have been the worst natural disaster in human history. **Figure 6.24** shows how often, on average, we expect Earth to be hit by objects of different sizes. Keep in mind that most impacts occur over oceans, since most of Earth's surface is water.

If we were to find an asteroid or a comet on a collision course with Earth, could we do anything about it? Many people have proposed schemes to save Earth by using nuclear weapons or other means to demolish or divert an incoming asteroid, but no one knows whether current technology is really up to the task. In the meantime, NASA is working to better understand the threat and to identify the vast majority of objects that could potentially impact Earth in the future.

 Think about it Study Figure 6.24. Based on the frequency of impacts large enough to cause serious damage, how much time and money do you think we should be spending to counter potential impact threats? Defend your opinion.

✳ THE PROCESS OF SCIENCE IN ACTION

6.3 Extinction of the Dinosaurs

The fossil record shows that dinosaurs were the dominant animals on Earth for more than 100 million years, until they abruptly went extinct about 65 million years ago. In fact, the death of the dinosaurs was only a small part of the biological devastation that seems to have occurred at that time: Up to 99% of all living plants and animals died, and up to 75% of all existing *species* were driven to extinction. This makes the event a clear example of a **mass extinction,** the rapid extinction of a large fraction of all living species. What could have caused the sudden die-off of the dinosaurs and other species? Today, the leading hypothesis invokes an impact, and the story of how scientists arrived at that hypothesis is this chapter's case study in the process of science in action.

Did an impact kill the dinosaurs?

The study of events that occurred on Earth in the past requires careful examination of rocks and fossils, along with modeling of processes that may have been responsible for what we observe. The case implicating an impact in the extinction of the dinosaurs has been built in just this way.

Impact Evidence In 1978, while analyzing geological samples collected in Italy, a scientific team led by father and son Luis and Walter Alvarez made a startling discovery. They found that a thin layer of dark sediments deposited about 65 million years ago—about the time the dinosaurs went extinct—was unusually rich in the element iridium. Iridium is a metal that is rare on Earth's surface but common in meteorites and therefore presumably common in asteroids and comets. Subsequent studies found the same iridium-rich layer in 65-million-year-old sediments around the world (**Figure 6.25**). Based on this evidence, the Alvarez team hypothesized that the extinction of the dinosaurs was caused by the impact of an asteroid or comet.

Further evidence for an impact comes from four other features found in the iridium-rich sediment layer: (1) unusually high abundances of several other metals, including osmium, gold, and platinum; (2) grains of *shocked quartz*, quartz crystals with a distinctive structure that indicates they experienced the high-pressure conditions of an impact; (3) spherical *rock droplets* of a type known to form when drops of molten rock cool and solidify in the air; and (4) soot. All these features point to an impact. The metal abundances look more like what we commonly find in meteorites than what we find elsewhere on Earth's surface. Shocked quartz is a characteristic of impact sites, including Meteor Crater in Arizona. The rock droplets presumably were made from molten rock splashed into the air by the force and heat of the impact. The soot probably came from vast forest fires ignited by impact debris. Some debris would have been blasted so high that it rose above the atmosphere and spread worldwide. It then would have plunged downward, with atmospheric friction heating it until it became a rain of hot, glowing rock.

More direct evidence of an impact comes from the presence of a large impact crater that appears to match the sediment layer in age (**Figure 6.26**). The size of the crater indicates that it was created by the impact of an asteroid or a comet measuring about 10 kilometers across.

The Extinction   Given the strong evidence that an impact occurred at the right time, how did it lead to extinction? Models suggest it could have happened as follows. On that fateful day some 65 million years ago, the asteroid or comet slammed into Mexico with the force of a hundred million hydrogen bombs (**Figure 6.27**). North America may have been devastated immediately. Hot debris from the impact rained around the rest of the world, igniting fires that killed many more living organisms. Longer term effects were even more severe. Dust and smoke remained in the atmosphere for weeks or months, blocking sunlight and causing temperatures to fall as if Earth were experiencing a harsh global winter. The reduced sunlight would have stopped photosynthesis for up to a year, killing large numbers of species throughout the food chain. Acid rain may have been another by-product, killing vegetation and acidifying lakes around the world. Chemical reactions

A layer rich in iridium and soot tells us a huge impact occurred at this point in geological (and biological) history.

Figure 6.25 | Around the world, sedimentary rock layers dating to 65 million years ago share the evidence of the impact of a comet or asteroid. Fossils of dinosaurs and many other species appear only in rocks below the iridium-rich layer.

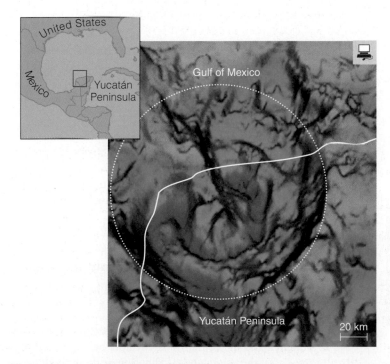

Figure 6.26 | This computer-generated image, based on measurements of small local variations in the strength of gravity, shows an impact crater about 200 kilometers across (dashed circle). The crater straddles the coast of Mexico's Yucatán Peninsula (see inset).

Figure 6.27 | This painting shows an asteroid or comet moments before its impact on Earth, some 65 million years ago. The impact probably caused the extinction of the dinosaurs.

in the atmosphere probably produced nitrous oxides and other compounds that dissolved in the oceans and killed marine organisms.

If these models are correct, the most astonishing fact may be that some species managed to survive. Among these were a few mammals, which may have survived in part because they lived in underground burrows and managed to store enough food to outlast the global winter that followed the impact. With the dinosaurs gone, these mammals became the new kings of the planet, evolving over the next 65 million years into numerous new species, ultimately including humans.

The Bottom Line There is still scientific dispute as to whether an impact was the sole cause of the extinction of the dinosaurs or just one of several causes. But the evidence leaves little doubt that a major impact coincided with the mass extinction of 65 million years ago, and we've constructed a plausible scenario for how the impact might have caused the extinction. This case illustrates a crucial point about how science progresses, because the idea that an impact could cause extinction was quite controversial when the Alvarez team first proposed it. After all, it was the first time anyone had suggested that an astronomical event might change the course of biological evolution. But the evidence was strong enough that scientists could not ignore it, so the ideas were published in scientific journals. Other scientists were then inspired to probe the ideas further. They searched for additional geological evidence that could support or refute the impact hypothesis, studied asteroid orbits and other past impacts to estimate the probability of such a major impact, and created models to predict the effects that an impact would have on the environment. These studies showed that such impacts were not only possible but inevitable over geological time scales. Combined with the extensive evidence of an impact and a model of the impact's devastating effects, an idea that was once greeted with great skepticism ultimately won widespread acceptance.

Summary of Key Concepts

6.1 Jovian Planets, Rings, and Moons

What are jovian planets like?

The jovian planets are far larger than the terrestrial worlds and are very different in nature. They have no solid surfaces and contain vast amounts of hydrogen, helium, and hydrogen compounds. They have strong weather in their atmospheres and layered interiors. They are each orbited by many moons and by rings that consist of countless individual particles.

Why are jovian moons so geologically active?

Even the largest jovian moons are barely larger than the Moon or Mercury, but many of them show evidence of substantial past or present geological activity. The source of this activity can probably be traced to the facts that *ice geology* can occur at lower temperatures than rock geology and that some of these moons have a source of heat—**tidal heating**—that is not important for the terrestrial worlds.

6.2 Asteroids, Comets, and the Impact Threat

Why are asteroids and comets grouped into three distinct regions?

Small bodies are found in three major groups: the rocky asteroids of the asteroid belt and the comets of the Kuiper belt and Oort cloud. In all three cases, the origin and distribution of these small bodies have been governed primarily by gravitational interactions between these bodies and the large jovian planets.

Do small bodies pose an impact threat to Earth?

Impacts have clearly occurred in Earth's past and will occur in the future. The level of threat depends on the size of the object: Impacts by smaller bodies cause less serious damage but occur more frequently. Large impacts can cause devastating damage but are quite rare.

6.3 Extinction of the Dinosaurs

Did an impact kill the dinosaurs?

It may not have been the sole cause, but a major impact clearly coincided with the **mass extinction** about 65 million years ago, in which the dinosaurs died out. Sediments from this era contain iridium and other evidence of an impact, and an impact crater of the right age lies along the coast of Mexico.

Investigations

Quick Quiz

Choose the best answer to each of the following; answers are in Appendix D. Explain your reasoning with one or more complete sentences.

1. In which list are the jovian planets in order of increasing distance from the Sun? (a) Jupiter, Saturn, Neptune, Uranus (b) Saturn, Jupiter, Uranus, Neptune (c) Jupiter, Saturn, Uranus, Neptune

2. What is Jupiter's Great Red Spot? (a) the largest mountain in the solar system (b) a very large cloud of ammonium hydrosulfide (c) a giant storm

3. The outer layers of most jovian moons are made primarily of (a) rock and metal. (b) ices. (c) hydrogen and helium.

4. Why is Io more volcanically active than our Moon? (a) Io is much larger. (b) Io has a higher concentration of radioactive elements. (c) Io has a different internal heat source.

5. What is unusual about Triton? (a) It orbits its planet backward. (b) It does not keep the same face toward its planet. (c) It is the only moon with rings.

6. Which moon shows evidence of rainfall and erosion by some liquid substance? (a) Europa (b) Titan (c) Ganymede

7. Saturn's many moons affect its rings primarily through (a) tidal forces. (b) orbital resonances. (c) magnetic field interactions.

8. The asteroid belt lies between the orbits of (a) Earth and Mars. (b) Mars and Jupiter. (c) Jupiter and Saturn.

9. Did a large terrestrial planet ever form in the region of the asteroid belt? (a) No, because there was never enough mass there. (b) No, because Jupiter prevented one from accreting. (c) Yes, but it was shattered by a giant impact.

10. Which objects have the most elliptical and tilted orbits? (a) asteroids in the asteroid belt (b) Kuiper belt comets (c) Oort cloud comets

11. About how often does a 1-kilometer object strike Earth? (a) every year (b) every million years (c) every billion years

12. Which of the following is *not* a key piece of evidence supporting the hypothesis that an impact caused the extinction of the dinosaurs? (a) rare metals found in 65-million-year-old sediments (b) radioactive isotopes found in 65-million-year-old dinosaur bones (c) an impact crater dating to 65 million years ago

Short-Answer/Essay Questions

Explain all answers clearly, with complete sentences and proper essay structure if needed. An asterisk (*) designates a quantitative problem, for which you should show all your work.

13. *Minor Ingredients Matter.* Suppose the jovian planet atmospheres were composed only of hydrogen and helium, with no hydrogen compounds at all. How would their atmospheres be different in terms of clouds, color, and weather? Explain.

14. *The New View of Titan.* What other planet or moon in the solar system does Titan most resemble, in your opinion? Summarize the similarities and differences in a few sentences.

15. *The Role of Jupiter.* Suppose that Jupiter had never existed. Describe at least three ways in which our solar system would be different, and clearly explain why.

16. *Asteroids vs. Comets.* Contrast the compositions and locations of comets and asteroids, and explain in your own words why they turned out differently.

17. *Oort Cloud vs. Kuiper Belt.* Explain in your own words how and why there are two different reservoirs of comets. Be sure to discuss where the two groups of comets formed and what kinds of orbits they travel.

18. *Observing Project: Jupiter's Moons.* Using binoculars or a small telescope, view the moons of Jupiter. Make a sketch of what you see, or take a photograph. Repeat your observations several times (nightly, if possible) over a period of a couple of weeks. Can you determine which moon is which? Can you measure the moons' orbital periods? Can you determine their approximate distances from Jupiter? Explain.

19. *Project: Dirty Snowballs.* Comets are said to resemble dirty snowballs in composition. If you have access to snow or ice, make a dirty snowball. (The ice chunks that form behind tires work well.) How much dirt does it take to darken snow? Find out by allowing your dirty snowball to melt in a container and measuring the approximate proportions of water and dirt afterward. What do your results tell you about comet composition?

*20. *Adding Up Asteroids.* It's estimated that there are a million asteroids measuring 1 kilometer across or larger. If a million such asteroids were combined into one object, how big would it be? Compare to the size of Earth. (*Hint:* Assume that each asteroid is a sphere with a radius of 0.5 km; the volume of a sphere is $\frac{4}{3}\pi r^3$.)

*21. *Jupiter's Mass.* Io orbits Jupiter at an average distance of 421,600 kilometers and with an orbital period of 1.77 days. Use these facts and Newton's version of Kepler's third law to calculate Jupiter's mass. (*Hint:* Be sure to convert the orbital distance to meters and the period to seconds so that you can use the value of the gravitational constant $G = 6.67 \times 10^{-11}$ m^3/kg s^2.)

7 Planets Around Other Stars

Learning Goals

7.1 Detecting Extrasolar Planets

- Why is it so difficult to detect planets around other stars?
- How do we detect planets around other stars?

7.2 Characteristics of Extrasolar Planets

- What have we learned about extrasolar planets?
- How do extrasolar planets compare with planets in our solar system?

✳ THE PROCESS OF SCIENCE IN ACTION

7.3 Revising the Nebular Theory

- Do extrasolar planets require us to modify our theory of solar system formation?

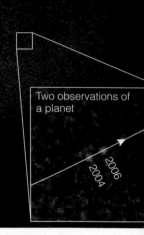

Two observations of a planet

2006
2004

The Hubble Space Telescope image on this page shows the first planet directly detected in visible light. The planet is shown twice, with one image taken two years after the other, revealing its orbital motion. Although this was the first planet imaged so directly, hundreds of other planets have been discovered around other stars through less direct techniques. These discoveries have profound implications. They demonstrate that planets are common in the universe, which increases the chance that we might someday find life elsewhere, perhaps even intelligent life. In addition, they give us many more worlds to study, a fact that enhances our ability to understand the planets of our own solar system, including Earth. In this chapter, we will investigate the rapidly developing science of planetary systems around distant stars.

7.1 Detecting Extrasolar Planets

The very idea of planets around other stars, or **extrasolar planets,** would have shattered the worldviews of many people throughout history. After all, cultures of the Western world long regarded Earth as the center of the universe, and most ancient cultures imagined the heavens to be a realm distinct from Earth.

The Copernican revolution, which taught us that Earth is a planet orbiting the Sun, opened up the possibility that planets might also orbit other stars. Still, until quite recently, no extrasolar planets were known. Our goal in this first section is to understand why the detection of extrasolar planets presents such an extraordinary technological challenge and how astronomers have begun to meet that challenge.

Why is it so difficult to detect planets around other stars?

We've known for centuries that other stars are distant suns, so it was natural to suspect that they would have their own planetary systems. The nebular theory of solar system formation, well established by the middle of the 20th century, made extrasolar planets seem even more likely. As we discussed in Chapter 4, the nebular theory explains our planetary system as a natural consequence of processes that accompanied the birth of our Sun. If the theory is correct, planets should be common throughout the universe. But prior to 1995, we lacked conclusive evidence of planets beyond our own solar system.

Why is it so difficult to detect extrasolar planets? You already know part of the answer if you think back to the scale model solar system discussed in Chapter 1. Recall that on a 1-to-10-billion scale, the Sun is the size of a grapefruit, Earth is a pinhead orbiting 15 meters away, and Jupiter is a marble orbiting 80 meters away. On the same scale, the distance to the nearest stars is equivalent to the distance across the United States. In other words, looking for an Earth-like planet orbiting the nearest star besides the Sun is like looking from San Francisco for a pinhead orbiting just 15 meters from a grapefruit in Washington, D.C. Seeing a Jupiter-like planet would be like seeing a marble about 80 meters from the grapefruit-size star.

Scale alone would make the task quite challenging, but it is further complicated by the fact that a Sun-like star would be a *billion times* brighter than the light reflected from any of its planets. Because even the best telescopes blur the light from stars at least a little, the glare of scattered starlight tends to overwhelm any small blips of planetary light.

In the early 1990s, these challenges made even some astronomers think that we were still decades away from finding extrasolar planets. But human ingenuity proved greater than the pessimists had guessed. Thanks to technological advances, clever planet-hunting strategies, and some unexpected differences between our solar system and others, we have begun to discover planets orbiting other stars. Although it is too soon to know for sure, it seems likely that *billions* of planetary systems inhabit our Milky Way Galaxy.

How do we detect planets around other stars?

The first clear-cut discovery of a planet around another Sun-like star—a star called 51 Pegasi—came in 1995. Hundreds of additional extrasolar planets have been discovered since that time, using several planet-finding strategies. If

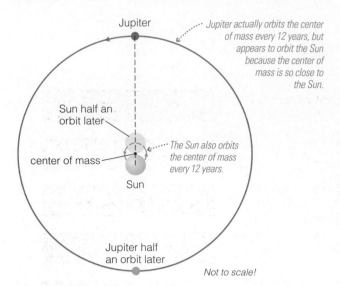

Jupiter

Jupiter actually orbits the center of mass every 12 years, but appears to orbit the Sun because the center of mass is so close to the Sun.

Sun half an orbit later

center of mass

The Sun also orbits the center of mass every 12 years.

Sun

Jupiter half an orbit later

Not to scale!

Figure 7.1 | This diagram shows how both the Sun and Jupiter actually orbit around their mutual center of mass, which lies very close to the Sun. The diagram is not to scale; the sizes of the Sun and its orbit are exaggerated about 100 times compared to the size shown for Jupiter's orbit, and Jupiter's size is exaggerated even more.

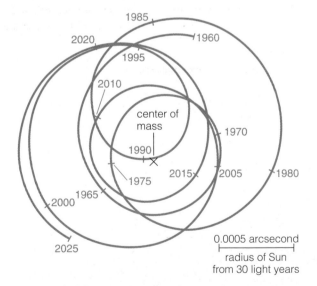

1985

2020   1960

1995

2010

center of mass

1970

1990

2015   2005   1980

1975

2000   1965

2025

0.0005 arcsecond

radius of Sun
from 30 light years

Figure 7.2 | This diagram shows the orbital path of the Sun from 1960 to 2025 around the center of mass of our solar system, as it would appear if viewed face-on from a distance of 30 light-years away. The Sun's complex motion reveals the gravitational effects of the planets (primarily Jupiter and Saturn). Notice that the entire range of motion during this period is only about 0.0015 arcsecond, almost 100 times smaller than the angular resolution of the Hubble Space Telescope.

we strip away the details, however, there are really only two basic ways to search for extrasolar planets:

1. *Directly.* Pictures or spectra of planets constitute direct evidence of their existence.

2. *Indirectly.* Precise measurements of a *star's* properties may indirectly reveal the effects of orbiting planets.

Direct detection is preferable in principle, because it can tell us far more about a planet's properties, but to date nearly all detections have been indirect.

Think about it Do a quick Web search on "extrasolar planets" to find the number of extrasolar planets currently known. How many have been found in the past year?

Gravitational Tugs Two indirect techniques—the *astrometric* and *Doppler techniques*—rely on observing stars in search of motion that we can attribute to gravitational tugs from orbiting planets. Although we usually think of a star as remaining still while planets orbit around it, that is only approximately correct. In reality, all objects in a star system, including the star itself, orbit the system's *center of mass,* which is essentially the balance point for all the mass of the solar system. The center of mass of our own solar system lies close to the Sun, because the Sun is far more massive than all the planets combined, but it is not exactly at the Sun's center.

We can see how this fact helps us to discover extrasolar planets by imagining the viewpoint of extraterrestrial astronomers observing our solar system from afar. Let's start by considering only the influence of Jupiter, the most massive planet in our solar system (**Figure 7.1**). The center of mass between the Sun and Jupiter lies just outside the Sun's visible surface, so what we usually think of as Jupiter's 12-year orbit around the Sun is really a 12-year orbit around this center of mass. Because the Sun and Jupiter are always on opposite sides of the center of mass (otherwise it wouldn't be a "center"), the Sun must orbit this point with the same 12-year period. The Sun's orbit traces out only a very small ellipse with each 12-year period, because the Sun's average orbital distance is barely larger than its own radius. Nevertheless, with sufficiently precise measurements, extraterrestrial astronomers could detect this orbital movement of the Sun and thereby deduce the existence of Jupiter without having ever seen the planet. They could even determine Jupiter's mass from the orbital characteristics of the Sun as it goes around the center of mass. A more massive planet located at the same distance would pull the center of mass farther from the Sun's center, giving the Sun a larger orbit and a faster orbital speed around the center of mass.

The other planets also exert gravitational tugs on the Sun, each adding a small additional effect to the effect of Jupiter. In principle, with sufficiently precise measurements of the Sun's orbital motion made over many decades, an extraterrestrial astronomer could deduce the existence of all the planets of our solar system (**Figure 7.2**). This is the essence of the **astrometric technique,** in which we make very precise measurements of stellar positions in the sky (*astrometric* means "measurement of the stars"). If a star "wobbles" gradually around its average position (the center of mass), we must be observing the influence of unseen planets. The primary difficulty with the astrometric technique is that we are looking for changes in position that are very small, even for nearby stars. In addition, the stellar motions are largest for massive planets orbiting far from their star, but the long orbital periods of such planets mean that it can take decades to notice the motion. As a result of these difficulties, the astrometric technique has been of only limited use to date, but astronomers hope it will prove successful with future space-based telescopes.

Tools of Science: The Doppler Effect

We can learn about the motions of distant objects relative to us by identifying shifts in their spectra caused by the **Doppler effect.** The Doppler effect is crucial not only to the Doppler technique for searching for extrasolar planets but also to many other areas of astronomy that we'll discuss later in the book.

You've probably noticed the Doppler effect on the *sound* of a train whistle near train tracks. If the train is stationary, the pitch of its whistle sounds the same no matter where you stand (**Figure 1a**). But if the train is moving, the pitch will sound higher when the train is coming toward you and lower when it's moving away from you. Just as the train passes by, you can hear the dramatic change from high to low pitch—a sort of "weeeeeeee–ooooooooooh" sound. To understand why, think about what happens to the sound waves coming from the train (**Figure 1b**). When the train is moving toward you, each pulse of a sound wave is emitted a little closer to you. The bunching up of the waves between you and the train gives them a shorter wavelength and higher frequency (pitch). After the train passes you, each pulse comes from farther away, stretching out the wavelengths and giving the sound a lower frequency.

The Doppler effect causes similar shifts in the wavelengths of light (**Figure 1c**). If an object is moving toward us, the light waves bunch up between us and the object, so its entire spectrum is shifted to shorter wavelengths. Because shorter wavelengths of visible light are bluer, the Doppler shift of an object coming toward us is called a *blueshift*. If an object is moving away from us, its light is shifted to longer wavelengths. We call this a *redshift*, because longer wavelengths of visible light are redder.

We look for Doppler shifts by studying *spectral lines* in an object's spectrum. (For a discussion of the origin of spectral lines, see Tools of Science, p. 129.) For example, suppose we recognize the pattern of hydrogen lines in the spectrum of a distant object (**Figure 2**). We know the rest wavelengths of the hydrogen lines—their wavelengths in stationary clouds of hydrogen gas—from laboratory experiments in which a tube of hydrogen gas is heated so that the wavelengths of the spectral lines can be measured. If the hydrogen lines from the object appear at longer wavelengths, then we know

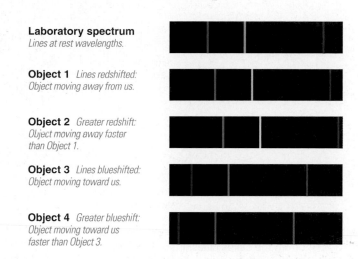

Laboratory spectrum
Lines at rest wavelengths.

Object 1   *Lines redshifted: Object moving away from us.*

Object 2   *Greater redshift: Object moving away faster than Object 1.*

Object 3   *Lines blueshifted: Object moving toward us.*

Object 4   *Greater blueshift: Object moving toward us faster than Object 3.*

Figure 2 | Spectral lines provide the crucial reference points for measuring Doppler shifts.

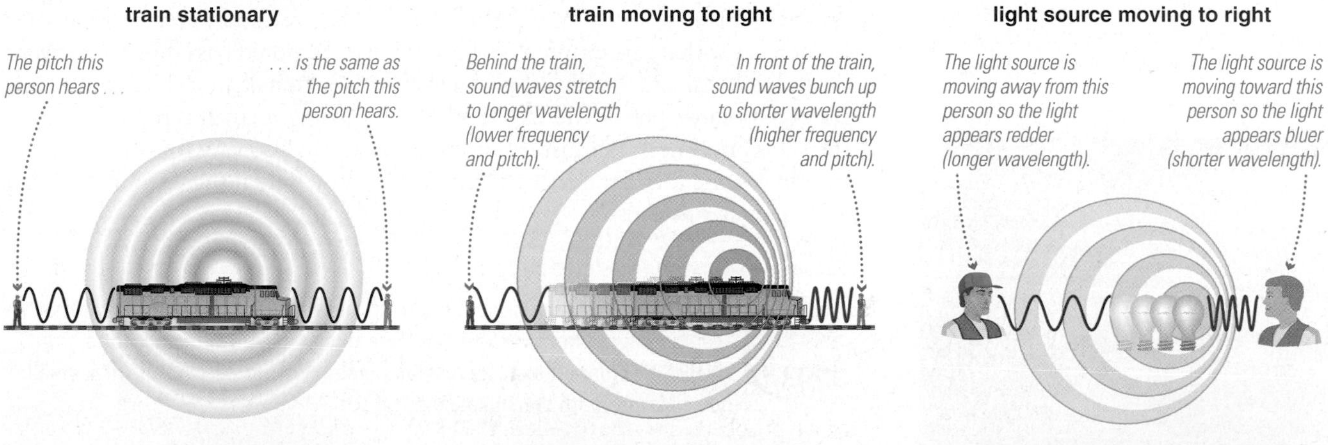

they are redshifted and the object is moving away from us. If the lines appear at shorter wavelengths, then we know they are blueshifted and the object is moving toward us. The larger the shift, the faster the object is moving. In fact, as long as the object is moving much slower than the speed of light, we can find its speed with a simple formula:

$$\frac{v_{\text{rad}}}{c} = \frac{\lambda_{\text{shift}} - \lambda_{\text{rest}}}{\lambda_{\text{rest}}}$$

where v_{rad} is the object's radial velocity, λ_{rest} is the rest wavelength of a particular spectral line, λ_{shift} is the shifted wavelength of the same line, and c is the speed of light. A positive answer means the object has a redshift and is moving away from us; a negative answer means it has a blueshift and is moving toward us.

train stationary

The pitch this person hears . . .   *. . . is the same as the pitch this person hears.*

a The whistle sounds the same no matter where you stand near a stationary train.

train moving to right

Behind the train, sound waves stretch to longer wavelength (lower frequency and pitch).   *In front of the train, sound waves bunch up to shorter wavelength (higher frequency and pitch).*

b For a moving train, the sound you hear depends on whether the train is moving toward you or away from you.

light source moving to right

The light source is moving away from this person so the light appears redder (longer wavelength).   *The light source is moving toward this person so the light appears bluer (shorter wavelength).*

c We get the same basic effect from a moving light source (although the shifts are usually too small to notice by eye).

Figure 1 | The Doppler effect. Each circle represents the crest of sound (or light) waves going in all directions from the source. For example, the circles from the train might represent waves emitted 0.001 second apart.

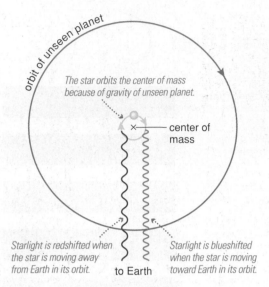

Figure 7.3 | The Doppler technique for discovering extrasolar planets. Light from the star is blueshifted as it comes toward us and redshifted as it moves away, with this pattern repeating for every orbit.

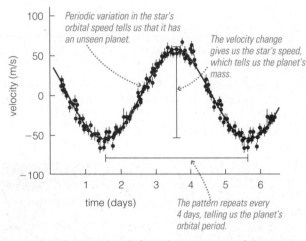

a A periodic Doppler shift in the spectrum of the star 51 Pegasi shows the presence of a large planet with an orbital period of about 4 days. Dots are actual data points; bars through dots represent measurement uncertainty.

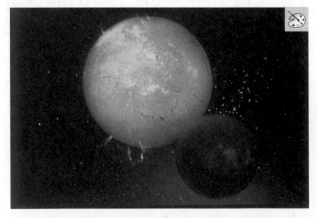

b Artist's conception of 51 Pegasi and its planet.

Figure 7.4 | Data from the first discovery of an extrasolar planet.

With the **Doppler technique,** we search for a star's orbital movement around the center of mass in a different way: by looking for changing *Doppler shifts* in a star's spectrum (see Tools of Science, p. 115). As long as a planet's orbit is *not* face-on to us, the planet's gravitational influence will cause its star to move alternately toward and away from us—and these motions cause alternating blueshifts and redshifts in the star's spectrum (**Figure 7.3**).

The 1995 discovery of a planet orbiting 51 Pegasi came when this star was found to have alternating blueshifts and redshifts with a period of about 4 days (**Figure 7.4a**). The 4-day period of the star's motion must be the orbital period of its planet. We can use this period with the star's mass and Newton's version of Kepler's third law (see Tools of Science, p. 96) to calculate the planet's orbital distance. The Doppler technique also allows us to estimate the planet's mass from the measured change in the star's velocity. (More technically, a precise estimate requires knowing the inclination of the orbit to our line of sight; if we do not know the inclination, we obtain a *lower limit* on the planet's mass.) The data in Figure 7.4a thereby enabled us to learn that the planet orbiting 51 Pegasi is similar to Jupiter in mass but orbits only about 0.05 AU from its star—so close that its surface temperature is probably over 1000 K. It is therefore an example of what we call a **hot Jupiter,** because it has a Jupiter-like mass but a much higher surface temperature (**Figure 7.4b**).

The Doppler technique has been used for the vast majority of planet discoveries to date. In some cases, Doppler data are good enough to tell us whether the star has more than one planet. Remember that if two or more planets exert a noticeable gravitational tug on their star, the Doppler data will show the combined effect of these tugs. As of 2009, at least 30 multiple-planet systems had been identified, including one with five planets. Keep in mind, however, that the Doppler technique is best suited to identifying massive planets that orbit relatively close to their star, because the star's orbital speed depends on the strength of the gravitational tug and gravity is strongest for massive planets with small orbital distances. The technique is much less useful for finding planets that orbit far from their stars or have low masses.

Transits and Eclipses A third indirect way of detecting distant planets relies on searching for slight changes in a star's brightness that occur when a planet passes in front of or behind it. If we were to examine a large sample of stars with planets, a small number of them (<1%) would by chance be aligned in such a way that one or more of the star's planets would pass directly between us and the star once each orbit. The result is a **transit,** in which the planet appears to move across the face of the star, causing a small temporary dip in the star's brightness. Because a star's brightness can also vary for other reasons, we can assume that a transiting planet is the cause only if the dimming repeats with a regular period.

Think about it What kind of planet is most likely to cause a transit across its star that we could observe from Earth: (a) a large planet close to its star, (b) a large planet far from its star, (c) a small planet close to its star, or (d) a small planet far from its star? Explain.

Figure 7.5 shows transit data for a planet orbiting the star HD209458. The transits occur every $3\frac{1}{2}$ days, telling us the planet's orbital period, and the 1.7% dip in the star's brightness tells us how the planet's radius compares to its star's radius. Half an orbit after a transit, the planet passes behind its star in what we call an **eclipse.** Observing an eclipse is much like observing a transit: In both cases, we measure the *combined* light from the star and planet,

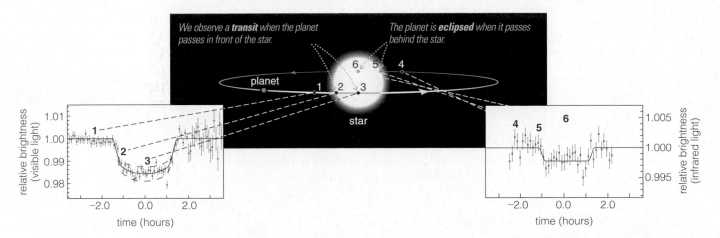

Figure 7.5 | This diagram shows the planet orbiting the star HD209458. The graphs show how the star's brightness changes during transits and eclipses, which each occur once during every $3\frac{1}{2}$ day orbit. During a transit, the star's brightness drops for about 2 hours by 1.7%, which tells us how the planet's radius compares to the radius of its star. During an eclipse, the infrared signal drops by 0.25%, which tells us about the planet's thermal emission.

so in principle there can be a dip in brightness whenever either object blocks light from the other. However, because planets generally emit only infrared light, the dips that occur during eclipses are usually measurable only at infrared wavelengths. For HD209458, the infrared brightness drops by about 0.25% during each eclipse, telling us that the planet emits 0.25% as much infrared radiation as the star. Using this information and the planet's radius measured during the transits, astronomers calculate the planet's temperature to be more than 1100 K.

The primary limitation of the transit and eclipse method is that it works only for the small fraction of planets whose orbits are nearly edge-on. But the method also has advantages, including the ability to take a spectrum of starlight transmitted through a planet's atmosphere. So far, astronomers have confirmed the existence of hydrogen, water, methane, and even a hint of sodium in the atmospheres of extrasolar planets. The transit method can also be used to search simultaneously for planets around vast numbers of stars and to detect much smaller planets than is possible with the Doppler technique. NASA's *Kepler* mission, launched in 2009, is designed to monitor some 100,000 stars for transits; if Earth-size planets are common, it should be able to detect dozens of them. A European Space Agency (ESA) spacecraft called *COROT* has already detected several transiting planets but may not be able to detect planets as small as Earth. In addition, telescopes as small as 4 inches in diameter have been used to discover transiting planets, so you can confirm for yourself with a backyard telescope some of the transits already detected.

Direct Detection The indirect planet-hunting techniques we have discussed so far have started a revolution in planetary science by demonstrating that our solar system is just one of many planetary systems. However, these indirect techniques tell us relatively little about the planets themselves, aside from their orbital properties and their masses or radii. To learn more about their nature, we need to obtain images or spectra of the planets themselves.

To date, astronomers have had only limited success, as the great distances and glare from stars make direct detection extremely difficult. **Figure 7.6** shows an infrared image of what is probably a jovian planet orbiting the star Beta Pictoris. The planet is so young that it is still glowing from the heat of formation. The chapter-opening photo shows the first confirmed direct detection of a planet with visible light. Astronomers are confident that the tiny dot really is a planet, because they observed it more than once and therefore could detect its orbital motion around its star. **Figure 7.7** summarizes the major planet detection techniques.

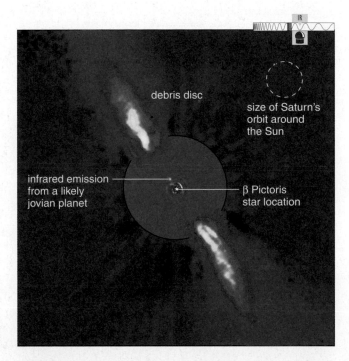

Figure 7.6 | This composite infrared image from the European Southern Observatory shows a probable jovian planet orbiting the star Beta Pictoris. The composite combines an image of the outer dust disk with a high-resolution image of the region closer to the star, revealing the planet. Both images were taken with the star itself blocked; the star's position has been added digitally.

The search for planets around other stars is one of the fastest growing and most exciting areas of astronomy. Although it has been only a little more than a decade since the first discoveries, known extrasolar planets already number well above 250. This figure summarizes major techniques that astronomers use to search for and study extrasolar planets.

① Gravitational Tugs: We can detect a planet by observing the small orbital motion of its star as both the star and its planet orbit their mutual center of mass. The star's orbital period is the same as that of its planet, and the star's orbital speed depends on the planet's distance and mass. Any additional planets around the star will produce additional features in the star's orbital motion.

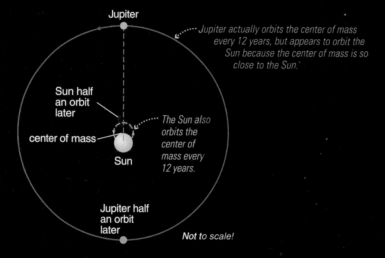

Jupiter

Jupiter actually orbits the center of mass every 12 years, but appears to orbit the Sun because the center of mass is so close to the Sun.

Sun half an orbit later

center of mass

Sun

The Sun also orbits the center of mass every 12 years.

Jupiter half an orbit later

Not to scale!

①a The Doppler Technique: As a star moves alternately toward and away from us around the center of mass, we can detect its motion by observing alternating Doppler shifts in the star's spectrum: a blueshift as the star approaches and a redshift as it recedes. This technique has revealed the vast majority of known extrasolar planets.

①b The Astrometric Technique: A star's orbit around the center of mass leads to tiny changes in the star's position in the sky. As we improve our ability to measure these tiny changes, we should discover many more extrasolar planets.

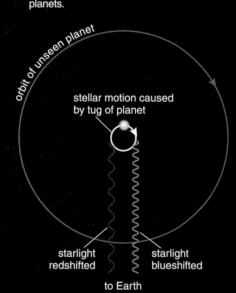

orbit of unseen planet

stellar motion caused by tug of planet

starlight redshifted

starlight blueshifted

to Earth

Current Doppler-shift measurements can detect an orbital velocity as small as 1 meter per second— walking speed.

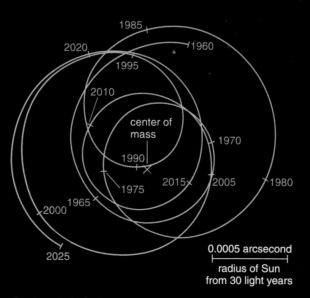

1985

2020

1960

1995

2010

center of mass

1990

1970

1975

2015   2005   1980

2000   1965

2025

0.0005 arcsecond

radius of Sun from 30 light years

The change in the Sun's apparent position, if seen from a distance of 10 light years, would be similar to the angular width of a human hair at a distance of 5 kilometers.

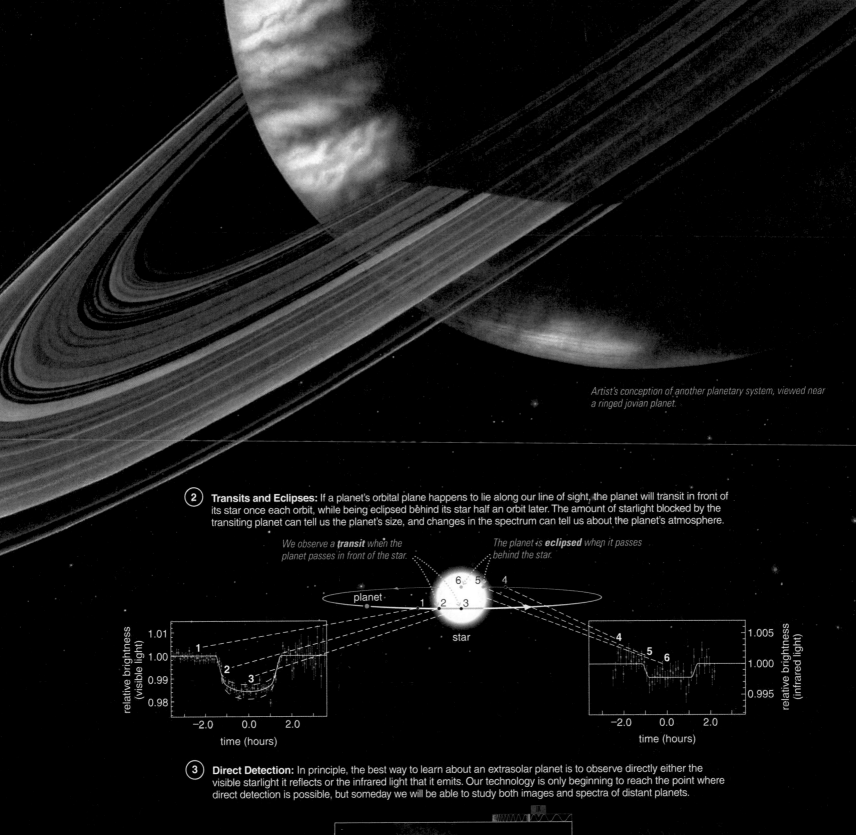

Artist's conception of another planetary system, viewed near a ringed jovian planet.

② **Transits and Eclipses:** If a planet's orbital plane happens to lie along our line of sight, the planet will transit in front of its star once each orbit, while being eclipsed behind its star half an orbit later. The amount of starlight blocked by the transiting planet can tell us the planet's size, and changes in the spectrum can tell us about the planet's atmosphere.

We observe a **transit** when the planet passes in front of the star.

The planet is **eclipsed** when it passes behind the star.

planet

star

③ **Direct Detection:** In principle, the best way to learn about an extrasolar planet is to observe directly either the visible starlight it reflects or the infrared light that it emits. Our technology is only beginning to reach the point where direct detection is possible, but someday we will be able to study both images and spectra of distant planets.

This infrared image shows a brown dwarf called 2M1207 (blue) . . .

. . . and what is probably a jovian planet (red) in orbit around it.

Think about it Go to the *Kepler* mission Web site. What is the status of the mission? How many planets has it discovered so far? Have these discoveries provided any major insights? Explain.

In the next few years, a new generation of large ground-based observatories may be able to provide even better images and spectra. Further down the line, both NASA and the European Space Agency hope to launch large orbiting telescopes with even greater capabilities. Within a couple of decades, we are likely to see the first crude images of Earth-size planets around other stars, and spectra of these worlds should allow us to search for signs of life-sustaining atmospheres and possibly of life itself.

Other Planet-Hunting Strategies The success of recent efforts to find extrasolar planets has led astronomers to think of many other possible ways of enhancing the search. One example is the Optical Gravitational Lensing Experiment (OGLE), a large survey of thousands of distant stars. Although it was not originally designed for planet detection, OGLE has already detected several planets by observing transits. It has also succeeded in detecting three planets by using *gravitational lensing,* an effect predicted by Einstein's general theory of relativity that occurs when one object's gravity bends or brightens the light of a more distant object. A different strategy is to look for the gravitational effects of unseen planets on the disks of dust that surround many stars, while another method involves searching for thermal emission from the impacts of accreting planetesimals. As we learn more about extrasolar planets, new search methods are sure to arise.

7.2 Characteristics of Extrasolar Planets

We have now discovered a large enough number of extrasolar planets that we can begin to search for patterns, trends, and groupings that might give us insight into how these planets compare to the planets of our own solar system and how they formed.

What have we learned about extrasolar planets?

The first step in looking for patterns and trends is to organize the existing information. We therefore begin by briefly summarizing the properties that we have so far identified of extrasolar planets.

Orbits **Figure 7.8** shows the orbits of the first 170 known extrasolar planets, all superimposed. Careful study of these orbits tell us a lot about how the layouts of other solar systems compare to that of our own. Despite how crowded the orbits are when viewed this way, at least two important facts should jump out at you. First, notice that only a handful of these planets have orbits that take them beyond about 5 AU, which is Jupiter's distance from our Sun. Most of the planets orbit very close to their host star; many of them orbit closer than Mercury orbits to the Sun in our solar system. Second, notice that many of the orbits are clearly elliptical, rather than nearly circular like the orbits of planets in our own solar system.

These data might at first seem to suggest that solar systems laid out like our own are quite rare. However, it is also possible that this result is a *selection effect,* arising from the fact that most of these planets have been detected with the Doppler technique. Recall that the Doppler technique is best suited to identifying massive planets that orbit relatively close to their

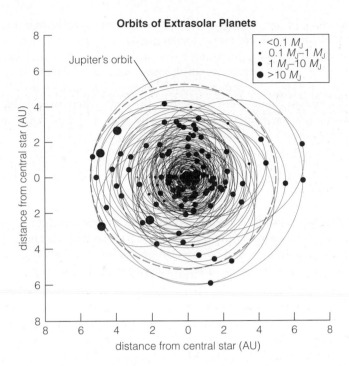

Orbits of Extrasolar Planets

Key:
· <0.1 M_J
· 0.1 M_J–1 M_J
● 1 M_J–10 M_J
● >10 M_J

Jupiter's orbit

distance from central star (AU)

distance from central star (AU)

Figure 7.8 | Orbital properties of 170 known extrasolar planets. This diagram shows all the orbits superimposed, as if all the planets orbited a single star. The sizes of the dots indicate approximate masses for the planets compared to Jupiter's mass, M_J (see key); the dots are located at the farthest point for each orbit.

star. Lower-mass planets are much more difficult to detect with this technique because of their weaker gravitational effects. Massive planets in more distant orbits are also difficult to detect, both because greater orbital distance means a weaker gravitational effect on the star and because long orbital periods can be identified only after many years of observation. We therefore say that the Doppler technique would tend to find, or *select,* massive planets in close-in orbits, even if such planets are comparatively rare. Already, some data for planets beyond the first 170 shown in Figure 7.8 suggest that there are more planets with orbits like those in our solar system than this figure indicates. However, until we have more data from other detection techniques, we will not really be sure whether close-in orbits like those in Figure 7.8 are common or rare.

Another important discovery has come from systems in which we have identified more than one planet. In many of these systems, the planets seem to have orbital resonances with each other; for example, one planet may have an orbital period that is exactly twice as long as that of another planet. Recall that orbital resonances play an important role in our own solar system among planetary rings, Jupiter's moons, and the asteroid and Kuiper belts. The data from extrasolar planets suggest that orbital resonances also help shape the overall layout of a planetary system.

Masses   The sizes of the dots in Figure 7.8 indicate the approximate masses of the planets. Again, we see what might at first seem to be a surprising trend: Most of the planets are more massive than Jupiter, and only a very few have much smaller masses. In fact, none of the planets discovered to date have masses as small as that of Earth (which has a mass of about 0.003 Jupiter mass). Again, this may simply be a selection effect that occurs because it is much easier to detect more massive planets. Scientists are therefore anxiously awaiting results from the *Kepler* mission, which should give us more meaningful statistics about the distribution of masses among extrasolar planets.

Sizes and Densities   The masses of known extrasolar planets suggest that most of them are jovian in nature, but mass alone cannot rule out the possibility of extra-large terrestrial planets—that is, very massive planets made of metal or rock. To distinguish between these possibilities, we need to know the sizes (diameters) of the planets, which we can use along with their masses to calculate their densities. We expect jovian planets to have large sizes and low densities, and terrestrial planets to have small sizes and higher densities. Unfortunately, the vast majority of known extrasolar planets have been detected by the Doppler technique, which gives us reasonable mass estimates but no information about size. However, in the cases where we also have size data from transits, the planets turn out to have sizes and densities consistent with what we expect for jovian planets. Again, the *Kepler* mission, by observing transits of many more planets, should give us important size data, which will improve our ability to make general statements about the sizes and densities of known extrasolar planets.

Compositions We have even less data about the composition of extrasolar planets, because we must obtain spectra. We have so far been able to obtain crude spectra in only a handful of cases. In one case (as of 2009), the spectral features identified water and methane, consistent with the idea that the planet is a jovian planet.

Think about it We now know that planets are common around other stars, but we do not yet know whether *Earth-like* planets are common. How would the discovery of Earth-like planets change our view of our place in the universe? Defend your opinion.

How do extrasolar planets compare with planets in our solar system?

Despite the limited data on extrasolar planets, we are already starting to answer key questions about other planetary systems. A particularly important question is whether planets in other star systems fit the same terrestrial and jovian categories as the planets in our solar system. So far the answer seems to be "possibly." The masses of the known extrasolar planets suggest that most of them are jovian in nature, a hypothesis supported by the limited data on their sizes, densities, and compositions. And while we have not yet detected any planets that we can confirm to be terrestrial in nature, this is probably because of the limitations of current technology for finding low-mass planets. There's some cause for optimism: Statistically, stars are more likely to have planets if they're rich in planet-forming ingredients other than hydrogen and helium, so the formation conditions for terrestrial planets look promising.

The data do, however, suggest some interesting variations on the terrestrial and jovian categories. A handful of planets have been discovered with masses only a few times that of Earth, suggesting that they might be "super-size" terrestrial planets, possibly even "waterworlds."

The most common variation is the numerous planets with masses suggesting that they are jovian but with orbits very close to their stars, making them *hot Jupiters* (see Figure 7.4b). These planets would probably have clouds much like those on Jupiter but of a different type. The temperatures on hot Jupiters would be far too high for gases such as ammonia or water vapor to condense into liquid droplets or ice flakes, as they do on our Jupiter. Instead, models suggest that a hot Jupiter with a temperature above about 1000 K would have clouds of rock dust containing common minerals.

We can even make educated guesses about a hot Jupiter's weather. Because a hot Jupiter is so close to its star, intense starlight would warm one side of the planet. As long as the planet rotates rapidly enough, the rotation and heat input should create patterns of planet-circling winds much like those on Jupiter. This would probably lead to a striped appearance much

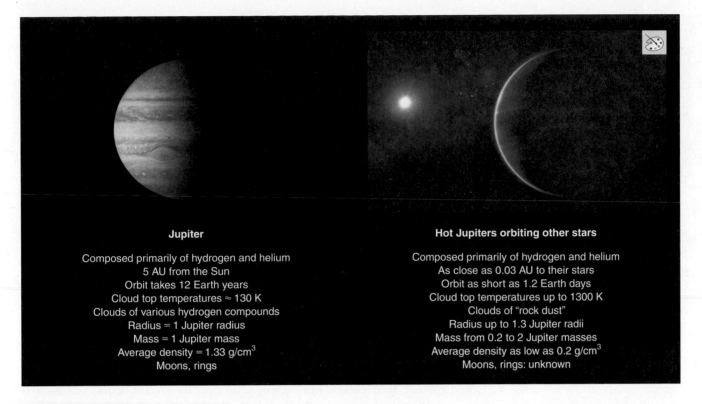

Jupiter

Composed primarily of hydrogen and helium
5 AU from the Sun
Orbit takes 12 Earth years
Cloud top temperatures ≈ 130 K
Clouds of various hydrogen compounds
Radius = 1 Jupiter radius
Mass = 1 Jupiter mass
Average density = 1.33 g/cm³
Moons, rings

Hot Jupiters orbiting other stars

Composed primarily of hydrogen and helium
As close as 0.03 AU to their stars
Orbit as short as 1.2 Earth days
Cloud top temperatures up to 1300 K
Clouds of "rock dust"
Radius up to 1.3 Jupiter radii
Mass from 0.2 to 2 Jupiter masses
Average density as low as 0.2 g/cm³
Moons, rings: unknown

like that of the real Jupiter. Moreover, hot Jupiters should be hot enough to glow faintly with visible light, so their stripes might be visible even on their night sides. **Figure 7.9** summarizes the similarities and differences expected between the real Jupiter and hot Jupiters.

Figure 7.9 | A summary of the expected similarities and differences between the real Jupiter and extrasolar hot Jupiters orbiting Sun-like stars.

✳ THE PROCESS OF SCIENCE IN ACTION

7.3 Revising the Nebular Theory

The discovery of extrasolar planets presents us with an opportunity to test our theory of solar system formation. Can our existing theory explain other planetary systems, or will we have to go back to the drawing board?

As we discussed in Chapter 4, the nebular theory holds that our solar system's planets formed as a natural consequence of processes that accompanied the formation of our Sun. If the theory is correct, then the same processes should accompany the births of other stars, so the nebular theory clearly predicts the existence of other planetary systems. In that sense, the recent discoveries of extrasolar planets mean that the theory has passed a major test, because its most basic prediction has been verified. However, extrasolar planets have also presented at least two significant challenges to this theory. We saw in Chapter 4 that the nebular theory predicts that jovian planets should form only in the cold outer regions of solar systems, while terrestrial planets should form closer in. The theory also predicts that planetary orbits should be nearly circular. Hot Jupiters and planets on highly elliptical orbits therefore seem to contradict the nebular theory.

Do extrasolar planets require us to modify our theory of solar system formation?

The nature of science demands that we question the validity of a theory whenever it is challenged by an observation or experiment [Section 3.2]. If the theory cannot explain the new observations or data, then we must revise or discard it.

The orbiting planet nudges particles in the disk . . .

. . . causing material to bunch up. These dense regions in turn tug on the planet, causing it to migrate inward.

Figure 7.10 | This figure shows a simulation of waves created by a planet embedded in a dusty disk of material surrounding its star. These waves may cause the planet to migrate inward.

The discovery of hot Jupiters has indeed caused scientists to revisit the nebular theory of solar system formation.

At first, scientists suspected that there might be something fundamentally wrong with the nebular theory, which led them to reexamine the theory in great detail. However, more than a decade of careful study did not turn up any obvious flaws in the basic theory. As a result, scientists now suspect that the basic predictions of the nebular theory are correct, and that jovian planets can be born only far from their stars and with nearly circular orbits. But how, then, can we explain hot Jupiters? And why are some extrasolar planet orbits so elliptical?

Most scientists now suspect that hot Jupiters are jovian planets that somehow migrated inward from their original orbits. In other words, they were born as ordinary jovian planets, just as the nebular theory predicts—on circular orbits in the cold outer regions of their solar systems—and only later moved inward to their current orbits. Scientists have come up with several possible ways in which this type of *planetary migration* might occur.

The most promising idea suggests that migration can be caused by waves passing through a gaseous disk (**Figure 7.10**). Recall that the nebular theory predicts that planets always form in a swirling disk of material around a young star. A massive planet's gravity and motion can disturb the otherwise evenly distributed disk material, generating waves that travel through the disk. As they pass by, the waves cause material to bunch up, and these clumps of material exert their own gravitational pull on the planet, robbing it of energy and causing it to migrate inward. Computer models confirm that waves in a nebula can cause young planets to spiral slowly toward their star. In our own solar system, this type of migration did not play a major role, probably because the solar wind cleared out the gas before it could have much effect. But planets may form earlier in other solar systems, allowing time for jovian planets to migrate substantially inward. It's also possible that a jovian planet could migrate inward as a result of multiple close encounters with much smaller planetesimals. Scientists are now applying this modified theory to our own solar system, and some evidence indicates that limited migration may have occurred in our solar system through this mechanism.

Astronomers have also developed new models to explain elliptical orbits through close encounters between young jovian planets. Such an encounter might send one planet out of the star system entirely, while the other is flung inward into a highly elliptical orbit. It's also possible that orbital resonances among jovian planets might cause their orbits to become more elliptical.

The bottom line is that discoveries of extrasolar planets have shown us that the general features of the nebular theory are probably correct, but that our original theory was incomplete. The original theory explains the formation of planets and the simple layout of a solar system such as ours, but it needs new features—such as planetary migration and gravitational encounters—to explain the differing layouts of other solar systems. We should not be too surprised by this fact, because scientific theories frequently need modification to accommodate new discoveries. For example, Newton's theory of gravity had to be modified by Einstein to account for effects observed in strong gravitational fields [Section 3.3], and the theory of atoms and subatomic particles has been modified numerous times as new discoveries have been made during the past century. Just as in those cases, modification of the nebular theory has opened new possibilities that we did not previously consider. We now recognize that solar systems can have a much wider range of arrangements than we had guessed before the discovery of extrasolar planets.

Summary of Key Concepts

7.1 Detecting Extrasolar Planets

Why is it so difficult to detect planets around other stars?

The great distances to stars and the fact that typical stars are a billion times brighter than the light reflected from any of their planets make it very difficult to detect **extrasolar planets.**

How do we detect planets around other stars?

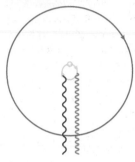

Nearly all known extrasolar planets have been discovered indirectly. We can detect a planet's gravitational effect on its star through the **astrometric technique,** which looks for small shifts in stellar position, or through the **Doppler technique,** which looks for the back-and-forth motion of stars revealed by Doppler shifts. We can also search for **transits** and **eclipses,** in which a system becomes slightly dimmer as a planet passes in front of or behind its star. Direct detection is only just beginning, but eventually it will allow us to obtain images and spectra of distant planets.

7.2 Characteristics of Extrasolar Planets

What have we learned about extrasolar planets?

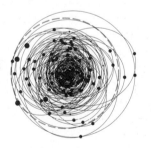

Most known extrasolar planets are much more massive than Earth. Many of them orbit surprisingly close to their stars and in highly elliptical orbits. We have limited information about their sizes and compositions, but current data are consistent with the idea that these planets are jovian in nature.

How do extrasolar planets compare with planets in our solar system?

The terrestrial and jovian categories probably still work for other solar systems, but with some variations. In particular, planets that are jovian in nature but orbit close to their stars, called **hot Jupiters,** probably differ from the jovian planets of our solar system in several ways, including their high temperatures and the composition of their clouds.

✳ THE PROCESS OF SCIENCE IN ACTION

7.3 Revising the Nebular Theory

Do extrasolar planets require us to modify our theory of solar system formation?

Our basic theory of solar system formation seems to be sound; we suspect that hot Jupiters probably were born with nearly circular orbits far from their stars but somehow migrated inward. Scientists are working to modify the nebular theory to allow for this type of migration and other surprising aspects of extrasolar planets.

Investigations

Quick Quiz

Choose the best answer to each of the following; answers are in Appendix D. Explain your reasoning with one or more complete sentences.

1. What method has detected the most extrasolar planets so far? (a) transit method (b) direct detection (c) Doppler technique
2. The extrasolar planets discovered so far most resemble (a) terrestrial planets. (b) jovian planets. (c) large icy worlds.
3. How many extrasolar planets have been detected? (a) between 10 and 100 (b) between 100 and 1000 (c) more than 1000
4. Which one of the following can the transit method tell us about a planet? (a) its mass (b) its size (c) the eccentricity of its orbit
5. Which method could detect a planet in an orbit that is face-on to Earth? (a) Doppler technique (b) transit method (c) astrometric technique
6. How is the planet orbiting 51 Pegasi different from Jupiter? (a) It's much closer to its star. (b) It has a much longer year. (c) It's much more massive.
7. Most known extrasolar planets are more massive than Jupiter because (a) smaller planets likely do not exist. (b) current detection methods are more sensitive to larger planets. (c) the Doppler technique usually overestimates planet masses.

8. Which detection method can be used to find planets with a backyard telescope? (a) Doppler technique (b) transit method (c) astrometric technique

9. Earth-size planets orbiting Sun-like stars (a) have already been discovered. (b) should be discovered in the next few years by ground-based telescopes. (c) should be discovered by 2015 by a space telescope.

10. Which detection method will the *Kepler* mission use? (a) Doppler technique (b) transit method (c) astrometric technique

11. How much do we know about the composition of extrasolar planets? (a) We don't yet have any composition information. (b) We have density information and spectra for a few planets. (c) We have composition information on the vast majority of planets.

12. What's the best explanation for the close-in orbits of hot Jupiters? (a) Hot Jupiters formed closer to their stars than Jupiter did. (b) Hot Jupiters formed farther out, like Jupiter, but then migrated inward. (c) The strong gravity of their stars pulled them in.

Short-Answer/Essay Questions

Explain all answers clearly, with complete sentences and proper essay structure if needed. An asterisk (*) designates a quantitative problem, for which you should show all your work.

13. *Why So Soon?* The detection of extrasolar planets came much sooner than astronomers expected. Was this a result of planets being different than expected or of technology improving faster than expected? Explain.

14. *Why Not Hubble?* Of the more than 200 extrasolar planets discovered, only one likely planet has ever been imaged by the Hubble Space Telescope. What limits Hubble's ability to image planets around other stars?

15. *Explaining the Doppler Technique.* In terms that an elementary school child would understand, explain how the Doppler technique works. It may help to use an analogy to explain the difficulty of direct detection and to describe the Doppler shift.

16. *Comparing Methods.* Make a list of the advantages and disadvantages of the Doppler and transit techniques. What kinds of planets are easiest to detect with each technique? Are there certain planets that cannot be detected at all with either method? What additional information can we obtain if we are able to detect a planet using both techniques? Explain.

17. *No Hot Jupiters Here.* How do we think hot Jupiters formed? Why didn't one form in our solar system?

18. *Orbital Resonances.* How may resonances be important in affecting extrasolar planet orbits? How are these effects similar to the effects of resonances in our solar system, and how are they different?

19. *A Year on HD209458b.* Imagine you're visiting the planet that orbits the star HD209458 (see Figure 7.5), hovering in the upper atmosphere in a suitable spacecraft. What would you see, and how would it be different from what you would see if you were floating in Jupiter's atmosphere? Consider factors like local conditions, clouds, how the Sun looks, and orbital motion.

20. *Lots of Big Planets.* Many of the extrasolar planets discovered so far are more massive than the most massive planet in our solar system. Does this mean our solar system is unusual? If so, how or why? If not, why not?

*21. *Planet Around 51 Pegasi.* The star 51 Pegasi has about the same mass as our Sun. A planet discovered around it has an orbital period of 4.23 days. The mass of the planet is estimated to be 0.6 times the mass of Jupiter. Use Kepler's third law to find the planet's average distance (semimajor axis) from its star. (*Hint:* Because the mass of 51 Pegasi is about the same as the mass of our Sun, you can use Kepler's third law in its original form, $p^2 = a^3$. Be sure to convert the period into years before using this equation.)

*22. *Doppler Calculations.* In hydrogen, the transition from level 2 to level 1 has a rest wavelength of 121.6 nm. Suppose you see this line at a wavelength of 120.5 nm in Star A and 121.2 nm in Star B. Calculate each star's speed, and state whether it is moving toward or away from us.

8 The Sun and Other Stars

Today, astronomy encompasses the study of the entire universe, but the root *astro* comes from the Greek for "star." Stars are therefore the namesakes of astronomy, and in this chapter we will begin our study of them. Most stars are so far away that even to our most powerful telescopes they appear only as points of light, and we must use a variety of observations and models to learn about their properties and life stories. One star, however, is close by and much easier to study: our Sun. We will begin by examining what we have learned about the nature of our star, shown in the photo as it appears to an ultraviolet telescope in space. Then we will turn our attention to the multitudes of other stars in the sky, comparing their properties with those of our Sun and seeing what these properties reveal about the lives of stars.

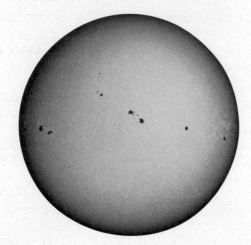

Figure 8.1 | This photo of the visible surface of the Sun shows several dark sunspots, each large enough to swallow our entire planet.

Table 8.1 Basic Properties of the Sun

Radius (R_{Sun})	696,000 km (about 109 times the radius of Earth)
Mass (M_{Sun})	2×10^{30} kg (about 300,000 times the mass of Earth)
Luminosity (L_{Sun})	3.8×10^{26} watts
Composition (by percentage of mass)	70% hydrogen, 28% helium, 2% heavier elements
Rotation rate	25 days (equator) to 30 days (poles)
Surface temperature	5800 K (average); 4000 K (sunspots)
Core temperature	15 million K

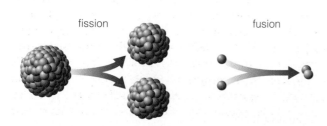

Figure 8.2 | Nuclear fission splits a nucleus into smaller nuclei, while nuclear fusion combines smaller nuclei into a larger nucleus.

8.1 Properties of the Sun

Our Sun is the nearest star, and it provides virtually all the light and heat on Earth. The source of the Sun's great power remained a mystery for most of human history, but today we understand not only how the Sun produces energy but also how the energy flow through the Sun determines its structure.

What is the Sun like?

The Sun is immense by any measure. Although it may look relatively small in our sky, from its distance and angular size we know that its radius is about 700,000 kilometers—more than 100 times the radius of Earth. Even **sunspots,** which appear as dark splotches on the Sun's surface, can be larger in size than Earth (**Figure 8.1**). We can calculate the Sun's mass by applying Newton's version of Kepler's third law (see Tools of Science, p. 96): It is about 2×10^{30} kilograms, which is 300,000 times the mass of Earth and 1000 times the combined mass of all the planets in our solar system.

The Sun's immense size is matched by its enormous output of energy. Its total power output, or **luminosity,** is an incredible 3.8×10^{26} watts. If we could somehow capture and store just 1 second's worth of the Sun's luminosity, it would be enough to meet current human energy demands for roughly the next 500,000 years. Of course, only a tiny fraction of the light radiated by the Sun reaches Earth, with the rest dispersing in other directions into space. Most of the Sun's energy is radiated as visible light, but the Sun also radiates light across most of the rest of the electromagnetic spectrum, including ultraviolet and X rays.

In terms of composition, the Sun is essentially a giant ball of hot gas or, more technically, of *plasma*—a gas in which many of the atoms are ionized because of the high temperature. Spectroscopy (see Tools of Science, p. 130) reveals that this plasma is composed of about 70% hydrogen and 28% helium (by mass), with all other elements making up only about 2%. Careful study of the Sun's spectrum also allows us to determine the Sun's surface temperature, which averages about 5800 K, except in sunspots, where the gas is a cooler 4000 K. The temperature must increase with depth within the Sun, and theoretical models tell us that the temperature reaches an astounding 15 million K at the Sun's center.

The entire Sun is rotating, though, unlike a spinning ball, not all parts of the Sun rotate at the same rate. The solar equator completes one rotation in about 25 days, and the rotation period increases with latitude to about 30 days near the solar poles. **Table 8.1** summarizes the basic properties of the Sun.

The Sun's enormous energy output comes from **nuclear fusion,** a process in which two or more atomic nuclei slam together hard enough that they stick and become one larger nucleus. Fusion reactions are quite different from the nuclear reactions used in nuclear reactors on Earth; those reactions release energy by splitting large nuclei—such as those of uranium or plutonium—into smaller ones in a process called **nuclear fission. Figure 8.2** summarizes the difference between fission and fusion.

Fusion occurs deep within the Sun, in the central region called the **core.** The 15-million-K plasma there is like a "soup" of hot gas, with bare, positively charged atomic nuclei (and negatively charged electrons) whizzing about at extremely high speeds. At any one time, some of these nuclei are on high-speed collision courses with each other. In most cases, electromagnetic forces deflect the nuclei, preventing actual collisions, because positive charges repel

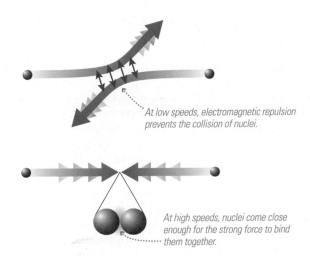

Figure 8.3 | Positively charged nuclei can fuse only if a high-speed collision brings them close enough for the strong force to come into play.

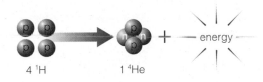

Figure 8.4 | In the Sun, four hydrogen nuclei (protons) fuse to make one helium nucleus (two protons and two neutrons).

Common Misconceptions

THE SUN IS NOT ON FIRE

We often say that the Sun is "burning," a term that conjures up images of a giant bonfire in the sky. However, the Sun does not burn the way a fire burns on Earth. Fires generate light through chemical changes that consume oxygen and produce a flame. The glow of the Sun has more in common with the glowing embers left over after the flames have burned out. Much like hot embers, the Sun's surface shines because it is hot enough to emit thermal radiation that includes visible light.

Hot embers quickly stop glowing as they cool, but the Sun keeps shining because its surface is kept hot by the energy rising from its core. Because this energy is generated by nuclear fusion, we sometimes say that it is the result of "nuclear burning"—a term intended to suggest nuclear changes in much the same way that "chemical burning" suggests chemical changes. Nevertheless, while we can say that the Sun undergoes nuclear burning in its core, it is not accurate to speak of any kind of burning on the Sun's surface, where light is produced primarily by thermal radiation.

one another. However, if nuclei collide with sufficient energy, they can fuse together to form a heavier nucleus (**Figure 8.3**).

Sticking positively charged nuclei together is not easy. The **strong force,** which binds protons and neutrons together in atomic nuclei, is the only force in nature that can overcome the electromagnetic repulsion between two positively charged nuclei. In contrast to gravitational and electromagnetic forces, which drop off gradually as the distances between particles increase (by an inverse square law), the strong force is more like glue or Velcro: It overpowers the electromagnetic force over very small distances but is insignificant when the distances between particles exceed the typical sizes of atomic nuclei. The key to nuclear fusion is to push the positively charged nuclei close enough together for the strong force to outmuscle electromagnetic repulsion. At the temperature of the Sun's core, hydrogen nuclei are moving fast enough that collisions can cause them to fuse together.

Fusion can combine only two nuclei at once, but a variety of different fusion reactions are always taking place simultaneously within the Sun. Overall, these fusion reactions end up transforming four individual hydrogen nuclei, which are protons, into a single helium nucleus containing two protons and two neutrons (**Figure 8.4**). Some of these reactions also produce tiny subatomic particles known as **neutrinos,** which are extremely lightweight and rarely interact with other matter.

Fusing hydrogen nuclei into helium nuclei generates energy because a helium nucleus has a mass slightly less (by about 0.7%) than the combined mass of four hydrogen nuclei. Therefore, when four hydrogen nuclei fuse into a helium nucleus, a little bit of mass disappears. This disappearing mass becomes energy in accord with Einstein's famous formula $E = mc^2$, which tells us that an amount of mass m can be transformed into an amount of energy E equal to the mass times the speed of light squared. Fusion in the Sun converts about 600 million tons of hydrogen into 596 million tons of helium every second, which means that 4 million tons of matter is turned into energy each second. Although this mass loss is enough to generate tremendous amounts of energy, it is insignificant compared to the Sun's total mass and does not affect the overall mass of the Sun in a measurable way.

The fact that the Sun is gradually transforming its core hydrogen into helium means that the Sun will eventually run out of hydrogen. We can estimate the Sun's overall lifetime by dividing the total amount of hydrogen that the Sun originally had in its core by the rate at which it is currently fusing this hydrogen. We find that the Sun's lifetime is about 10 billion years. Since the Sun is now about $4\frac{1}{2}$ billion years old, it will continue to shine for about another 5 billion years. In the next chapter, we'll discuss what will happen to the Sun after it uses up its core supply of hydrogen.

How does energy escape from the Sun?

Energy released by nuclear fusion in the Sun's core must somehow find its way to the surface, where it is radiated into space in the form of photons. Although we cannot look directly inside the Sun, we can investigate how solar energy escapes by using mathematical models of the Sun's interior. These models agree well with our observations of the Sun, successfully accounting for its size, surface temperature, and luminosity. This agreement gives us confidence in our understanding of what the solar interior is like. Let's follow the journey of solar energy from the solar core until its release into space.

The Sun's Interior   Nearly all the energy released by fusion in the core takes the form of gamma-ray photons. Although these photons travel at the speed of light, the path they take through the Sun's interior zigzags so much

Tools of Science: Spectroscopy

Spectroscopy is the study of light that has been spread out into a spectrum by a device such as a prism. **Figure 1** shows the three basic kinds of spectra that we observe: (1) a **continuous spectrum** contains smooth light across a broad range of wavelengths; (2) an **emission line spectrum** has bright lines on a dark background; and (3) an **absorption line spectrum** has dark lines on a continuous background.

Spectra provide a wealth of information about the objects we study. We have previously discussed how a *thermal radiation* spectrum—which is one type of continuous spectrum—tells us an object's surface temperature (see Tools of Science, p. 80) and how *Doppler shifts* in the wavelengths of spectral lines tell us the speed of an object toward or away from us (see Tools of Science, p. 115). We learn even more from the spectral lines themselves.

The laws of quantum mechanics (see Tools of Science, p. 148) dictate that each different kind of atom, ion, or molecule produces

spectral lines at a unique set of wavelengths—in essence, its "chemical fingerprint." We can therefore determine the composition of a distant object by identifying and comparing all the different "fingerprints" in its spectrum. It does not matter whether the lines are absorption lines or emission lines; the wavelengths of the lines are what tell us an object's composition.

We can illustrate how the spectrum of a star like the Sun reveals its composition by putting these ideas together. The hot Sun emits thermal radiation, much like the hot light bulb in Figure 1, which produces a continuous spectrum. The low-density, uppermost layers of the Sun act like a cloud of gas that the continuous spectrum must pass through on its way to Earth. Atoms in these layers absorb light at specific wavelengths, producing an absorption line spectrum. The wavelengths of those lines tell us the kinds of atoms that are present, and a more detailed analysis tells us the relative proportions of those elements.

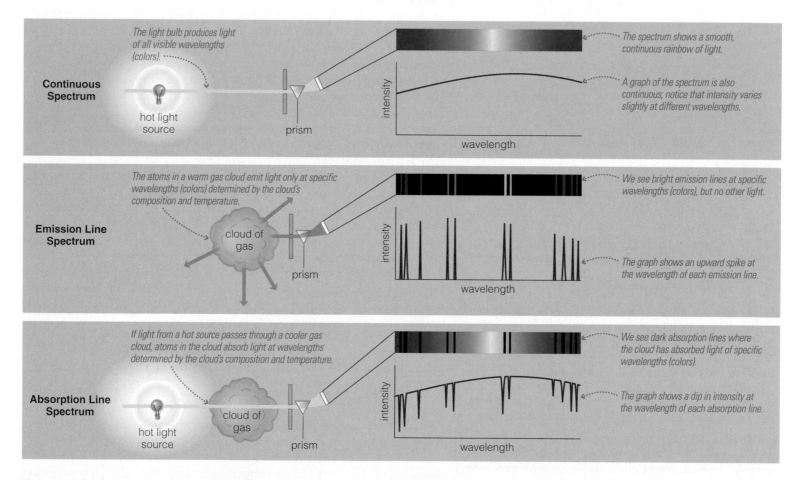

Figure 1 | Examples of conditions under which we see the three basic types of spectra.

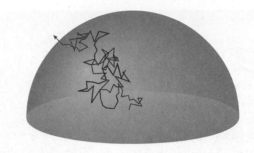

that it takes them a very long time to make any outward progress. Deep in the solar interior, the plasma is so dense that a photon can travel only a fraction of a millimeter in any one direction before it interacts with an electron. Each time a photon collides with an electron, the photon gets deflected into a new and random direction. The photon thereby bounces around the dense interior in a haphazard way (sometimes called a *random walk*) and only very gradually works its way outward (**Figure 8.5**). The layer directly above the core is called the **radiation zone** because energy in the form of electromagnetic radiation is moving through it primarily by way of these randomly bouncing photons.

Figure 8.5 | A photon in the solar interior bounces randomly among electrons, slowly working its way outward.

At the top of the radiation zone, where the temperature has dropped to about 2 million K, the solar plasma absorbs photons more readily (rather than just bouncing them around). This absorption creates the conditions needed for convection [Section 5.1] and hence marks the bottom of the Sun's **convection zone.** Recall that convection occurs because hot gas is less dense than cool gas. Like hot-air balloons, hot bubbles of plasma rise upward through the convection zone, while cooler plasma near the top slides around the rising bubbles and sinks. The rising of hot plasma and sinking of cool plasma form a cycle that transports energy outward from the base of the convection zone to the solar surface, or **photosphere.** This convecting gas explains the mottled appearance of the photosphere that we see in close-up photographs (**Figure 8.6**).

Radiation and convection within the Sun transport fusion energy very slowly by human standards—the journey of energy from the core to the photosphere takes several hundred thousand years. However, a small fraction of the fusion energy goes into the neutrinos produced by hydrogen fusion. Neutrinos escape from the Sun quickly because they can pass through the solar interior without interacting with other particles. Traveling at nearly the speed of light, they reach us about 8 minutes after they leave the core. Although neutrinos are extremely difficult to capture, sensitive neutrino detectors have measured the flow of neutrinos from fusion in the Sun. These measurements agree with predictions made by our models of the solar interior, indicating that we really do know what is going on inside the Sun, even though we cannot see beneath the photosphere.

The Solar Atmosphere

Once energy from the solar interior reaches the photosphere, it can escape directly into space as thermal radiation, because the gas layers above the photosphere are transparent to visible light. The photosphere is therefore the visible surface of the Sun. The transparent layer just above the photosphere, in which the temperature rises to over 10,000 K, is called the **chromosphere.** Above that is the **corona,** in which temperatures reach more than 1 million K. Gas at the top of the corona is so hot that it emits X rays, and some of it escapes the Sun in an outward flow of charged particles known as the **solar wind. Figure 8.7** schematically shows all the layers of the Sun.

As energy liberated by nuclear fusion rises to the photosphere by convection, it helps to stimulate a wide variety of phenomena including *sunspots* on the photosphere, gigantic loops of glowing gas that connect pairs of sunspots, and huge explosions known as *solar flares*. At the root of all these features are magnetic fields generated by the plasma circulating through the convection zone. Magnetic fields cause sunspots and other phenomena in the solar atmosphere by restricting the directions in which charged solar plasma particles can flow. Sunspots occur where magnetic fields prevent hot plasma from mixing with cooler plasma in the photosphere, allowing the temperature of the plasma in sunspots to drop as low as 4000 K, significantly cooler than the 5800 K plasma that surrounds them. This lower temperature explains why sunspots appear darker than the rest of the solar surface. The

Bright spots appear on Sun's surface where hot gas is rising . . .   . . . then the gas sinks after it has cooled off.

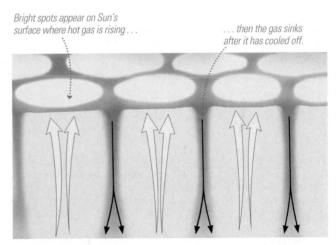

a This diagram shows convection beneath the Sun's surface: hot gas (yellow arrows) rises while cooler gas (black arrows) descends around it.

Hot gas is rising here . . .   . . . and cooler gas is sinking here.   VIS

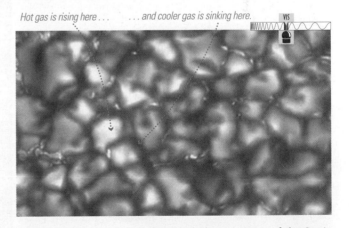

b This photograph shows the mottled appearance of the Sun's photosphere. The bright spots, each about 1000 kilometers across, correspond to the rising plumes of hot gas in the diagram in part (a).

Figure 8.6 | The Sun's photosphere churns with rising hot gas and falling cool gas as a result of underlying convection.

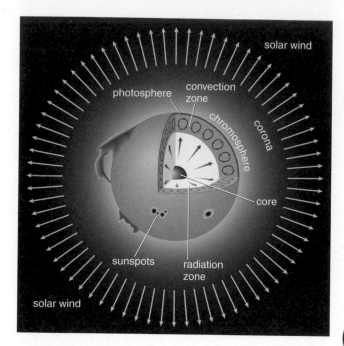

Figure 8.7 | The basic structure of the Sun.

Figure 8.8 | Strong magnetic fields keep sunspots cooler than the surrounding photosphere, while magnetic loops can arch from the sunspots to great heights above the Sun's surface.

a Pairs of sunspots are connected by tightly wound magnetic field lines.

loops connecting sunspot pairs are lines of a magnetic field that has trapped glowing gas high in the corona (**Figure 8.8**). Flares occur when pent-up magnetic field energy is suddenly released into the solar atmosphere. Particularly strong solar flares can affect Earth because they feed energetic bursts of charged particles into the solar wind, which can end up producing the beautiful lights of the *aurora* [Section 5.1] as these bursts interact with Earth's upper atmosphere.

The phenomena we observe in the solar atmosphere change over relatively short periods of time and therefore constitute what we call **solar activity** (or *solar weather*). The most notable pattern of solar activity is the **sunspot cycle**—a cycle in which the average number of sunspots gradually rises and falls (**Figure 8.9**). At the time of *solar maximum*, when sunspots are most numerous, we may see dozens of sunspots on the Sun at one time. In contrast, we may see few or no sunspots at the time of *solar minimum*. The length of time between maximums is 11 years on average, but we have observed it to be as short as 7 years and as long as 15 years. There have also been some periods during which no sunspots have been seen for many decades.

8.2 Properties of Other Stars

As we turn our attention from the Sun to the other stars in the sky, it is natural to wonder what life would be like if Earth were orbiting another star. How bright would it be in daytime? What color starlight would bathe Earth's surface? Would the star live long enough for life to develop? In this section, we'll touch on some of the answers to these questions as we study the properties of stars.

How do we measure the properties of stars?

Much of what we know about stars other than our Sun comes from measurements of just three basic properties: luminosity, surface temperature, and mass. We have already seen how the Sun measures up in these categories. Let's now examine how we go about determining these properties in other stars.

b This X-ray photo (from NASA's *TRACE* mission) shows hot gas trapped within looped magnetic field lines.

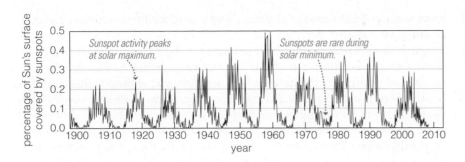

Figure 8.9 | This graph shows how the number of sunspots on the Sun changes with time. The vertical axis shows the percentage of the Sun's surface covered by sunspots. The cycle has a period of approximately 11 years.

Measuring Luminosity If you go outside on any clear night, you'll immediately see that stars differ in brightness. Some stars are so bright that we can use them to identify constellations. Others are so dim that our naked eyes cannot see them at all. However, these differences in brightness do not by themselves tell us anything about the luminosities of the stars—that is, about how much light they are actually generating—because distance affects brightness. For example, the stars Procyon and Betelgeuse appear about equally bright in our sky, but Betelgeuse actually emits about 5000 times as much light as Procyon. It has about the same brightness in our sky because it is much farther away.

Because two similar-looking stars can be generating very different amounts of light, we need to distinguish clearly between a star's brightness in our sky and the actual amount of light that it emits into space (**Figure 8.10**):

- When we talk about how bright stars look in our sky, we are talking about **apparent brightness**—the brightness of a star as it appears to our eyes.

- When we talk about how bright stars are in an absolute sense, regardless of their distance, we are talking about **luminosity**—the total amount of power that a star emits into space.

The apparent brightness of a star (or any other light source) depends on both its luminosity and its distance. More specifically, apparent brightness obeys an *inverse square law* with distance (**Figure 8.11**), much like the inverse square law that describes the force of gravity [Section 3.3]. For example, if we viewed the Sun from twice Earth's distance, it would appear dimmer by a factor of $2^2 = 4$. If we viewed it from 10 times Earth's distance, it would appear $10^2 = 100$ times dimmer. We can express this relationship with a simple formula called the **inverse square law for light**:

$$\text{apparent brightness} = \frac{\text{luminosity}}{4\pi \times (\text{distance})^2}$$

Because the standard units of luminosity are watts, the units of apparent brightness are *watts per square meter*. (The 4π in the formula comes from the fact that the surface area of a sphere is given by $4\pi \times (\text{radius})^2$.)

Think about it Suppose Star A is four times as luminous as Star B. How will their apparent brightnesses compare if they are both the same distance from Earth? How will their apparent brightnesses compare if Star A is twice as far from Earth as Star B? Explain.

In principle, we can always determine a star's apparent brightness by carefully measuring with a telescope the amount of light we receive from the star per square meter. The inverse square law for light therefore provides our primary way of measuring stellar luminosities, since we can calculate luminosity

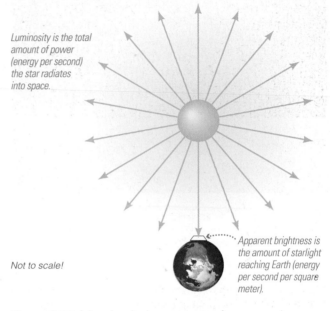

Luminosity is the total amount of power (energy per second) the star radiates into space.

Not to scale!

Apparent brightness is the amount of starlight reaching Earth (energy per second per square meter).

Figure 8.10 | Luminosity is a measure of power, and apparent brightness is a measure of power per unit area.

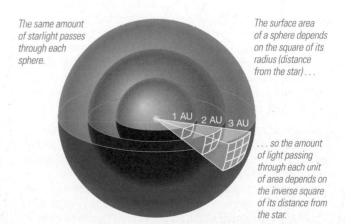

The same amount of starlight passes through each sphere.

The surface area of a sphere depends on the square of its radius (distance from the star)...

1 AU 2 AU 3 AU

...so the amount of light passing through each unit of area depends on the inverse square of its distance from the star.

Figure 8.11 | The inverse square law for light: The apparent brightness of a star declines with the square of its distance.

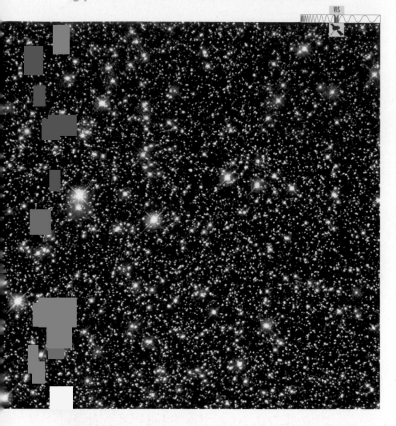

Figure 8.12 | This Hubble Space Telescope photo shows a wide variety of stars that differ in color and brightness. Most of the stars in this photo are at roughly the same distance, about 2000 light-years from the center of our galaxy. Clouds of gas and dust obscure our view of visible light from most of our galaxy's central regions, but a gap in the clouds allows us to see the stars in this photo.

Common Misconceptions

PHOTOS OF STARS

Photographs of stars, star clusters, and galaxies convey a great deal of information, but they also contain a few artifacts that are not real. For example, different stars seem to have different sizes in photographs such as Figure 8.12, but stars are so far away that they should all appear as mere points of light. The sizes are an artifact of how our instruments record light. Bright stars tend to be overexposed in photographs, making them appear larger than dimmer stars. Overexposure also explains why the centers of globular clusters and galaxies usually look like big blobs in photographs: The central regions of these objects contain many more stars than the outskirts, and the combined light of so many stars tends to blur together.

Spikes around bright stars in photographs, often making the pattern of a cross with a star at the center, are another artifact. You can see these spikes around many of the brightest stars in Figure 8.12. These spikes are not real but rather are created by the interaction of starlight with the supports holding the secondary mirror in the telescope [Section 3.1]. The spikes generally occur only with point sources of light like stars, and not with larger objects like galaxies.

from apparent brightness and distance. The most difficult part of this process is measuring a star's distance.

The most direct way to measure a star's distance is with *stellar parallax,* the small annual shifts in a star's apparent position caused by Earth's motion around the Sun [Section 2.3]. Astronomers measure stellar parallax by comparing observations of a nearby star made 6 months apart (see Figure 2.18). The nearby star appears to shift against the background of more distant stars because we are observing it from opposite points of Earth's orbit. Measuring the precise amount of the star's annual shift due to parallax and then using some geometry enables us to calculate the star's distance [Section 12.1].

Once we have determined a star's luminosity from its apparent brightness and distance, we usually state the result in comparison to the Sun's luminosity, which we write as L_{Sun} for short. For example, Proxima Centauri, the nearest of the three stars in the Alpha Centauri system and the nearest star besides our Sun, is only about 0.0006 times as luminous as the Sun, or $0.0006L_{Sun}$. Betelgeuse, the bright left-shoulder star of Orion, has a luminosity of $38,000L_{Sun}$, meaning that it is 38,000 times as luminous as the Sun. Studies of the luminosities of many stars have taught us that stars have a wide range of luminosities, with our Sun somewhere in the middle. The dimmest stars have luminosities 1/10,000 that of the Sun ($10^{-4} L_{Sun}$) and are extremely common, while the brightest stars are about 1 million times as luminous as the Sun ($10^6 L_{Sun}$) and are very rare.

Measuring Surface Temperature Surface temperature is the next fundamental property we can measure. You might wonder why we emphasize *surface* temperature rather than interior temperature. The answer is that only surface temperature is directly measurable; interior temperatures must be inferred from mathematical models of stellar interiors.

Measuring a star's surface temperature is somewhat easier than measuring its luminosity, because the star's distance doesn't affect the measurement. We determine surface temperature from either the star's color or its spectrum. **Figure 8.12** shows that stars come in a variety of colors. They come in different colors because they emit thermal radiation, and a thermal radiation spectrum depends only on the surface temperature of the object that emits it (see Tools of Science, p. 80). For example, the Sun's 5800 K surface temperature causes it to emit most strongly in the middle of the visible portion of the spectrum, which is why the Sun looks yellow or white in color. A cooler star, such as Betelgeuse (surface temperature 3400 K), looks red because it emits much more red light than blue light. A hotter star, such as Sirius (surface temperature 9400 K), emits more blue light than red light and therefore has a bluish color.

We can measure surface temperature more precisely by carefully studying a star's spectral lines. Stars with spectral lines of highly ionized elements must be fairly hot, because it takes a high temperature to ionize atoms. Stars with spectral lines of molecules must be relatively cool, because molecules break apart into individual atoms unless they are at relatively cool temperatures. The types of spectral lines present in a star's spectrum therefore provide a direct measure of the star's surface temperature.

Astronomers classify stars according to surface temperature by assigning a **spectral type** determined from the spectral lines present in a star's spectrum. The hottest (bluest) stars are called spectral type O, followed in order of declining surface temperature by spectral types B, A, F, G, K, and M (**Table 8.2**). The traditional mnemonic for remembering this sequence, OBAFGKM, is "Oh Be A Fine Girl/Guy, Kiss Me!" Astronomers sometimes subdivide the spectral types by following the letter with a number from 0 through 9. For example, the Sun's spectral type is G2, which means it is slightly hotter than a G3 star but slightly cooler than a G1 star.

Table 8.2 The Spectral Sequence

Spectral Type	Example(s)	Temperature Range	Key Absorption Line Features	Brightest Wavelength (color)	Typical Spectrum
O	Stars of Orion's Belt	>30,000 K	Lines of ionized helium, weak hydrogen lines	<97 nm (ultraviolet)*	
B	Rigel	30,000 K– 10,000 K	Lines of neutral helium, moderate hydrogen lines	97–290 nm (ultraviolet)*	
A	Sirius	10,000 K– 7500 K	Very strong hydrogen lines	290–390 nm (violet)*	
F	Polaris	7500 K– 6000 K	Moderate hydrogen lines, moderate lines of ionized calcium	390–480 nm (blue)*	
G	Sun, Alpha Centauri A	6000 K– 5000 K	Weak hydrogen lines, strong lines of ionized calcium	480–580 nm (yellow)	
K	Arcturus	5000 K– 3500 K	Lines of neutral and singly ionized metals, some molecules	580–830 nm (red)	
M	Betelgeuse, Proxima Centauri	<3500 K[†]	Strong molecular lines	>830 nm (infrared)	

*All stars above 6000 K look more or less white to the human eye because they emit plenty of radiation at all visible wavelengths.
[†]Two new spectral types, L and T, have been proposed for starlike objects with surface temperatures even cooler than those of M stars. However, unlike other stars, most of these objects are not fueled by nuclear fusion.

The range of surface temperatures for stars is much narrower than the range of luminosities. The coolest stars (spectral type M) have surface temperatures as low as 3000 K; the hottest stars (spectral type O) have surface temperatures as great as 50,000 K. Cool red stars are much more common than hot blue stars.

Measuring Mass Stellar mass is generally more difficult to measure than surface temperature or luminosity. The most dependable method for determining a star's mass relies on Newton's version of Kepler's third law (see Tools of Science, p. 96). Recall that this law can be applied only when we can observe one object orbiting another, and it requires that we measure both the orbital period and the average orbital distance of the orbiting object. For stars, these requirements generally mean that we can apply the law to measure masses only in **binary star systems** in which two stars orbit each other.

Binary star systems in which one star occasionally eclipses the other are particularly useful for mass measurements, because we know that the stars are orbiting in the same plane as our line of sight (**Figure 8.13**). Doppler shift measurements therefore tell us the true orbital velocities of the stars, because they are moving directly toward or directly away from us during part of the orbit. Combining these velocity measurements with the orbital period, which is simply the time between eclipses of one of the stars, gives the average orbital distance. Using the orbital period and an orbital distance, we can calculate the masses of the stars with Newton's version of Kepler's third law.

Careful observations of many different binary star systems containing stars of many different masses have helped to establish the overall mass range

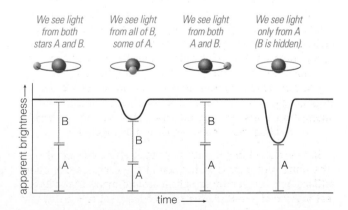

Figure 8.13 | If the stars in a binary star system orbit in the same plane as our line of sight, they will occasionally eclipse each other, causing a temporary drop in the apparent brightness we measure for the system. These types of binary systems are especially useful for measuring stellar masses.

for stars. This range extends from as little as 0.08 times the mass of the Sun ($0.08M_{Sun}$) to about 150 times the mass of the Sun ($150M_{Sun}$). We'll discuss the reasons for this mass range in Chapter 9.

What patterns do we find in the properties of stars?

We have seen that stars come in a wide range of luminosities, surface temperatures, and masses. But are these characteristics randomly distributed among stars, or can we find patterns that provide clues to what determines the properties of stars?

Before reading any further, take another look at Figure 8.12 and think about how you would classify these stars. Almost all of them are at nearly the same distance from Earth, so we can compare their true luminosities by looking at their apparent brightnesses in the photograph. If you look closely, you'll notice a couple of important patterns:

- Most of the brightest stars are reddish in color.

- Aside from those relatively few bright red stars, there's a general trend to the luminosities and colors among all the rest of the stars: The brighter ones are blue or white, the more modest ones are yellowish like our Sun, and the dimmest ones are barely visible specks of red.

Keeping in mind that colors tell us about surface temperature—blue is hotter and red is cooler—you can see that these patterns must be telling us about relationships between surface temperature and luminosity. Careful study of these relationships reveals that stars fall into three major groups.

The Main Sequence Although the very bright red stars stand out in Figure 8.12, the majority of the stars in the photograph follow the trend in which redder color—which means lower surface temperature—goes with lower luminosity, while bluer color—which means higher surface temperature—goes with higher luminosity. We'll see in the next chapter that the stars that follow this trend all have an important property in common: Like our Sun, these stars are generating energy by fusing hydrogen in their cores. This group is called the **main sequence,** for reasons that should become clear in the next section. Because these stars follow a clear trend in which luminosity depends on surface temperature, and surface temperature is specified by spectral type, we can infer the luminosity of a main-sequence star simply by determining its spectral type.

Observations of binary systems containing stars belonging to the main sequence have shown that their luminosities and surface temperatures are closely related to their masses. Stars of spectral type G, like our Sun, have masses near $1M_{Sun}$, luminosities near $1L_{Sun}$, and surface temperatures near 5800 K, placing them approximately in the middle of the range of each property. In comparison, the coolest stars in the main sequence—those of spectral type M—have masses less than $0.3M_{Sun}$, luminosities less than $0.01L_{Sun}$, and surface temperatures less than 3500 K. At the other extreme, hot stars of spectral type O have masses greater than about $20M_{Sun}$, luminosities exceeding $30,000L_{Sun}$, and surface temperatures hotter than 30,000 K.

We can estimate the lifetimes of these stars from their masses and luminosities, just as we did for the Sun. Recall that the Sun has enough hydrogen fuel in its core to continue radiating its current luminosity for 10 billion years. Now consider a star of spectral type B, with a mass of $10M_{Sun}$ and a luminosity of $10,000L_{Sun}$. It starts with 10 times the fuel of the Sun but consumes this fuel at a rate 10,000 times faster. Its lifetime must therefore be only about 1/1000 as long as the Sun's lifetime, or about 10 million years. Cosmically speaking, this is a remarkably short time, one reason why massive stars are so rare: Most of the massive stars that were ever born have long since died.

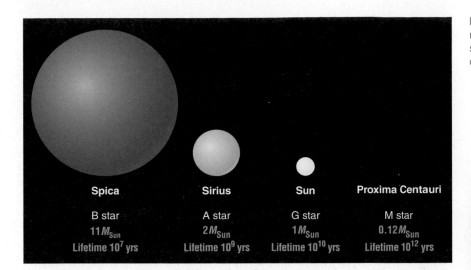

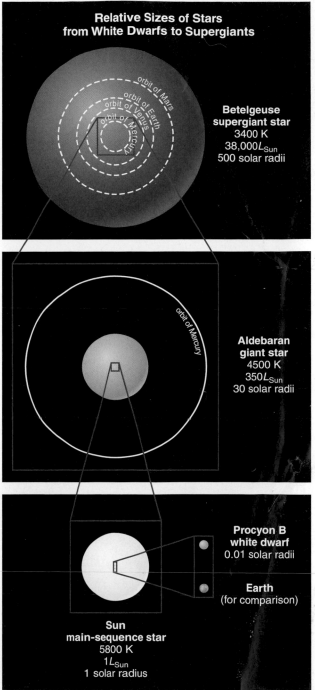

Figure 8.14 | Four typical hydrogen-burning stars from the main sequence, shown to scale. Note that the more massive stars are hotter and more luminous than the less massive ones, but have shorter lifetimes.

On the other end of the scale, a 0.3-solar-mass star emits a luminosity just 0.01 times that of the Sun and consequently lives roughly $0.3/0.01 = 30$ times as long as the Sun, or 300 billion years. In a universe that is now about 14 billion years old, even the most ancient of these small, dim, red stars of spectral type M still survive and will continue to shine faintly for hundreds of billions of years to come. **Figure 8.14** compares these properties for a sample of four stars from the main sequence.

Giants and Supergiants   The bright red stars in Figure 8.12 do not follow the trends we find for the main-sequence stars. In fact, because these stars are redder than the Sun, their surfaces must be cooler than that of the Sun, even though they are far more luminous. How can relatively cool stars have such high luminosities? Remember that a star's surface temperature determines the amount of light it emits per unit surface area: Hotter stars emit much more light per unit surface area than cooler stars. For example, a blue star emits far more total light than a red star of the same size. Therefore, a star that is red and cool can be very luminous only if it has a very large surface area.

We conclude that these very bright red stars must be far larger in size than our Sun. These large stars are known as either **giants** or **supergiants,** depending on their size. Giants have radii that can be 10 to 100 times that of the Sun, and supergiants can be even larger. For example, Aldebaran is a giant star more than 10 times as large in radius as our Sun. Betelgeuse, the left shoulder in the constellation Orion, is an enormous supergiant with a radius roughly 500 times that of the Sun, equivalent to more than twice the Earth–Sun distance (**Figure 8.15**). As we'll discuss in Chapter 9, we now know that giant and supergiant stars have run out of hydrogen fuel in their cores and are nearing the ends of their lives.

White Dwarfs   The third major group of stars consists of **white dwarfs.** These stars are so dim that none can be seen in Figure 8.12. A typical white dwarf, such as Procyon B (the last star shown in Figure 8.15), has a luminosity of only about $0.001L_{Sun}$ but a surface temperature hotter than that of the Sun. For such a hot star to be so dim, it must be much smaller in size than the Sun. In fact, most white dwarfs are similar in radius to Earth, even though they are similar in mass to the Sun. The matter in white dwarfs must be compressed to an extremely high density, unlike anything found on Earth.

We'll see in Chapter 9 that white dwarfs are small and hot because they are the remaining embers of giants that have run out of usable fuel and blown off their outer layers. They are hot because they are essentially exposed stellar cores, but they are dim because they lack an ongoing energy source and radiate

Figure 8.15 | The relative sizes of stars. A supergiant like Betelgeuse would fill the inner solar system. A giant like Aldebaran would fill the inner third of Mercury's orbit. The Sun is a hundred times larger in radius than a white dwarf, which is roughly the same size as Earth.

only their leftover heat into space. We'll discuss the nature of white dwarfs and other stellar corpses in Chapter 10.

Complete Stellar Classification We have discussed two key ways of classifying stars. First, we can classify a star by spectral type, which essentially tells us its surface temperature. The hottest stars have spectral type O, and temperatures decline in the order of the spectral sequence OBAFGKM. Second, we have found that stars fall into three major groupings: a *main sequence* consisting of stars that burn hydrogen in their cores, for which higher surface temperature goes with higher luminosity; giants and supergiants, which are extraordinarily luminous and large in size; and white dwarfs, which are extremely dim and small in size. The complete classification of a star therefore consists of two parts: (1) its spectral type, stated as one of the letters OBAFGKM, and (2) whether it is a star that burns hydrogen in its core, a giant or supergiant, or a white dwarf. For example, we can say that our Sun is a spectral type G hydrogen-burning star, while Betelgeuse is a spectral type M supergiant.

✳ THE PROCESS OF SCIENCE IN ACTION

8.3 Visualizing Patterns Among Stars

Major discoveries in science sometimes come about when someone happens to look at the data in just the right way. The patterns we have just discussed were discovered in this way about a century ago. This discovery of the patterns among stars is our case study in the process of science for this chapter, and it demonstrates the importance of visualizing data, and particularly the use of graphs, in scientific discovery.

How did we discover the patterns in stellar properties?

Our story of discovery begins at the end of the 19th century, when Edward Pickering was director of the Harvard College Observatory. He was interested in studying and classifying stellar spectra, but this was tedious and time-consuming work. He therefore hired numerous assistants, whom he called "computers." Most of his computers were women who had studied physics or astronomy at women's colleges such as Wellesley and Radcliffe (**Figure 8.16**). At the time, institutions like Harvard did not permit women to hold faculty positions, so Pickering's project offered a rare opportunity for these women to continue their work in astronomy.

Improving the Spectral Types One of the first computers was Williamina Fleming. Following Pickering's suggestions, Fleming classified stellar spectra according to the strength of the spectral lines of hydrogen: type A for the strongest hydrogen lines, type B for slightly weaker hydrogen lines, and so on, to type O for stars with the weakest hydrogen lines. As more stellar spectra were obtained and the spectra were studied in greater detail, it became clear that this classification scheme based solely on hydrogen lines was inadequate. Ultimately, the task of finding a better classification scheme fell to Annie Jump Cannon, who joined Pickering's team in 1896. Building on the work of Fleming and another of Pickering's computers, Antonia Maury, Cannon soon realized that the spectral classes fell into a natural order—but not the alphabetical order determined by hydrogen lines alone. Moreover, she found that some of the original classes overlapped others and could be eliminated. Cannon discovered that the natural sequence consisted

Figure 8.16 | Women astronomers pose with Edward Pickering at Harvard College Observatory in 1913. Annie Jump Cannon is fifth from the left in the back row.

of just a few of Pickering's original classes in the order OBAFGKM, which we now know to represent a sequence of surface temperatures. This breakthrough laid the foundation for our modern understanding of stellar properties.

Hertzsprung-Russell Diagrams Around the same time that Cannon was reorganizing the spectral types, Danish astronomer Ejnar Hertzsprung and American astronomer Henry Norris Russell pioneered a new approach to analyzing the relationship between spectral type and luminosity. Building on the work of Cannon and others, they independently decided to make graphs of stellar properties by plotting stellar luminosity on one axis and spectral type on the other. These graphs ultimately revealed fundamental patterns among the properties of stars and are now called **Hertzsprung-Russell (H-R) diagrams.** These diagrams remain among the most important tools in astronomical research and are central to our study of stars.

Figure 8.17 (p. 140) displays a modern example of an H-R diagram. It is a graph with surface temperature on one axis and luminosity on the other:

- The horizontal axis represents a star's surface temperature, which corresponds to both its spectral type and its color. Temperature *decreases* from left to right because Hertzsprung and Russell based their diagrams on the spectral sequence OBAFGKM.

- The vertical axis represents a star's luminosity, in units of the Sun's luminosity (L_{Sun}). Stellar luminosities span a wide range, so each tick mark represents a luminosity 10 times as large as that of the prior tick mark.

All you need to know to plot a star on an H-R diagram is (1) its luminosity and (2) its surface temperature or spectral type. Each location on the diagram represents a unique combination of spectral type and luminosity. For example, the dot representing the Sun in Figure 8.17 corresponds to the Sun's surface temperature, 5800 K, and its luminosity, $1L_{Sun}$. Because luminosity increases upward on the diagram and surface temperature increases leftward, stars near the upper left are hot and luminous. Similarly, stars near the upper right are cool and luminous, stars near the lower right are cool and dim, and stars near the lower left are hot and dim.

Think about it Explain how the colors of the stars in Figure 8.17 help indicate stellar surface temperature. Do these colors tell us anything about interior temperatures? Why or why not?

Soon after Hertzsprung and Russell started plotting stars according to luminosity and spectral type, it became clear that stars cluster into three different regions of an H-R diagram:

- Most stars fall along the prominent streak known as the *main sequence,* which runs from the upper left to the lower right on the H-R diagram. All these stars are generating energy like the Sun, by fusing hydrogen into helium in their cores.

- *Giants and supergiants* are found to the upper right of the main sequence because they are very luminous and tend to have relatively cool surfaces.

- *White dwarfs* are found to the lower left of the main sequence because they are less luminous than main-sequence stars but have relatively high surface temperatures.

The H-R diagram also helps us visualize relationships among stellar radii. If two stars have the same surface temperature, one can be more luminous than the other only if it is larger in size. Stellar radii therefore increase as we go

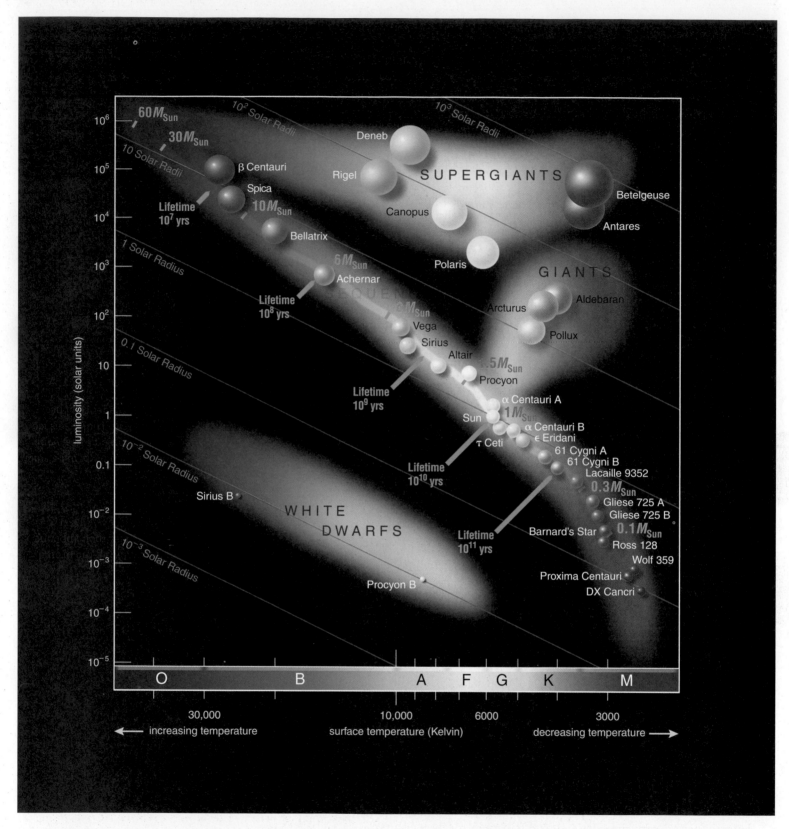

Figure 8.17 | An H-R diagram, one of astronomy's most important tools, shows a star's surface temperature or spectral type along the horizontal axis and its luminosity along the vertical axis. Several of the brightest stars in the sky are plotted here, along with a few of those closest to Earth. They are not drawn to scale—the diagonal lines, labeled in solar radii, indicate how large they are compared to the Sun. The lifetime and mass labels apply only to main-sequence stars.

from the high-temperature, low-luminosity corner on the lower left of the H-R diagram to the low-temperature, high-luminosity corner on the upper right. Diagonal lines in Figure 8.17 indicate the radii of the stars in the graph.

Mass and Lifetime Along the Main Sequence   We have already noted that luminosity and surface temperature are both closely related to mass for main-sequence stars. Purple labels in Figure 8.17 indicate the masses of stars along the main sequence and show that *stellar masses decrease downward along the main sequence.* To make the labels easier to see, **Figure 8.18** repeats the same data but shows only the main sequence rather than the entire H-R diagram. At the upper end of the main sequence are the hot, luminous O stars, with masses as high as 150 times that of the Sun ($150M_{Sun}$). On the lower end are the cool, dim M stars, with as little as 0.08 times the mass of the Sun ($0.08M_{Sun}$).

From this close relationship between mass and luminosity, we can also determine a hydrogen-burning star's lifetime. Green labels in Figures 8.17 and 8.18 indicate the lifetimes of main-sequence stars. Like masses, stellar lifetimes vary in an orderly way as we move along the main sequence: Massive stars near the upper end of the main sequence have *shorter* lives than less massive stars near the lower end.

The Bottom Line   An H-R diagram visually represents an enormous amount of information about stars. A single point representing a star on the diagram tells us the star's luminosity, surface temperature, spectral type, color, and radius. If that point is on the main sequence, it also tells us the mass and life-time of the star. The first major discovery made with H-R diagrams was of the distinct patterns among these properties, but that was just the beginning. In the next chapter, we will investigate *why* the luminosity and surface temperature of a main-sequence star depend on its mass and why main-sequence stars become giants and supergiants after they run out of core hydrogen. As our investigation progresses, we will continue to make use of these remarkable diagrams.

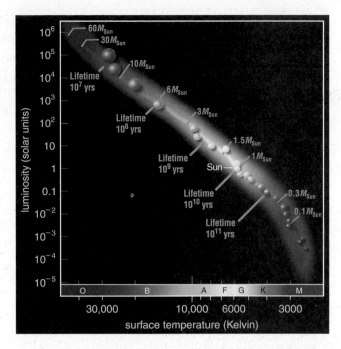

Figure 8.18 | The main sequence from Figure 8.17 is isolated here so that you can more easily see how masses and lifetimes vary along it. Notice that more massive hydrogen-burning stars are brighter and hotter but have shorter lifetimes. (Stellar masses are given in units of solar masses: $1M_{Sun} = 2 \times 10^{30}$ kg.)

Summary of Key Concepts

8.1 Properties of the Sun

What is the Sun like?

The Sun is a giant ball of plasma that generates energy by **nuclear fusion** of hydrogen into helium in its **core,** where the temperature is 15 million K.

How does energy escape the Sun?

Energy released by fusion in the Sun's core travels first outward through the **radiation zone** and then through the **convection zone** to the **photosphere,** from which it escapes into space in the form of photons. Convection beneath the photosphere helps to generate magnetic fields that produce **sunspots** in the photosphere and other phenomena higher up in the **chromosphere** and **corona.**

8.2 Properties of Other Stars

How do we measure the properties of stars?

We can determine a star's luminosity by measuring its **apparent brightness** and distance and then using the **inverse square law for light.** We can determine a star's surface temperature from either its color or its **spectral type.** We can measure the

masses of stars in **binary star systems** by determining the orbital period and average orbital distance of the stars and then applying Newton's version of Kepler's third law.

What patterns do we find in the properties of stars?

Most stars follow the same general trend, in which high-mass stars are very luminous and blue, with high surface temperatures, while lower-mass stars are less luminous and redder, with cooler surface temperatures. This group is known as the **main sequence.** The exceptions are **giants** and **supergiants,** which are much larger in radius and more luminous than normal stars, and **white dwarfs,** which are much smaller in radius and tend to have hotter than normal surface temperatures.

☀ THE PROCESS OF SCIENCE IN ACTION

8.3 Visualizing Patterns Among Stars

How did we discover the patterns in stellar properties?

Plotting the properties of stars on a **Hertzsprung-Russell (H-R) diagram,** which is a graph of surface temperature (or spectral type)

versus luminosity, revealed the relationships between stellar properties. Most stars fall along the *main sequence* of the diagram, which runs diagonally from the upper left to the lower right. Giants and supergiants appear to the upper right of the main sequence, while white dwarfs appear to the lower left.

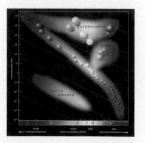

Investigations

Quick Quiz

Choose the best answer to each of the following; answers are in Appendix D. Explain your reasoning with one or more complete sentences.

1. Which of these groups of particles has the greatest mass? (a) a helium nucleus with two protons and two neutrons (b) four electrons (c) four individual protons

2. Scientists estimate the central temperature of the Sun using (a) probes that measure changes in Earth's atmosphere. (b) mathematical models of the Sun. (c) telescopic observations of the Sun's interior.

3. At the center of the Sun, fusion converts hydrogen into (a) plasma. (b) radiation and elements like carbon and nitrogen. (c) helium and energy.

4. Solar energy leaves the core of the Sun in the form of (a) photons. (b) rising hot gas. (c) sound waves.

5. Which of these layers of the Sun is coolest? (a) photosphere (b) chromosphere (c) corona

6. Why do sunspots appear darker than their surroundings? (a) They are cooler than their surroundings. (b) They block some of the sunlight from the photosphere. (c) They do not emit any light.

7. What do we need to measure in order to determine a star's luminosity? (a) apparent brightness and mass (b) apparent brightness and temperature (c) apparent brightness and distance

8. What two pieces of information would you need in order to measure the masses of stars in an eclipsing binary system? (a) the time between eclipses and the average distance between the stars (b) the period of the binary system and its distance from the Sun (c) the velocities of the stars and the Doppler shifts of their absorption lines

9. Which of these stars has the coolest surface temperature? (a) an A star (b) an F star (c) a K star

10. Which of these stars is the most massive? (a) a main-sequence A star (b) a main-sequence G star (c) a main-sequence M star

11. Which of these stars has the longest lifetime? (a) a main-sequence A star (b) a main-sequence G star (c) a main-sequence M star

12. Which of these stars has the largest radius? (a) a supergiant A star (b) a giant K star (c) a supergiant M star

Short-Answer/Essay Questions

Explain all answers clearly, with complete sentences and proper essay structure if needed. An asterisk (*) designates a quantitative problem, for which you should show all your work.

13. *A Really Strong Force.* How would the interior temperature of the Sun be different if the strong force that binds nuclei together were 10 times stronger?

14. *Measuring the Sun's Rotation.* Suppose you take pictures of the Sun several days in a row, noting the changing position of a particular sunspot. Now, suppose a friend of yours standing on Pluto does the same thing. Would you and your friend measure the same rate of movement for the sunspot? Explain.

15. *Covered with Sunspots.* Describe what the Sun would look like from Earth if the entire photosphere were the same temperature as a sunspot.

16. *Inside the Sun.* Describe how scientists determine what the interior of the Sun is like. Why have we not yet sent a probe into the Sun to obtain direct information about the interior?

17. *Interpreting the H-R Diagram.* Using the information in Figure 8.17, describe how Proxima Centauri differs from Sirius.

18. *Parallax from Jupiter.* Suppose you could travel to Jupiter and observe changes in positions of nearby stars during one orbit of Jupiter around the Sun. Describe how those changes would be different from what we measure from Earth. How would your ability to measure the distances to stars be different from the vantage point of Jupiter?

19. *An Expanding Star.* Describe what would happen to the surface temperature of a star if its radius doubled in size with no change in luminosity.

20. *Colors of Eclipsing Binaries.* Figure 8.13 shows an eclipsing binary system consisting of a small blue star and a larger red star. Explain why the decrease in apparent brightness of the combined system is greater when the blue star is eclipsed than when the red star is eclipsed.

*21. *The Lifetime of the Sun.* The total mass of the Sun is about 2×10^{30} kg, of which about 75% was hydrogen when the Sun formed. However, only about 13% of this hydrogen ever becomes available for fusion in the core. The rest remains in layers of the Sun where the temperature is too low for fusion.
   a. Use the given data to calculate the total mass of hydrogen available for fusion over the lifetime of the Sun.
   b. The Sun fuses about 600 billion kilograms of hydrogen each second. Based on your result from part (a), calculate how long the Sun's initial supply of hydrogen can last. Give your answer in both seconds and years.
   c. Given that our solar system is now about 4.6 billion years old, when will the Sun run out of hydrogen for fusion?

*22. *The Inverse Square Law for Light.* Earth is about 150 million kilometers from the Sun, and the apparent brightness of the Sun in our sky is about 1300 watts/m². Using these two facts and the inverse square law for light, determine the apparent brightness that we would measure for the Sun *if* we were located at the following positions.
   a. Half Earth's distance from the Sun
   b. Twice Earth's distance from the Sun
   c. Five times Earth's distance from the Sun

9 Stellar Lives

Learning Goals

9.1 Lives in Balance
- Why do stars shine so steadily?
- Why do a star's properties depend on its mass?

9.2 Star Death
- What will happen when our Sun runs out of fuel?
- How do high-mass stars end their lives?

✳ THE PROCESS OF SCIENCE IN ACTION

9.3 Testing Stellar Models with Star Clusters
- What do star clusters reveal about the lives of stars?

This photo shows the Orion Nebula, a huge interstellar cloud that acts as a stellar nursery, giving birth to thousands of stars over a period of a few million years. These stars then live out their lives, with fates dictated by their masses. Low-mass stars shine steadily for billions of years, sharing an ultimate fate similar to that of our Sun. High-mass stars shine so brightly that they burn out in just a few million years, dying in spectacular explosions that scatter their ashes into space. In this chapter, we'll study the how and why of star births, lives, and deaths. Along the way, we'll learn about the future of our own Sun and see how high-mass stars have produced the "star stuff" from which we and our planet are made.

9.1 Lives in Balance

The story of a star's life is in many ways the story of a battle between two opposing forces: pressure and gravity. A star is born when gravity becomes strong enough to overcome pressure within a gas cloud and compress it until nuclear fusion begins. A star lives while the energy generated by fusion can keep pressure and gravity in steady balance. A star dies as it exhausts its fuel for fusion and goes out of balance again. In this section, we examine how a star achieves the steady state of balance that fusion maintains.

Why do stars shine so steadily?

Most stars shine steadily for most of their lives. Annual changes in our Sun's light output are less than 0.1%, and models indicate that the Sun's luminosity has changed by less than 30% over the last $4\frac{1}{2}$ billion years. This type of steady energy output has been critical for life on Earth, and perhaps for life on planets around other stars. But what keeps stars shining so steadily? To answer this question, let's first look at how stars are born.

Star Birth   Stars form out of interstellar gas, within clouds in which the inward pull of gravity becomes stronger than the outward push of gas pressure. Two things can help gravity win out over pressure and start the contraction of a cloud of gas: (1) high density, because packing the gas particles closer together makes the gravitational forces between them stronger, and (2) low temperature, because lowering a cloud's temperature reduces the gas pressure. We therefore expect star-forming clouds to be colder and denser than most other interstellar gas clouds.

Observations confirm this idea: Stars are indeed born within the coldest and densest clouds, called **molecular clouds** (**Figure 9.1**) because their low temperatures allow hydrogen atoms to pair up to form hydrogen molecules. They typically have temperatures of only 10–30 K. And while their densities are still low enough that they would qualify as superb vacuums by earthly standards, these molecular clouds are hundreds to thousands of times as dense as other regions of interstellar space. The molecular clouds that give birth to stars also tend to be quite large—typically containing enough material to form

Figure 9.1 | A star-forming molecular cloud in the constellation Scorpius. The region pictured here is about 50 light-years across.

Newborn stars produce white patches in the cloud where starlight illuminates surrounding gas.

The cloud looks dark where dust particles block the light from more distant stars.

thousands of stars—because more total mass also helps gravity overcome gas pressure. The gas in these large clouds tends to form smaller clumps as they collapse, with each clump forming an individual star system. That is why stars are generally born in clusters. Our own Sun was presumably born in a cluster, but after $4\frac{1}{2}$ billion years the Sun's siblings have dispersed and mingled with the other stars in our galaxy.

We have already seen what happens to a clump of gas that gives birth to an individual star system, such as the *solar nebula* that gave birth to our Sun [Section 4.2]. Gravity causes the cloud to shrink in size—a process called **gravitational contraction.** This contraction causes the gas temperature to rise, especially in the center of the cloud, because it converts gravitational potential energy into thermal energy. Gravitational contraction proceeds rapidly at first, because photons from the cloud quickly carry away the thermal energy generated by gravitational contraction, thereby preventing a buildup of gas pressure that could counteract gravity. Eventually, however, the central region of the cloud becomes so dense that photons cannot easily escape, so the pressure builds and contraction slows. The cloud center is now a **protostar** that shines with energy generated by ongoing gravitational contraction. It is not yet a true star because its core is not yet hot enough to sustain nuclear fusion.

Despite the energy losses from its surface, a protostar continues to heat up, because much of the energy released by gravitational contraction remains trapped inside it. This process of ongoing shrinkage and heating continues until the central temperature and density grow high enough to sustain nuclear fusion. Gravitational contraction finally stops when the energy production from fusion in the core matches the energy losses from the star's surface. The thermal energy content of the star then remains constant, enabling pressure and gravity to come into steady balance. This is why we say that the star is "born" when its core becomes hot enough for sustained nuclear fusion.

Two Kinds of Balance Looking back at the story of star birth, we can see that a star becomes a steady source of light when it achieves two kinds of balance. First, the outward push of internal gas pressure must balance the inward pull of gravity to prevent the star from expanding or contracting. A stack of acrobats provides a simple example of this kind of balance, which we will call **gravitational equilibrium** (Figure 9.2). The bottom person supports the weight (due to gravity) of everybody above him, so his arms must push upward with enough pressure to support all this weight. At each higher level, the overlying weight is less, so it's a little easier for each additional person to hold up the rest of the stack.

Gravitational equilibrium works much the same within a star like the Sun, except the outward push against gravity comes from internal gas pressure rather than an acrobat's arms. The internal pressure precisely balances gravity at every point within the star, thereby keeping it stable in size (Figure 9.3). Because the weight of overlying layers is greater as we look deeper below the surface, the pressure increases with depth.

 Think about it Earth's atmosphere is also in gravitational equilibrium, with the pressure in lower layers supporting the weight of upper layers. Use this idea to explain why the air gets thinner at higher altitudes.

The second kind of balance is the **energy balance** between the rate at which fusion releases energy in the star's core and the rate at which the star's surface radiates this energy into space. Energy balance is important because without it, the balance between pressure and gravity does not remain steady. If fusion in the core does not replace the energy radiated from the surface, thereby keeping the total thermal energy content constant, then gravitational contraction will cause the core to shrink and force its temperature to rise.

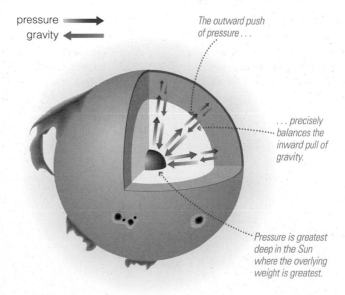

Figure 9.2 | An acrobat stack demonstrates gravitational equilibrium: The lowest person supports the most weight and feels the greatest pressure; the overlying weight and underlying pressure decrease for those higher up.

pressure →
gravity ←

The outward push of pressure . . .

. . . precisely balances the inward pull of gravity.

Pressure is greatest deep in the Sun where the overlying weight is greatest.

Figure 9.3 | Gravitational equilibrium in the Sun. At each point inside, the pressure pushing outward balances the weight of the overlying layers. This balance works the same way in other stars.

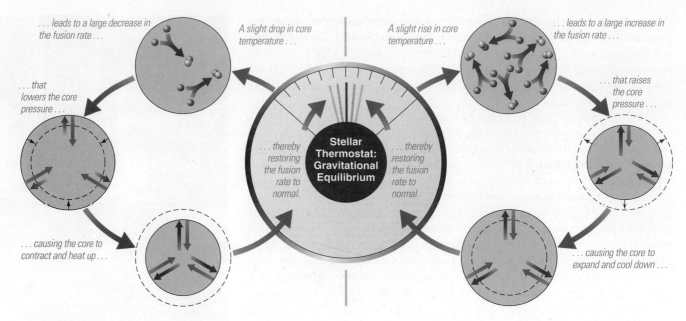

... leads to a large decrease in the fusion rate ...

A slight drop in core temperature ...

A slight rise in core temperature ...

... leads to a large increase in the fusion rate ...

... that lowers the core pressure ...

Stellar Thermostat: Gravitational Equilibrium

... that raises the core pressure ...

... thereby restoring the fusion rate to normal.

... thereby restoring the fusion rate to normal.

... causing the core to contract and heat up ...

... causing the core to expand and cool down ...

Figure 9.4 | The stellar thermostat. Gravitational equilibrium regulates a star's core temperature. Everything is in balance if the amount of energy leaving the core equals the amount of energy produced by fusion. A rise in core temperature causes the core to expand, lowering its temperature to its original value. A decrease in core temperature causes the core to contract, also restoring the original core temperature.

The Stellar Thermostat The interplay between these two kinds of balance enables stars to shine steadily for millions or billions of years. The key lies with a self-regulating **stellar thermostat** that quickly offsets any small change in temperature so that balance is restored (**Figure 9.4**). To see how it works, let's use our Sun as an example.

Suppose that for some reason the Sun's core temperature rose very slightly. Protons in the core would collide with more energy, enabling more fusion reactions to happen. In fact, the rate of fusion is very sensitive to temperature, so the small temperature increase would cause the fusion rate to soar, generating lots of extra energy. Because energy moves slowly through the Sun's interior, this extra energy would be bottled up in the core, causing an increase in the core pressure. The push of this pressure would temporarily exceed the pull of gravity, causing the core to expand and cool. This cooling, in turn, would cause the fusion rate to drop back down until the core returned to its original size and temperature, restoring both gravitational equilibrium and energy balance.

A slight drop in the Sun's core temperature would trigger an opposite chain of events. The reduced core temperature would lead to a decrease in the rate of nuclear fusion, causing a drop in pressure and contraction of the core. As the core shrank, its temperature would rise until the fusion rate returned to normal and restored the core to its original size and temperature.

To summarize, a star's luminosity will remain steady as long as hydrogen fusion continues in its core and keeps the star in both gravitational equilibrium and energy balance. A star like the Sun can shine steadily for about 10 billion years in this balanced state, because that is how long its core supply of hydrogen can last.

Why do a star's properties depend on its mass?

Now that we have seen how the Sun remains balanced, we are ready to look at why the luminosity, surface temperature, mass, and lifetime of a main-sequence star are all so closely related. The stars on the main sequence of an H-R diagram are steadily fusing hydrogen into helium in their cores, just like the Sun, meaning that each one has achieved both gravitational equilibrium and energy balance. However, the core temperature at which a star achieves these two kinds of balance depends on its mass. Because mass determines a star's balancing point, it determines all the rest of a star's major properties and also determines which objects can become stars and which cannot.

Mass and the Main Sequence To see why a star's state of balance depends on its mass, consider what happens inside a protostar more massive than the Sun. Because this protostar has more mass than the Sun, gravitational contraction converts more gravitational potential energy into thermal energy during its protostar stage. Fusion therefore begins earlier in the process of contraction and brings this higher-mass protostar into balance with a higher core temperature and a greater luminosity. The luminosity of this newly formed star can be much greater than the Sun's because even a slightly higher core temperature leads to a much higher rate of fusion, enabling the star to achieve energy balance with both a greater radius and a hotter surface temperature.

All the trends we see with mass along the main sequence reflect the properties stars have when they first come into balance. Stars with less mass than the Sun need to contract more before fusion begins and come into balance with a lower core temperature, lower luminosity, cooler surface, and smaller radius. They will also live much longer than the Sun because their lower fusion rates consume the fuel for fusion more slowly. Stars with more mass than the Sun contract less before fusion begins and come into balance with a higher core temperature, greater luminosity, hotter surface, and larger radius. They have much shorter lifetimes than the Sun because they consume their fuel so rapidly.

Limits on Stellar Masses The interplay between mass and the two kinds of balance in stars also explains why stars can exist only within a relatively narrow range of masses, from about $0.08 M_{Sun}$ to about $150 M_{Sun}$. On the high-mass end, models indicate that stars much more massive than $150 M_{Sun}$ can't exist because they cannot remain in balance: The enormous energy output from fusion in their cores would generate so much pressure and such a great luminosity that it would drive their outer layers into space. Observations confirm this idea: Stars much more massive than $150 M_{Sun}$ haven't been found.

On the low-mass end, calculations show that a protostar cannot become hot enough for nuclear fusion unless its mass is at least $0.08 M_{Sun}$, equivalent to about 80 times the mass of Jupiter. If its mass is less than this, internal pressure will halt its gravitational contraction before its core reaches the 10-million-K threshold needed for efficient hydrogen fusion. In that case, the protostar will stabilize in size as a "failed star" known as a **brown dwarf** (**Figure 9.5**). Because they do not sustain steady fusion in their cores, brown dwarfs never achieve the energy balance that would allow them to shine steadily. Instead, they slowly cool with time as they radiate away their internal thermal energy. Brown dwarfs occupy a fuzzy gap between what we call a planet and what we call a star. They radiate primarily in the infrared and are far dimmer than normal stars, making them extremely difficult to detect.

Pressure in Brown Dwarfs The source of the pressure that stops gravity from squeezing a brown dwarf's core to the point at which it could sustain fusion is quite different from the type of pressure we encounter in most situations. Ordinary gas pressure is often called **thermal pressure** because it is closely linked to temperature: Raising the temperature increases the particle speeds and thereby raises the thermal pressure. Thermal pressure is not what ultimately balances gravity in a brown dwarf. Instead, the crush of gravity in a brown dwarf is halted by something called **degeneracy pressure,** a type of pressure that does not depend on temperature at all. It depends instead on the laws of *quantum mechanics* that also give rise to distinct energy levels in atoms (see Tools of Science, p. 148).

Figure 9.5 | Brown dwarfs are "failed stars" that have masses lower than the $0.08 M_{Sun}$ required to sustain nuclear fusion. This artist's conception shows a brown dwarf orbited by a planet (to its left) in a system with multiple stars. The reddish color approximates how a brown dwarf would appear to human eyes. The bands are shown because we expect brown dwarfs to look more like jovian planets than like stars.

In much the same way that restrictions on electrons in atoms allow them to occupy only particular energy levels, the quantum laws restrict how closely electrons can be packed in a gas. We can see how this fact leads to degeneracy pressure with a simple analogy. Imagine an auditorium in which the laws of quantum mechanics dictate the spacing between chairs and people represent electrons (**Figure 9.6**). As though playing a game of musical chairs, the people are always moving from seat to seat, just as electrons must remain constantly in motion. Most objects are like auditoriums with many more available chairs than people, so the people (electrons) can easily find chairs as they move about. However, the cores of protostars with masses below $0.08M_{Sun}$ are like much smaller auditoriums with so few chairs that the people (electrons) fill nearly all of them. Because there are virtually no open seats, the people (electrons) can't all squeeze into a smaller section of the auditorium.

This resistance to squeezing is the origin of degeneracy pressure. If the people were really like electrons, quantum laws would also require them to

Tools of Science: Quantum Laws and Astronomy

Although astronomy is primarily concerned with very large things, the behavior of large objects sometimes depends on the laws that govern their constituent particles. The branch of physics that deals with the behavior of atoms and subatomic particles is called *quantum mechanics*, so we say that these particles obey *quantum laws*.

Perhaps the most important application of quantum laws in astronomy arises in spectroscopy (see Tools of Science, p. 130). Consider a hydrogen atom, which has one proton and one electron. It's tempting to think of the electron as a tiny particle that orbits the proton, but that would imply that the electron could have any orbit with any amount of energy. This is not the case; quantum laws dictate that the electron can have only particular amounts of energy, and not other energies in between. As an analogy, suppose you're washing windows on a building. If you stand on an adjustable platform, you can stop at any height above the ground. But if you use a ladder, you can stand only at the particular heights of the rungs on the ladder, and not at other heights in between. The possible energies of electrons in atoms are like the specific heights on a ladder: Only a few particular energy levels are possible (**Figure 1**).

This fact explains why atoms produce emission and absorption lines with particular wavelengths. An electron can change its energy level only if it gains or loses the particular amount of energy that separates a pair of levels. Therefore, an atom can absorb only photons with the particular energies needed for electrons to rise from one energy level to another; because the energy of a photon depends on its wavelength (shorter wavelength means higher energy), the atom absorbs only particular wavelengths of light. An atom emits photons when an electron falls between energy levels, and therefore can emit only photons with the same set of particular wavelengths that it can absorb.

Another important quantum effect in astronomy is the degeneracy pressure discussed in this chapter, which arises from the combination of two quantum laws: (1) the uncertainty principle, which places a limit on the certainty with which we can know the combination of a subatomic particle's location and velocity, and (2) the exclusion principle, which places a limit on how closely electrons and other subatomic particles can

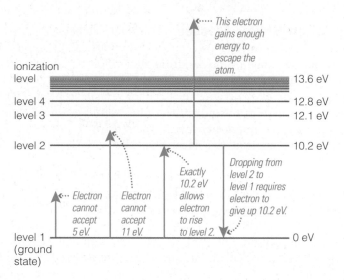

Figure 1 | Energy levels for the electron in a hydrogen atom, with energies given in units of electron-volts, or eV. Electrons can move between energy levels only if they gain or lose the particular energies that separate the levels.

be packed together. In a brown dwarf or white dwarf, electrons are packed in such a way that the exclusion principle limits how much more they can be packed, and this resistance to further packing creates degeneracy pressure. The uncertainty principle comes into play because the electrons are very close together, particularly in white dwarfs. This implies very little uncertainty in their locations, so their velocities must be highly uncertain, which means some of them must move with very high speeds. Because no particle can exceed the speed of light, there is also a limit to how much electron degeneracy pressure can resist the crush of gravity. Once that limit is surpassed, as is the case for a stellar core during a supernova, the electrons must combine with protons to make neutrons. The neutrons can then provide degeneracy pressure of their own—but only to a point, as we will discuss in the next chapter.

move faster and faster to find open seats as they were squeezed into a smaller section. However, their speeds would have nothing to do with temperature. Degeneracy pressure and the particle motion that goes with it arise *only* because of the restrictions on where the particles can go, which is why temperature does not affect them.

9.2 Star Death

We have seen that a star will shine steadily as long as it can maintain balance between gravity and pressure and between the rates at which energy is generated in its core and released into space at its surface. Hydrogen fusion in the core can maintain this state of balance as long as the core contains enough hydrogen. But fusion cannot continue forever. In this section, we will investigate what happens when core hydrogen runs out and a star can no longer remain in balance.

What will happen when our Sun runs out of fuel?

Like everything else in a star's life, its fate when it runs out of core hydrogen depends on its mass. Scientists have been able to learn about the end stages of stellar lives through a combination of theoretical modeling and observations of stars that appear to be in these late stages. These studies show that stars divide into two major groups by mass: Low-mass stars, like our Sun, end their lives relatively quietly, while high-mass stars (more than $8M_{Sun}$) die in titanic explosions. Our Sun is a fairly typical low-mass star, so we will use it as a prototype to investigate the final life stages of low-mass stars before moving on to the more dramatic deaths of high-mass stars.

Red Giant Stage When the Sun's core hydrogen is finally depleted, nuclear fusion in the core will temporarily cease. With no fusion to balance the thermal energy losses from the star's surface, the core will once again be out of balance, just as it was when the Sun was a protostar, and gravitational contraction of the core will resume. Somewhat surprisingly, the Sun's outer layers will expand outward during this time. Over a period of about a billion years—or about 10% of its main-sequence lifetime—the Sun will slowly grow in size and luminosity to become a **red giant.** At the peak of its red giant phase, the Sun will be more than 100 times larger in radius and more than 1000 times brighter in luminosity than it is today.

To understand why the Sun will expand even while its core is shrinking, we need to think about the composition of the core at the end of the Sun's main-sequence life. After the core exhausts its hydrogen, it will be made almost entirely of helium, because helium is the product of hydrogen fusion. However, the gas surrounding the core will still contain plenty of fresh hydrogen that has never previously undergone fusion. Because gravity shrinks both the *inert* (nonburning) helium core and the surrounding *shell* of hydrogen, the hydrogen shell will soon become hot enough for **hydrogen shell burning**— hydrogen fusion in a shell around the core (**Figure 9.7**). In fact, the shell will become so hot that hydrogen shell burning will proceed at a much higher rate than core hydrogen fusion does today. The higher fusion rate will generate enough energy to dramatically increase the Sun's luminosity and enough pressure to push the surrounding layers of gas outward, expanding the Sun's size.

Helium Burning Gravitational contraction of the inert helium core will continue—with the outer layers of the red giant Sun continuing to grow larger and more luminous—until the core temperature reaches 100 million K. At that point, **helium fusion** will ignite through a reaction that converts three helium nuclei into one carbon nucleus (**Figure 9.8**). Energy is released because the

a When there are many more available places (chairs) than particles (people), a particle is unlikely to try to occupy the same place as another particle. The only pressure comes from the temperature-related motion of the particles.

b When the number of particles (people) approaches the number of available places (chairs), finding an available place requires that the particles move faster than they would otherwise. The extra motion creates degeneracy pressure.

Figure 9.6 | An auditorium analogy to explain degeneracy pressure. Chairs represent available places (quantum states) for electrons, and people who must keep moving from chair to chair represent electrons.

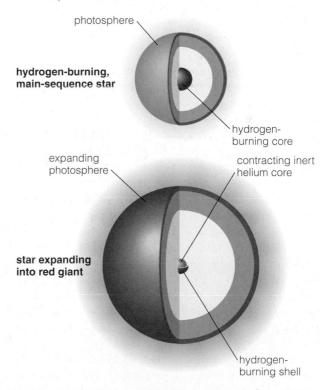

photosphere

hydrogen-burning, main-sequence star

hydrogen-burning core

expanding photosphere

contracting inert helium core

star expanding into red giant

hydrogen-burning shell

Figure 9.7 | After a star like the Sun ends its main-sequence life, its inert helium core contracts while hydrogen shell burning begins. The high rate of fusion in the hydrogen shell forces the star's upper layers to expand outward.

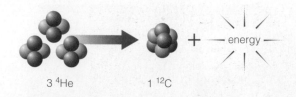

Figure 9.8 | Helium fusion converts three helium nuclei into one carbon nucleus, releasing energy in the process.

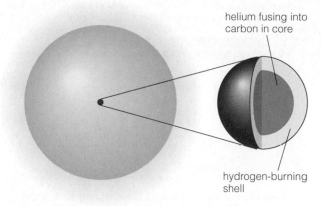

helium fusing into carbon in core

hydrogen-burning shell

Figure 9.9 | Core structure of the Sun after helium burning begins. Helium fusion will cause the core and hydrogen-burning shell to expand and slightly cool, thereby reducing the overall energy generation rate compared to the rate during the red giant stage. The outer layers will then shrink back, making the helium-burning Sun smaller than the Sun was during its red giant stage.

carbon-12 nucleus has a slightly lower mass than the three helium-4 nuclei, and the lost mass becomes energy in accord with $E = mc^2$. The temperature required for helium fusion is much greater than that for hydrogen fusion because helium nuclei, which contain two protons, have a greater positive charge than hydrogen nuclei, which contain just one proton. The greater charge means that helium nuclei repel one another more strongly than hydrogen nuclei, and therefore must slam into one another at much higher speeds in order to fuse [Section 8.1].

Once helium fusion begins, thermal pressure in the core will temporarily become stronger than gravity, forcing the core to expand. This core expansion will push the hydrogen-burning shell outward, lowering its temperature and its fusion rate. As a result, even though core helium fusion *and* hydrogen shell burning will be taking place simultaneously (**Figure 9.9**), the Sun's total energy production will fall from the peak it reached during the red giant stage. This reduction in energy production means that the Sun, now a *helium-burning star*, will be somewhat smaller in size, lower in luminosity, and yellower in color than it was as a red giant. With fusion once again operating in the core, the helium-burning Sun will regain the same sort of balance it had as a main-sequence star.

Last Gasps   It is only a matter of time until a helium-burning star fuses all its core helium into carbon. In the Sun, the core helium will run out after about 100 million years—only about 1% as long as the Sun's 10-billion-year hydrogen-burning lifetime. When the core helium is exhausted, fusion will again cease. The core, now made of the carbon produced by helium fusion, will begin to shrink once again under the crush of gravity.

The exhaustion of core helium will cause the Sun to expand once again, just as it did in becoming a red giant. This time, the trigger for the expansion will be helium fusion in a shell around the inert carbon core. Meanwhile, the hydrogen shell will still burn atop the helium layer. The Sun will have become a *double shell–burning star*. Both shells will contract along with the inert core, driving their temperatures and fusion rates so high that the Sun will expand to an even greater size and luminosity than in its first red giant stage.

The furious burning in the helium and hydrogen shells cannot last long—a few million years or less. The only hope of extending the Sun's life will then lie with its carbon core, but this is a false hope for a star like the Sun because of degeneracy pressure. Carbon fusion is possible only at temperatures above about 600 million K, and degeneracy pressure will halt the contraction of the Sun's inert carbon core before it ever gets that hot, in the same way that it halts the contraction of a brown dwarf. With the carbon core unable to undergo fusion and provide a new source of energy, the Sun will finally have reached the end of its life.

 Think about it   Suppose the universe contained only low-mass stars. Would elements heavier than carbon exist? Why or why not?

The Sun's last act will be to eject its outer layers into space, creating a huge shell of gas expanding away from the inert carbon core. The exposed core will still be very hot and will therefore emit intense ultraviolet radiation. This radiation will ionize the gas in the expanding shell, making it glow brightly as a **planetary nebula.** We see many examples of planetary nebulae around other low-mass stars that have recently died in this same way (**Figure 9.10**). The glow of a planetary nebula fades as the exposed core cools and the ejected gas disperses into space, disappearing within a million years or less.

The fate of the leftover carbon core is even more interesting. Degeneracy pressure will halt its collapse when it is about the size of Earth, but it will still have much of the Sun's original mass. It will be quite hot, but its small size will mean it is very dim. In other words, it will be the type of object that we call a

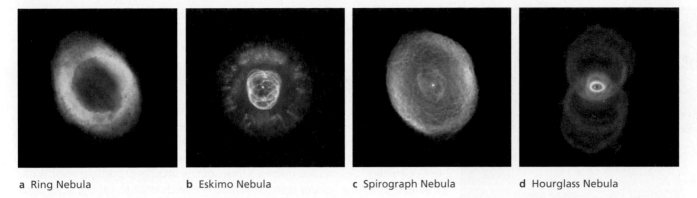

a Ring Nebula   b Eskimo Nebula   c Spirograph Nebula   d Hourglass Nebula

Figure 9.10 | Hubble Space Telescope photos of planetary nebulae, which form when low-mass stars cast off their outer layers of gas in their final death throes. The central white dots are the remaining hot cores of the stars that ejected the gas. These hot cores ionize and energize the shells of gas that surround them. As the nebula gas disperses into space, the hot core remains as a white dwarf.

white dwarf [Section 8.2]. White dwarfs are small in radius but high in mass because they are the compressed, exposed cores of dead stars, supported against the crush of gravity by degeneracy pressure.

The Sun's Life on an H-R Diagram   Mathematical models enable astronomers to determine how the Sun's surface temperature and luminosity should change as it evolves. Plotting the results on an H-R diagram produces a **life track** (also called an *evolutionary track*) that shows the Sun's luminosity and surface temperature at each point in its life (**Figure 9.11**).

Notice that the Sun's life track goes generally upward as it becomes a red giant, because the Sun will expand in size and luminosity while its inert helium core is contracting. The track also goes slightly to the right during this stage, because its surface temperature falls a bit. At the tip of the red giant stage, helium fusion begins and the Sun contracts in size, as shown by the track going down and left as it becomes a helium-burning star. The life track again turns upward as the Sun enters its second red giant phase, this time with energy generated by fusion in shells of both helium and hydrogen. As the Sun ejects its planetary nebula, the dashed curve indicates that we are shifting from plotting the surface temperature of a red giant to plotting the surface temperature of the

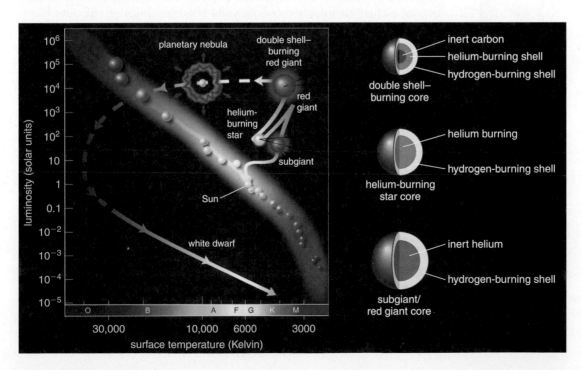

Figure 9.11 | The life track of our Sun from the time it first becomes a hydrogen-burning, main-sequence star to the time it dies as a white dwarf. Core structure is shown at key stages.

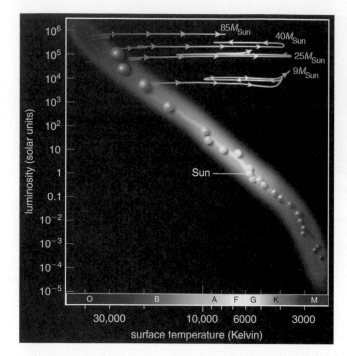

Figure 9.12 | Life tracks on the H-R diagram from main-sequence star to red supergiant for selected high-mass stars. Labels on the tracks give the star's mass at the beginning of its main-sequence life. Because of the strong wind from such a star, its mass can be considerably smaller when it leaves the main sequence. (Based on models from A. Maeder and G. Meynet.)

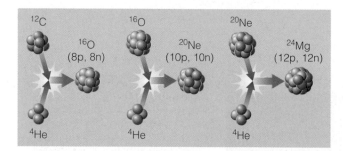

a Many reactions proceed through *helium capture*, in which fusion joins a helium nucleus to some other nucleus.

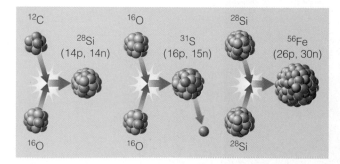

b At extremely high temperatures, fusion of even heavier nuclei can occur.

Figure 9.13 | A few of the many nuclear reactions that occur in the final stages of a high-mass star's life.

exposed core left behind. The curve becomes solid again near the lower left, indicating that this core is a hot white dwarf. From that point, the curve continues downward and to the right as the white dwarf cools and fades.

How do high-mass stars end their lives?

We now turn our attention to the end stages in the lives of high-mass stars. At first, these end stages are very similar to those of a low-mass star like the Sun, except they proceed much more rapidly. For example, a 25-solar-mass star lasts only a few million years as a hydrogen-burning main-sequence star before it goes out of balance. As its core hydrogen runs out, a hydrogen-burning shell forms around a shrinking helium core, generating so much energy that the star's outer layers expand outward until the star becomes a **supergiant.** Gravitational contraction of the helium core continues until it becomes hot enough to fuse helium into carbon. The high-mass star fuses helium into carbon so rapidly that it is left with an inert carbon core after no more than a few hundred thousand years.

After that, the life stages of this 25-solar-mass star become quite different from those of the Sun. Gravitational contraction simply continues after carbon fusion ends, the inert carbon core shrinks, and the core pressure, temperature, and density all rise. The shrinking core gets hotter and hotter, and it soon becomes hot enough for carbon fusion. As we'll discuss shortly, the core and shells go through several more phases of fusion of increasingly heavy elements—producing the "star stuff" that makes our lives possible—while the star's outer layers continue to swell in size.

Despite the dramatic events taking place in its interior, the high-mass star's outer appearance changes slowly. As each stage of core fusion ceases, the surrounding shell burning intensifies and further inflates the star's outer layers. Each time the core flares up, the outer layers contract somewhat, but the star's overall luminosity remains about the same. The result is that the star's life track zigzags across the top of the H-R diagram (**Figure 9.12**). In the most massive stars, the core changes happen so quickly that the outer layers don't have time to respond, and the star progresses steadily toward becoming a red supergiant.

One of these massive red supergiant stars happens to be relatively nearby: Betelgeuse, the upper-left shoulder of Orion. Its radius is over 500 solar radii, or more than twice the distance from the Sun to Earth. We have no way of knowing exactly what stage of nuclear burning is now taking place in its core. Betelgeuse may have a few thousand years of nuclear burning still ahead, or it may right now be in its final stages of life. If the latter is the case, then sometime soon we will witness one of the most dramatic events that ever occurs in the universe: a supernova.

Advanced Nuclear Burning A high-mass star like Betelgeuse can make elements heavier than carbon because degeneracy pressure is never able to halt the contraction of its core. The crush of gravity within the star is simply too strong. Each time the core exhausts one source of fuel for fusion, it contracts and heats up until it can fuse even larger nuclei, while fusion of lighter elements continues in multiple shells around the core. The complete set of nuclear reactions in a high-mass star's final stages of life is quite complex, and many different reactions may take place simultaneously, making elements like oxygen, silicon, and sulfur (**Figure 9.13**). Near the end, the star's central region resembles the inside of an onion, with layer upon layer of shells burning different elements (**Figure 9.14**).

The core continues shrinking, heating, and fusing new elements until iron starts to pile up. If iron were like the other elements in prior stages of nuclear burning, core contraction would stop when iron fusion ignited. However, iron is unique among the elements: It is the one element from which it is *not* possible to generate any kind of nuclear energy.

To understand why iron is unique, remember that only two basic processes can release nuclear energy: *fusion* of light elements into heavier ones

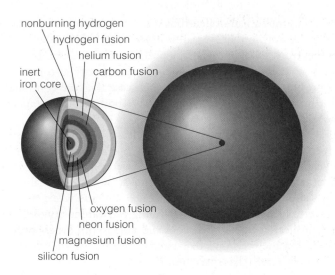

nonburning hydrogen
hydrogen fusion
helium fusion
carbon fusion
inert iron core
oxygen fusion
neon fusion
magnesium fusion
silicon fusion

Figure 9.14 | The multiple layers of nuclear burning in the core of a high-mass star during the final days of its life.

and *fission* of heavy elements into not-so-heavy ones (see Figure 8.2). Recall that hydrogen fusion converts four protons (hydrogen nuclei) into a helium nucleus that consists of two protons and two neutrons, which means that the total number of *nuclear particles* (protons and neutrons combined) does not change. However, this fusion reaction generates energy (in accord with $E = mc^2$) because the *mass* of the helium nucleus is less than the combined mass of the four hydrogen nuclei that fused to create it—despite the fact that the *number* of nuclear particles is unchanged.

In other words, fusing hydrogen into helium generates energy because helium has a lower *mass per nuclear particle* than hydrogen. Similarly, fusing three helium-4 nuclei into one carbon-12 nucleus generates energy because carbon has a lower mass per nuclear particle than helium, which means that some mass disappears and becomes energy in this fusion reaction. This decrease in mass per nuclear particle from hydrogen to helium to carbon is part of a general trend shown in **Figure 9.15**.

The mass per nuclear particle tends to decrease as we go from light elements to iron, which means that fusion of light nuclei into heavier nuclei generates energy. This trend reverses beyond iron: The mass per nuclear particle tends to *increase* as we look to still heavier elements. As a result, elements heavier than iron can generate nuclear energy only through fission into lighter elements.

Iron has the lowest mass per nuclear particle of all nuclei and therefore cannot release energy by either fusion or fission. Once the matter in a stellar core turns to iron, it can generate no more energy. For the iron core, the only hope of resisting the crush of gravity lies with degeneracy pressure, but iron keeps piling up until even degeneracy pressure cannot support the core. What ensues is the ultimate nuclear waste catastrophe: The star explodes, scattering all of its newly made elements into interstellar space.

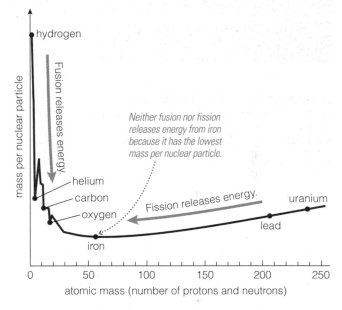

Fusion releases energy.

Neither fusion nor fission releases energy from iron because it has the lowest mass per nuclear particle.

Fission releases energy.

hydrogen
helium
carbon
oxygen
iron
lead
uranium

mass per nuclear particle

atomic mass (number of protons and neutrons)

Figure 9.15 | Overall, the average mass per nuclear particle declines from hydrogen to iron and then increases. Selected nuclei are labeled to provide reference points. (This graph shows only the most general trends. A more detailed graph would show numerous up-and-down bumps superimposed on the general trends. The vertical scale is arbitrary, but shows the general idea.)

 Think about it How would the universe be different if hydrogen, rather than iron, had the lowest mass per nuclear particle? Why?

The Supernova Explosion Degeneracy pressure cannot support the inert iron core because the immense gravity of a high-mass star pushes electrons past their quantum mechanical limit. Once they get too close together, they can no longer exist freely. In an instant, the electrons disappear by combining with protons to form neutrons (**Figure 9.16**), releasing the tiny subatomic particles known as *neutrinos* [Section 8.1] in the process. With the degeneracy pressure gone, gravity has free rein. In a fraction of a second, an iron core with a mass comparable to that of our Sun and a size larger than that

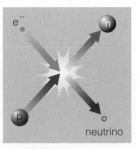

e⁻
n
p
neutrino

Figure 9.16 | During the final, catastrophic collapse of a high-mass stellar core, electrons and protons combine to form neutrons, accompanied by the release of neutrinos.

of Earth collapses into a ball of neutrons just a few kilometers across, making the type of stellar corpse that we call a *neutron star*. The collapse halts only because the neutrons have a degeneracy pressure of their own. In some cases, the remaining core may be massive enough that gravity also overcomes the degeneracy pressure of the neutrons, in which case the core continues to collapse until it becomes a *black hole*. We'll discuss both neutron stars and black holes in the next chapter.

The gravitational collapse of the core releases an enormous amount of energy—more than a hundred times what the Sun will radiate over its entire 10-billion-year lifetime. This energy drives the outer layers of the star off into space in a titanic explosion called a **supernova.** The heat of the explosion makes the gas shine with dazzling brilliance. For about a week, the supernova blazes as powerfully as 10 billion Suns, rivaling the luminosity of a moderate-size galaxy. The ejected gases slowly cool and fade in brightness over the next several months, but they continue to expand outward until they eventually mix with other gases in interstellar space (**Figure 9.17**).

Star Stuff   Our high-mass star has died, but the variety of elements produced in the star's nuclear furnace are now scattered throughout the gas clouds of interstellar space. Millions or billions of years later, a new round of star formation may incorporate this supernova debris into a new generation of stars. Some of the heavy elements that came from this supernova will become the building blocks of new planets—and perhaps even new life forms—around those newborn stars. In fact, the building blocks of Earth and of our very own bodies came from supernovae in the distant past.

We are confident that the majority of the heavy elements within the solar nebula that made our planet came from earlier supernovae, because our observations of those elements match the predictions of supernova models.

Figure 9.17 | This Hubble Space Telescope photograph shows the Crab Nebula, the remnant of the supernova observed in A.D. 1054.

Figure 9.18 shows the measured abundances of elements in our solar system. Hydrogen and helium are by far the most abundant, because all other elements (except lithium) have been produced only by stars. Even-numbered elements are more common than adjacent odd-numbered elements, because advanced fusion reactions typically fuse helium nuclei (atomic number = 2) with other even-numbered elements. Abundances drop off even further beyond iron, because the heavier elements are produced only in small quantities during or after supernova explosions. The telltale patterns we see among these abundances therefore show that we ourselves are "star stuff," built from the debris of stars that exploded long ago.

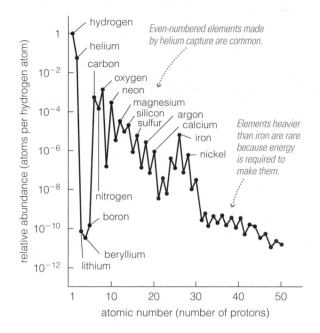

Figure 9.18 | This graph shows the observed relative abundances of elements in our solar system compared to the abundance of hydrogen (given as 1). For example, the abundance of nitrogen is about 10^{-4}, which means that there are about $10^{-4} = 0.0001$ times as many nitrogen atoms in the solar system as hydrogen atoms.

✳ THE PROCESS OF SCIENCE IN ACTION

9.3 Testing Stellar Models with Star Clusters

Now that you have heard the life stories of stars, you might be wondering how scientists can be so confident in their models of the Sun and other stars. After all, the lives of stars are so much longer than human lifetimes that all our observations of stars amount to only very brief snapshots of their lives. The answer, as always in science, is that those models make very specific predictions about our observations of stars that have been confirmed through repeated testing.

Observations of star clusters have been particularly useful in testing stellar models for two key reasons:

1. All the stars in a cluster lie at about the same distance from Earth, meaning that the apparent brightness of each star directly tells how its luminosity compares with that of other stars in the cluster.

2. All the stars in a cluster formed at about the same time, from the same large molecular cloud, meaning that they are all now about the same age.

We will therefore devote this chapter's exploration of the process of science in action to investigating how star clusters enable us to test our models of stellar lives.

What do star clusters reveal about the lives of stars?

Star clusters come in two basic types: modest-size **open clusters** and densely packed **globular clusters.** The two types differ not only in how densely they are packed with stars but also in age. Open clusters, such as the Pleiades (**Figure 9.19**), typically contain a few hundred to a few thousand relatively young stars in a region about 30 light-years across.

Globular clusters are much more densely packed and contain some of the oldest stars in the universe. A globular cluster can contain more than a million stars in a ball-shaped region no more than about 150 light-years across. Its center may have 10,000 stars packed into a space just a few light-years across (**Figure 9.20**). The view from a planet in a globular cluster would be marvelous, with thousands of stars lying closer than Alpha Centauri is to the Sun.

The value of star clusters for testing our models becomes apparent when we plot their stars on an H-R diagram. **Figure 9.21** shows an

Figure 9.19 | The Pleiades is a nearby open cluster of stars visible in the constellation Taurus. The Pleiades is often called the *Seven Sisters*, although only six of the cluster's several thousand stars are easily visible to the naked eye. The region shown is about 11 light-years across.

Figure 9.20 | The globular cluster M80 is more than 12 billion years old. The prominent reddish stars in this Hubble Space Telescope photo are red giant stars nearing the ends of their lives. The central region pictured here is about 15 light-years across.

H-R diagram for the Pleiades. Most of the stars fall along the main sequence, with one important exception: The Pleiades' stars trail away to the right of the main sequence at the upper end. That is, the hot, short-lived stars of spectral type O are missing from the main sequence. This fact demonstrates that hot, massive stars do indeed have shorter lifetimes than cooler, less massive stars. The Pleiades is old enough for its main-sequence O stars to have already ended their hydrogen-burning lives, but not quite old enough for all of its stars of spectral type B to have exhausted their core hydrogen.

The precise point on the H-R diagram at which the Pleiades' main sequence diverges from the main sequence is called its **main-sequence turnoff point.** Theoretical models indicate that the main-sequence lifetime of stars at this point on the main sequence is roughly 100 million years, so we conclude that this must be the age of the Pleiades. Stars with lifetimes longer than 100 million years are still fusing hydrogen in their cores and hence remain main-sequence stars.

The models also predict what should happen over the next few billion years. First, the B stars in the Pleiades ought to die out, followed by the A stars and the F stars. If we could make an H-R diagram for the Pleiades every few million years, we would expect to see the main sequence gradually grow shorter. This is exactly what we see in **Figure 9.22**, which shows the Pleiades along with stars from three other open clusters plotted on the same H-R diagram. Two of the clusters are older than the Pleiades: One has no more B stars and the other has no more A stars. The remaining cluster is much younger than the Pleiades, with a main-sequence turnoff point indicating an age of only 14 million years. Only its most massive O stars have died out, with some of these stars now found in the upper-right part of the H-R diagram, indicating that they have become supergiants, just as predicted by our models of high-mass stars.

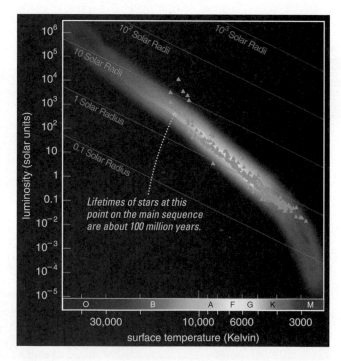

Figure 9.21 | An H-R diagram for the stars of the Pleiades. Triangles represent individual stars. The Pleiades cluster is missing its upper main-sequence stars, indicating that these stars have already ended their hydrogen-burning lives. The main-sequence turnoff point at about spectral type B6 tells us that the Pleiades is about 100 million years old.

Suppose a star cluster is precisely 10 billion years old. Where would you expect to find its main-sequence turnoff point? Would you expect this cluster to have any main-sequence stars of spectral type A? Would you expect it to have main-sequence stars of spectral type K? Explain. (*Hint:* What is the lifetime of our Sun?)

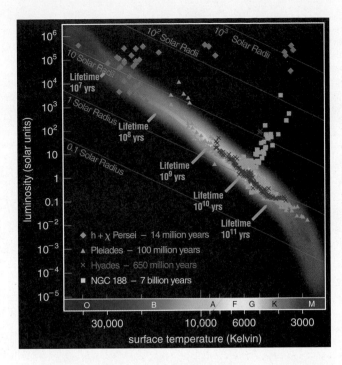

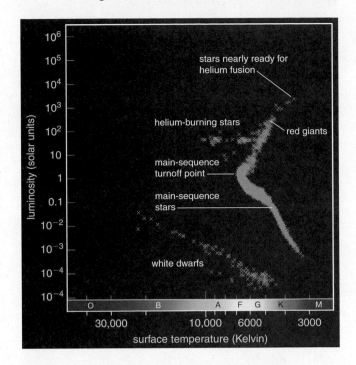

Figure 9.22 | This H-R diagram shows stars from four clusters. Their differing main-sequence turnoff points indicate different ages.

Figure 9.23 | This H-R diagram shows stars from the globular cluster M4, with an age of around 13 billion years. These stars are currently going through many of the same stages that models predict our Sun will go through near the end of its life.

Eventually, we would expect the H-R diagram of the Pleiades to look something like the one in **Figure 9.23**, which shows the stars of a 13-billion-year-old globular cluster. Within this diagram, you can see stars in many of the different life stages that the Sun will go through. Stars along the lower right remain on the main sequence, meaning that they are still fusing hydrogen in their cores. Just above and to the right of the main-sequence turnoff point, we see stars that have just begun their expansion into red giants as their cores have shut down and hydrogen shell burning has begun. The upward line of stars to the right of the turnoff point shows that red giants gradually grow in radius and luminosity. At the tip of this line are the red giants on the verge of helium fusion. Stars that have already begun helium fusion appear below and to the left of the red giants, because they are somewhat smaller, hotter, and less luminous. Finally, to the lower left of the main sequence we see stars that ended their lives and became white dwarfs.

Comparisons between diagrams like Figure 9.23 and computer models of stars give us great confidence that these models realistically reflect the lives of stars. You can do a quick check for yourself by comparing Figure 9.23 with Figure 9.11, which shows model predictions for the Sun's life track. The match is not exact, because these stars are not exactly the same as the Sun, but all the general features are indeed similar. Astronomers can make more detailed comparisons by computing models for entire star clusters and plotting the results on an H-R diagram. These efforts both confirm the stories of stellar lives that we have told in this chapter and continue to help us improve our stellar models. **Figure 9.24** summarizes the life stories of both low- and high-mass stars, as determined from our star models.

All stars spend most of their time as main-sequence stars and then change dramatically near the ends of their lives. This figure illustrates the life stages of a high-mass star and a low-mass star and shows how long a star spends in each stage of life. Notice that the lifetime of a low-mass star is far longer than that of a high-mass star.

1 Protostar: A star system forms when a cloud of interstellar gas collapses under gravity.

2 Blue main-sequence star: In the core of a high-mass star, four hydrogen nuclei fuse into a single helium nucleus by the series of reactions known as the CNO cycle.

3 Red supergiant: After core hydrogen is exhausted, the core shrinks and heats. Hydrogen shell burning begins around the inert helium core, causing the star to expand into a red supergiant.

4 Helium-burning supergiant: Helium fusion begins when the core temperature becomes hot enough to fuse helium into carbon. The core then expands, slowing the rate of hydrogen shell burning and allowing the star's outer layers to shrink.

LIFE OF A HIGH-MASS STAR ($25M_{Sun}$)

Protostar

Blue main-sequence star

Red supergiant

Helium-burning supergiant

Main sequence lifetime: 5 million years

Endpoint: Neutron star or black hole

Main sequence lifetime: 10 billion years

Protostar

Yellow main-sequence star

LIFE OF A LOW-MASS STAR ($1M_{Sun}$)

1 Protostar: A star system forms when a cloud of interstellar gas collapses under gravity.

2 Yellow main-sequence star: In the core of a low-mass star, four hydrogen nuclei fuse into a single helium nucleus by the series of reactions known as the proton-proton chain.

3 Red giant star: After core hydrogen is exhausted, the core shrinks and heats. Hydrogen shell burning begins around the inert helium core, causing the star to expand into a red giant.

5 **Multiple shell–burning supergiant:** After the core runs out of helium, it shrinks and heats until fusion of heavier elements begins. Late in life, the star fuses many different elements in a series of shells while iron collects in the core.

6 **Supernova:** Iron cannot provide fusion energy, so it accumulates in the core until degeneracy pressure can no longer support it. Then the core collapses, leading to the catastrophic explosion of the star.

7 **Neutron star or black hole:** The core collapse forms a ball of neutrons, which may remain as a neutron star or collapse further to make a black hole.

Multiple shell–burning supergiant

Supernova

Later stages: < 1 million years

TIME

Later stages: 1 billion years

Endpoint: White dwarf

Red giant star   Helium-burning star   Double shell–burning red giant   Planetary nebula

4 **Helium-burning star:** Helium fusion begins when the core becomes hot enough to fuse helium into carbon. The core then expands, slowing the rate of hydrogen shell burning and allowing the star's outer layers to shrink.

5 **Double shell–burning red giant:** Helium shell burning begins around the inert carbon core after the core helium is exhausted. The star then enters its second red giant phase, with fusion in both a hydrogen shell and a helium shell.

6 **Planetary nebula:** The dying star expels its outer layers in a planetary nebula, leaving behind the exposed inert core.

7 **White dwarf:** The remaining white dwarf is made primarily of carbon and oxygen because the core of the low-mass star never grows hot enough to produce heavier elements.

Summary of Key Concepts

9.1 Lives In Balance

Why do stars shine so steadily?

A star is born when **gravitational contraction** causes a cloud of interstellar gas to contract until hydrogen fusion ignites in its core. It can then shine steadily as long as it remains in two kinds of balance: (1) **gravitational equilibrium,** in which the outward push of pressure balances the inward pull of gravity, and (2) **energy balance,** in which the energy produced by fusion in the core matches the energy lost in the form of radiation from its surface.

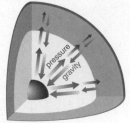

Why do a star's properties depend on its mass?

A star's mass determines the core temperature at which fusion brings it into balance. Stars less massive than the Sun come into balance with a cooler core temperature, lower luminosity, and smaller radius. Stars more massive than the Sun come into balance with a hotter core temperature, greater luminosity, and larger radius. Energy balance is possible for stars between $150M_{Sun}$ and $0.08M_{Sun}$. Below this mass, **degeneracy pressure** prevents gravity from making the star hot enough for efficient hydrogen fusion, and the object becomes a "failed star" known as a **brown dwarf.**

9.2 Star Death

What will happen when our Sun runs out of fuel?

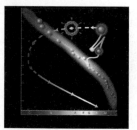

When hydrogen fusion stops in the core, the Sun will go out of balance. The core will then start contracting, and **hydrogen shell burning** will begin around it, causing the Sun to become a **red giant.** When the core becomes hot enough for **helium fusion,** making carbon in the core, it will temporarily return to a state of balance. After the core runs out of helium, fusion of hydrogen and helium will continue in shells around a contracting carbon core, but degeneracy pressure will stop the contraction before carbon fusion begins. During this time, the star will increase in size and luminosity until it ejects its outer layers in a **planetary nebula,** leaving behind a **white dwarf** supported by degeneracy pressure.

How do high-mass stars end their lives?

After a high-mass star (more than $8M_{Sun}$) runs out of core hydrogen, it goes through hydrogen shell burning and helium fusion stages similar to those of the Sun, and expands into a **supergiant.** However, fusion does not stop with carbon, because degeneracy pressure cannot hold off the crushing power of gravity within a high-mass star. Its core continues to contract, heat up, and produce even heavier elements until it starts to make iron. Because iron fusion cannot release energy, iron continues to pile up in the core until gravity causes the core to collapse and the star explodes as a **supernova,** dispersing the heavy elements necessary for life into the galaxy.

✳ THE PROCESS OF SCIENCE IN ACTION

9.3 Testing Stellar Models with Star Clusters

What do star clusters reveal about the lives of stars?

Star clusters come in two types. **Open clusters** are loose collections of up to several thousand stars and are generally younger than our Sun. **Globular clusters** are more densely packed, contain hundreds of thousands of stars, and are generally older than our Sun. Because all the stars in a cluster were born about the same time, we can determine a star cluster's age from 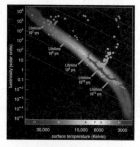 its **main-sequence turnoff point.** Comparing clusters of different ages shows how the properties of stars change as they age, enabling us to test our models of stellar lives.

Investigations

Quick Quiz

Choose the best answer to each of the following; answers are in Appendix D. Explain your reasoning with one or more complete sentences.

1. What slows down the contraction of a star-forming cloud when it makes a protostar? (a) production of fusion energy (b) magnetic fields (c) trapping of thermal energy inside the protostar

2. Which of these stars has the hottest core? (a) a white main-sequence star (b) an orange main-sequence star (c) a red main-sequence star

3. Which of these stars has the hottest core? (a) a blue main-sequence star (b) a red supergiant (c) a red main-sequence star

4. Which of these stars does not have fusion occurring in its core? (a) a red giant (b) a red main-sequence star (c) a blue main-sequence star

5. What happens to a low-mass star after helium fusion begins? (a) Its luminosity goes up. (b) Its luminosity goes down. (c) Its luminosity stays the same.

6. What would stars be like if hydrogen had the smallest mass per nuclear particle? (a) Stars would be brighter. (b) All stars would be red giants. (c) Nuclear fusion would not occur in stars of any mass.

7. What would stars be like if carbon had the smallest mass per nuclear particle? (a) Supernovae would be more common. (b) Supernovae would never occur. (c) High-mass stars would be hotter.

8. What would you be most likely to find if you returned to the solar system in 10 billion years? (a) a neutron star (b) a white dwarf (c) a black hole

9. In which of the following objects does degeneracy pressure fail to stop gravitational contraction? (a) a brown dwarf (b) a white dwarf (c) the core of a high-mass star

10. What happens to the core of a high-mass star after it runs out of hydrogen? (a) It shrinks and heats up. (b) It shrinks and cools down. (c) Helium fusion begins right away.

11. Which of these star clusters is youngest? (a) a cluster whose brightest main-sequence stars are white (b) a cluster whose brightest stars are red (c) a cluster containing stars of all colors

12. Which of these star clusters is oldest? (a) a cluster whose brightest main-sequence stars are white (b) a cluster whose brightest main-sequence stars are yellow (c) a cluster containing stars of all colors

Short-Answer/Essay Questions

Explain all answers clearly, with complete sentences and proper essay structure if needed. An asterisk (*) designates a quantitative problem, for which you should show all your work.

13. *Homes to Civilization?* We do not yet know how many stars have Earth-like planets, nor do we know the likelihood that such planets harbor advanced civilizations like our own. However, some stars can probably be ruled out as candidates for advanced civilizations. For example, given that it took a few billion years for humans to evolve on Earth, it seems unlikely that advanced life would have had time to evolve around a star that is only a few million years old. For each of the following stars, decide whether you think it is possible that it could harbor an advanced civilization. Explain your reasoning in one or two paragraphs.
   a. A $10M_{Sun}$ main-sequence star
   b. A $1.5M_{Sun}$ main-sequence star
   c. A $1.5M_{Sun}$ red giant
   d. A $1M_{Sun}$ helium-burning star
   e. A red supergiant

14. *Brown Dwarfs.* How are brown dwarfs like jovian planets? In what ways are brown dwarfs like stars?

15. *Rare Elements.* Lithium, beryllium, and boron are elements with atomic numbers 3, 4, and 5, respectively. Despite their being three of the five simplest elements, Figure 9.18 shows that they are rare compared to many heavier elements. Suggest a reason for their rarity. (*Hint:* Consider the process by which helium fuses into carbon.)

16. *Future Skies.* As a red giant, the Sun will have an angular size in Earth's sky of about 30°. What will sunset and sunrise be like? About how long will they take? Do you think the color of the sky will be different from what it is today? Explain.

17. *Research: Historical Supernovae.* Historical accounts exist of supernovae in the years 1006, 1054, 1572, and 1604. Choose one of these supernovae and learn more about historical records of the event. Did the supernova influence human history in any way? Write a two- to three-page summary of your research findings.

*18. *Supernova Betelgeuse.* The distance from Earth to the red supergiant Betelgeuse is approximately 427 light-years. If it were to explode as a supernova, it would be one of the brightest stars in the sky. Right now, the brightest star other than the Sun is Sirius, with a luminosity of $26L_{Sun}$ and a distance of 8.6 light-years. How much brighter than Sirius would the Betelgeuse supernova be in our sky if it reached a maximum luminosity of $10^{10}L_{Sun}$?

10 The Bizarre Stellar Graveyard

Welcome to the afterworld of stars, the fascinating domain of white dwarfs, neutron stars, and black holes. The photograph above shows a neutron star, and we'll see in this chapter that dead stars like this one behave in unusual ways. With fusion no longer providing energy to keep these dead stars in balance, their only hope of staving off the crushing power of gravity lies in the quantum mechanical effect of degeneracy pressure. But even this strange form of pressure cannot save the most massive stellar cores, which collapse to become *black holes* that drastically curve the space around them and make time appear to grind to a halt. Prepare to be amazed by the eerie inhabitants of the stellar graveyard!

10.1 White Dwarfs and Neutron Stars

In Chapter 9, we saw that stars of different masses leave different types of stellar corpses. Low-mass stars like the Sun leave behind white dwarfs when they die. Higher-mass stars die in the titanic explosions known as supernovae, leaving behind neutron stars or black holes. We will explore the stellar graveyard in order of mass, starting with white dwarfs and neutron stars.

What are white dwarfs?

A **white dwarf** is essentially the exposed core of a low-mass star that has died and shed its outer layers in a planetary nebula [Section 9.2]. It is quite hot when it first forms, because it was recently the inside of a star, but slowly cools with time. White dwarfs are stellar in mass but small in size (radius) [Section 8.2], which is why they are generally quite dim compared to stars like the Sun. However, the hottest white dwarfs can shine brightly in high-energy ultraviolet and X-ray light (**Figure 10.1**).

A white dwarf's starlike mass and small size make gravity very strong near its surface. If gravity were unopposed, it would crush the white dwarf to an even smaller size, so some sort of pressure must be pushing back equally hard to keep the white dwarf stable. As we discussed in Chapter 9, the source is *degeneracy pressure*—a type of pressure that arises when subatomic particles are packed as closely as the laws of quantum mechanics allow. More specifically, the degeneracy pressure in white dwarfs arises from closely packed electrons, so we call it **electron degeneracy pressure.** A white dwarf therefore exists in a state of balance in which the outward push of electron degeneracy pressure matches the inward crush of gravity.

White Dwarf Composition, Density, and Size Because a white dwarf is the core left over after a star has ceased nuclear burning, its composition reflects the products of the star's final nuclear-burning stage. The white dwarf left behind by a $1M_{Sun}$ star will be made mostly of carbon, since stars like the Sun fuse helium into carbon in their final stage of life.

Despite its ordinary-sounding composition, a scoop of matter from a white dwarf would be unlike anything seen on Earth. A typical white dwarf has a mass similar to that of the Sun compressed into an object the size of Earth. If you recall that Earth is smaller than a typical sunspot, you can imagine that packing the entire mass of the Sun into the volume of Earth is no small feat. The density of a white dwarf is so high that a teaspoon of its material would weigh as much as a small truck if you could bring that material to Earth.

More massive white dwarfs are actually smaller in size than less massive ones. For example, a $1.3M_{Sun}$ white dwarf is half the diameter of a $1.0M_{Sun}$ white dwarf (**Figure 10.2**). The more massive white dwarf is smaller because its stronger gravity compresses its matter to a greater density. According to the laws of quantum mechanics, the electrons in a white dwarf respond to this compression by moving faster, which makes the degeneracy pressure strong enough to resist the greater force of gravity. The most massive white dwarfs are therefore the smallest.

The White Dwarf Limit The fact that electron speeds are higher in more massive white dwarfs leads to a fundamental limit on the maximum mass of a white dwarf. Theoretical calculations show that electron speeds would reach the speed of light in a white dwarf with a mass of about 1.4 times the mass of the Sun ($1.4M_{Sun}$). Because neither electrons nor anything else can travel faster than the speed of light, no white dwarf can have a mass greater than this $1.4M_{Sun}$ **white dwarf limit** (also called the *Chandrasekhar limit*, after its discoverer).

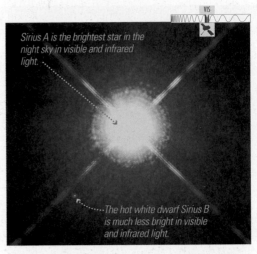

Sirius A is the brightest star in the night sky in visible and infrared light.

The hot white dwarf Sirius B is much less bright in visible and infrared light.

a Sirius as seen in infrared light by the Hubble Space Telescope.

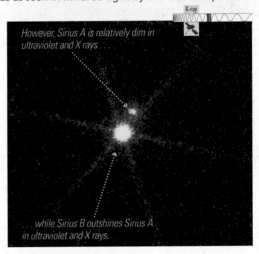

However, Sirius A is relatively dim in ultraviolet and X rays . . .

. . . while Sirius B outshines Sirius A in ultraviolet and X rays.

b Sirius as seen by the Chandra X-Ray Telescope.

Figure 10.1 | The binary star system Sirius consists of a main-sequence star, Sirius A, that is the brightest star in the night sky and a white dwarf, Sirius B. The hot white dwarf in the binary actually outshines its companion in ultraviolet and X-ray light. (The spikes emanating from the stars are not real. They are artifacts created by telescope optics.)

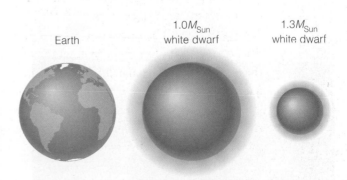

Earth

$1.0M_{Sun}$ white dwarf

$1.3M_{Sun}$ white dwarf

Figure 10.2 | More massive white dwarfs are actually *smaller* (and thus denser) than less massive white dwarfs. Earth is shown for scale.

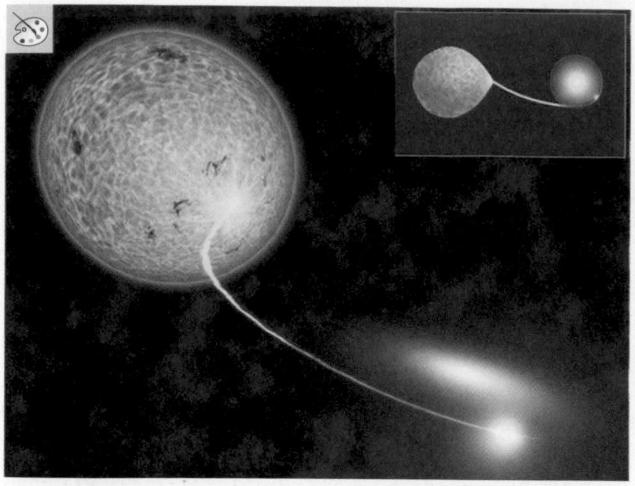

Figure 10.3 | This artist's conception shows how mass spilling from a companion star (left) forms an accretion disk (right) around a white dwarf (at the center of the disk). The inset shows how the system looks from above rather than from the side.

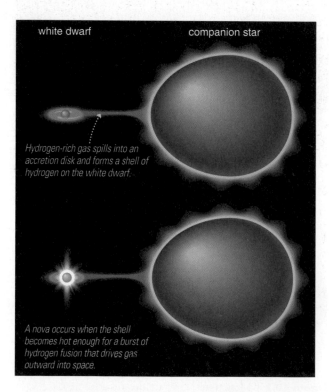

white dwarf   companion star

Hydrogen-rich gas spills into an accretion disk and forms a shell of hydrogen on the white dwarf.

A nova occurs when the shell becomes hot enough for a burst of hydrogen fusion that drives gas outward into space.

Figure 10.4 | A nova occurs when hydrogen fusion ignites on the surface of a white dwarf in a binary star system.

Strong observational evidence supports this theoretical limit on the mass of a white dwarf. Many known white dwarfs are members of binary systems, and hence we can measure their masses [Section 8.2]. In every observed case, the white dwarfs have masses below $1.4M_{Sun}$, just as theoretical models predict.

White Dwarfs and Accretion   Left to itself, a single white dwarf will never again shine as brightly as the star it once was. With no source of fuel for fusion, it will cool over time into a cold black dwarf. Its size will never change, because electron degeneracy pressure will forever keep it stable against the crush of gravity. However, the situation can be quite different for a white dwarf in a binary system in which the two stars orbit close together.

A white dwarf in a binary system can gradually gain mass if its companion is a main-sequence or giant star (**Figure 10.3**). When a clump of mass first spills over from the companion star to the white dwarf, it has some small orbital velocity. The law of conservation of angular momentum (see Tools of Science, p. 62) dictates that the clump must orbit faster and faster as it falls toward the white dwarf's surface. The infalling matter therefore forms a whirlpool-like disk around the white dwarf. Because the process in which matter collects into a larger body is called *accretion*, this rapidly rotating disk is called an **accretion disk.**

Accretion can provide a "dead" white dwarf with a new energy source. The inward-spiraling gas in the accretion disk becomes very hot as its gravitational potential energy is converted into thermal energy, causing it to shine with intense ultraviolet or X-ray radiation. More dramatic events can occur as hydrogen gas from the companion star accumulates on the surface of the white dwarf.

Novae   The hydrogen spilling toward the white dwarf from its companion gradually spirals inward through the accretion disk and eventually falls onto the surface of the white dwarf. The white dwarf's strong gravity compresses this hydrogen gas into a thin surface layer. Both the pressure and the temperature rise as the layer builds up with more accreting gas. When the temperature at the bottom of the layer reaches about 10 million K, hydrogen fusion suddenly ignites.

The white dwarf blazes back to life as its hydrogen layer burns. This thermonuclear flash causes the binary system to shine for a few glorious weeks as a **nova** (**Figure 10.4**). A nova is far less luminous than a supernova, but still can shine as brightly as 100,000 Suns. It generates heat and pressure, ejecting most of the material that has accreted onto the white dwarf. This material expands outward, creating a remnant that sometimes remains visible years after the nova explosion. Accretion resumes after a nova explosion subsides, so the entire process can repeat itself. In some cases, novae have been observed to repeat after just a few decades. More commonly, thousands of years pass between nova explosions.

White Dwarf Supernovae   Each time a nova occurs, the white dwarf ejects some of its mass. Each time a nova subsides, the white dwarf begins to accrete matter again. Theoretical models cannot yet tell us whether the net result should be a gradual increase or decrease in the white dwarf's mass. Nevertheless, observations show that in at least some cases accreting white dwarfs in binary systems continue to gain mass as time passes. If a white dwarf gains enough mass, it can one day approach the $1.4M_{Sun}$ white dwarf limit. This day is the white dwarf's last.

Remember that white dwarfs left behind by stars like our Sun consist largely of carbon. As a white dwarf's mass approaches $1.4M_{Sun}$, its temperature rises enough for carbon fusion to begin. Once carbon fusion gets started, it ignites almost instantly throughout the white dwarf, releasing an enormous amount of thermal energy and causing a huge jump in internal pressure. When this happens, the white dwarf explodes completely in what we will call a **white dwarf supernova.**

According to our understanding of novae and white dwarf supernovae, can either of these events ever occur with a white dwarf that is *not* a member of a binary star system? Explain.

The "carbon bomb" explosion that causes a white dwarf supernova is quite different from the explosion resulting from the collapse of an iron core in a high-mass star [Section 9.2], which we will call a **massive star supernova.**[*] Astronomers can distinguish between the two types of supernova by studying their light. Both types shine brilliantly, with peak luminosities about 10 billion times that of the Sun ($10^{10} L_{\text{Sun}}$), but the luminosity of a white dwarf supernova fades quickly during the first few weeks and then declines more gradually, while the decline in brightness of a massive star supernova is often more complicated (**Figure 10.5**). In addition, spectra of white dwarf supernovae always lack hydrogen lines because white dwarfs contain very little hydrogen, while spectra of massive star supernovae generally contain prominent hydrogen lines because massive stars usually have plenty of hydrogen in their outer layers when they explode.

What are neutron stars?

White dwarfs with densities of 5 tons per teaspoon may seem incredible, but neutron stars are stranger still. The possibility that neutron stars might exist was first proposed in the 1930s, but many astronomers thought it preposterous that nature would make anything so bizarre. Nevertheless, a vast amount of evidence now makes it clear that neutron stars really exist. Let's explore the evidence by first examining the properties of neutron stars. Then we will discuss their initial discovery and how they reveal their presence in binary star systems.

Neutron Star Properties A **neutron star** is essentially a ball of neutrons created by the collapse of a high-mass star's iron core in a massive star supernova (**Figure 10.6**). Typically just 10 kilometers in radius yet more massive than the Sun, neutron stars are like giant atomic nuclei made almost entirely of neutrons and held together by gravity. Like white dwarfs, neutron stars resist the crush of gravity with the degeneracy pressure that arises when particles are packed as closely as nature allows. In the case of neutron stars, however, it is neutrons rather than electrons that are closely packed, so **neutron degeneracy pressure** supports them against the crush of gravity.

The force of gravity at the surface of a neutron star is awe-inspiring. A rocket ship trying to escape from a neutron star's surface would need to move at about half the speed of light. If you foolishly chose to visit a neutron star's surface, gravity would immediately squash you into a microscopically thin puddle of subatomic particles.

Things would be only slightly less troubling if a bit of neutron star could somehow come to visit you. A paper clip with the density of neutron star material would outweigh Mount Everest. If such a paper clip magically appeared in your hand, you could not prevent it from falling. Down it would plunge, passing through Earth like a rock falling through air. It would gain speed until it reached Earth's center, and its momentum would carry it onward until it slowed to a stop on the other side of our planet. Then it would fall back down again. If it came in from space, each plunge of the neutron star material would drill a different hole through the rotating Earth. In the words of astronomer Carl Sagan, the inside of Earth would "look briefly like Swiss cheese" (until the melted rock flowed to fill

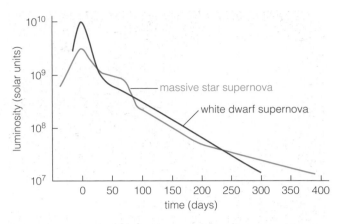

Figure 10.5 | The curves on this graph show how the luminosities of two different supernovae fade with time. The white dwarf supernova fades quickly at first and then more gradually a few weeks after its peak, while the massive star supernova fades in a more complicated pattern. (Notice that the luminosity scale on the *y*-axis is expressed in powers of 10.)

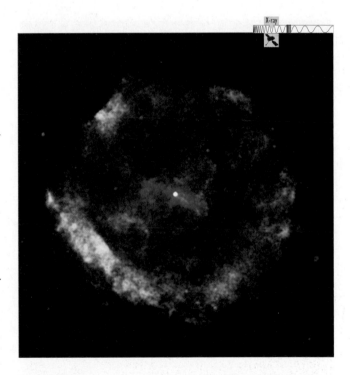

Figure 10.6 | This X-ray image from the Chandra X-Ray Observatory shows the supernova remnant G11.2-03, the remains of a supernova observed by Chinese astronomers in A.D. 386. The white dot at the center represents X rays from the neutron star left behind by the supernova. The different colors correspond to different X-ray wavelength bands. The region pictured is about 23 light-years across.

[*]Observationally, astronomers classify supernovae as *Type II* if their spectra show hydrogen lines, and *Type I* otherwise. All Type II supernovae are assumed to be massive star supernovae. However, a Type I supernova can be either a white dwarf supernova or a massive star supernova in which the star blew away all its hydrogen before exploding. Type I supernovae appear in three classes whose light curves differ, called *Type Ia, Type Ib,* and *Type Ic.* Only Type Ia supernovae are thought to be white dwarf supernovae.

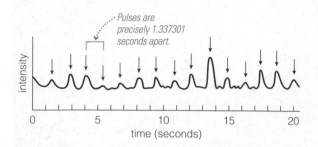

Figure 10.7 | About 20 seconds of data from the first pulsar, discovered by Jocelyn Bell in 1967.

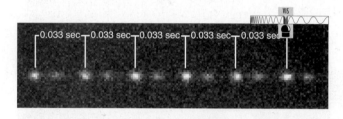

Figure 10.8 | This time-lapse image of the pulsar at the center of the Crab Nebula, a supernova remnant, shows its main pulse recurring every 0.033 second. The fainter pulses are thought to come from the pulsar's other lighthouse-like beam. (Photo from the Very Large Telescope of the European Southern Observatory.)

in the holes) by the time friction finally brought the piece of neutron star to rest at Earth's center.

In the unfortunate event that an *entire* neutron star came to visit you, it would not fall at all. Because it would be only about 10 kilometers in radius, the neutron star would probably fit in your hometown. However, since it would be 300,000 times as massive as Earth, the neutron star's immense surface gravity would quickly destroy your hometown and the rest of civilization. By the time the dust settled, the former Earth would be squashed into a shell no thicker than your thumb on the surface of the neutron star.

Pulsars The first observational evidence for neutron stars came in 1967, when a graduate student named Jocelyn Bell discovered a strange source of radio waves. Bell had helped her adviser, Anthony Hewish, build a radio telescope ideal for discovering fluctuating sources of radio waves. She was trying to interpret the flood of data pouring out of this instrument in October 1967 when she noticed a peculiar signal. After ruling out other possibilities, she concluded that *pulses* of radio waves were arriving from somewhere near the direction of the constellation Cygnus at precise 1.337301-second intervals (**Figure 10.7**).

The pulses coming from Cygnus were very surprising because no known astronomical object pulsated so regularly. In fact, the pulsations came at such precise intervals that they were nearly as reliable for measuring time as the most precise human-made clocks. For a while, the mysterious source of the radio waves was dubbed "LGM" for Little Green Men—only half-jokingly. Today we refer to such rapidly pulsing radio sources as **pulsars.**

The mystery of what exactly pulsars are was solved in 1968, when astronomers found pulsars at the center of two supernova remnants, the Vela Nebula and the Crab Nebula (**Figure 10.8**). This evidence showed that pulsars must be neutron stars left behind by supernova explosions.

Pulsars "pulse" because the neutron star is spinning rapidly as a result of the conservation of angular momentum: As an iron core collapses into a neutron star, its rotation rate must increase as it shrinks in size. The collapse also bunches the magnetic field lines running through the core far more tightly, greatly amplifying the strength of the magnetic field. As a result, a neutron star's magnetic field can be a trillion times as strong as Earth's. This intense magnetic field directs beams of radiation out along its magnetic poles. If a neutron star's magnetic poles are not aligned with its rotation axis, each beam of radiation sweeps round and round like a lighthouse beam, so we see a pulse of light each time the beam sweeps past Earth (**Figure 10.9**).

The pulsation rate of a pulsar changes very slowly with time. The continual twirling of a pulsar's magnetic field generates electromagnetic radiation that carries away energy and angular momentum, causing the neutron star's rotation rate to slow gradually. The pulsar in the Crab Nebula, for example, currently spins about 30 times per second. Two thousand years from now, it will spin less than half as fast. Eventually, a pulsar's spin slows so much and its magnetic field becomes so weak that we can no longer detect it. In addition, some spinning neutron stars may be oriented so that their beams do not sweep past our location. Therefore, we have the following rule: *All pulsars are neutron stars, but not all neutron stars are pulsars.*

 Think about it Suppose we observe no pulses of radiation from a neutron star. Is it possible that a civilization in some other star system would see this neutron star as a pulsar? Explain.

We know that pulsars must be neutron stars because no other massive object could spin so fast. A white dwarf, for example, can spin no faster than about once per second. A faster spin would tear it apart because gravity would no longer be strong enough to hold it together. Pulsars have been discovered that rotate as fast as 625 times per second. Only an object as small and dense as

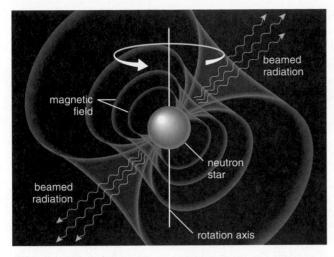

a A pulsar is a rotating neutron star that beams radiation along its magnetic axis.

b If the magnetic axis is not aligned with the rotation axis, the pulsar's beams sweep through space like lighthouse beams. Each time one of the pulsar's beams sweeps across Earth, we see a pulse of radiation.

Figure 10.9 | Radiation from a rotating neutron star can appear to pulse like the beams from a lighthouse.

a neutron star has strong enough gravity at its surface to spin so fast without breaking apart.

Neutron Stars in Binary Systems   Like white dwarfs, neutron stars in close binary systems can produce brilliant bursts of fusion as gas overflowing from a companion star creates a hot, swirling accretion disk. However, in the neutron star's extremely strong gravitational field, infalling matter releases an amazing amount of gravitational potential energy. Dropping a brick onto a neutron star would liberate as much energy as an atomic bomb.

The huge amount of energy released by infalling matter from the companion star makes a neutron star's accretion disk much hotter and more luminous than the accretion disk around a white dwarf. The high temperatures in the inner regions of the accretion disk cause it to radiate powerfully in X rays. Some close binaries with neutron stars emit 100,000 times as much energy in X rays as our Sun emits in all wavelengths of light combined, so we call them **X-ray binaries.**

Like accreting white dwarfs that occasionally erupt into novae, accreting neutron stars sporadically erupt with a pronounced spike in luminosity

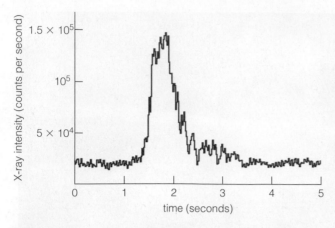

Figure 10.10 | Light curve of an X-ray burst. In this particular burst, the X-ray luminosity of the neutron star spiked to over 6 times its usual brightness in a matter of seconds.

(**Figure 10.10**). Because these eruptions release their energy primarily in the form of X rays, we call them **X-ray bursts,** and the systems that produce them are known as **X-ray bursters.** Much like novae, X-ray bursts result from the sudden ignition of nuclear fusion. However, while novae occur when hydrogen fusion ignites on the surface of a white dwarf in a close binary system, X-ray bursts arise from the ignition of helium fusion on the neutron star in a close binary system.

10.2 Black Holes

The story of stellar corpses would be strange enough if it ended with white dwarfs and neutron stars, but it doesn't. Sometimes, the gravity of a stellar corpse becomes so strong that nothing can prevent it from collapsing under its own weight. The stellar corpse collapses without end, crushing itself out of existence and forming perhaps the most bizarre type of object in the universe: a *black hole.*

What are black holes?

The basic idea behind a black hole originated near the end of the 18th century. Newton's laws of motion and gravity were well known by that time, and scientists were starting to think about the force of gravity at the surface of an object with a large amount of mass but a small radius. In particular, they wondered whether the gravity of such an object could be so strong that nothing, not even light, could escape its gravitational pull.

Escape Velocity   To illustrate the conditions required for gravity to keep something from escaping, we'll use the example of a spacecraft orbiting Earth. Firing its rockets to give it more kinetic energy will increase the size of its orbit. If it gains enough kinetic energy, it can end up in an unbound orbit that allows it to escape Earth completely (**Figure 10.11**). Although it might make more sense to say that the spacecraft achieves "escape energy," we instead say that it achieves **escape velocity,** because the velocity it needs in order to escape does not depend on the spacecraft's mass. The escape velocity from Earth's surface is about 40,000 km/hr, or 11 km/s. This is the minimum speed required to escape Earth's gravity for any object that starts near Earth's surface.

Now think about what would happen if we could make Earth smaller while keeping its mass the same. Making Earth more compact would raise the escape velocity, because the strength of gravity at Earth's surface depends on the inverse square of its radius [Section 3.3]. If we could squeeze Earth down to about the size of a golf ball, its escape velocity would reach the speed of light. Scientists of the 18th century assumed that the light emitted by such an object would behave like a rock thrown upward, eventually slowing to a stop and falling back down.

Einstein's theory of relativity (see Tools of Science, p. 169) shows that extremely compact, massive objects are considerably stranger than people imagined two centuries ago, but their basic idea was correct: It is indeed possible for an object's gravity to be so strong that not even light can escape. Physicist John Wheeler named such objects "black holes" in 1967, long after they were first proposed. They are called "black" because no light can escape from their surfaces. They are called "holes" because they are like holes in the observable universe in the following sense: If you enter a black hole, you leave the region of the universe that we can observe with a telescope and can never return.

Black Hole Formation   A black hole forms when a massive object is somehow squeezed down to a much smaller size. We have already seen how extreme squeezing can happen inside a massive star when its iron core reaches the white dwarf limit of $1.4M_{Sun}$. Electron degeneracy pressure cannot resist

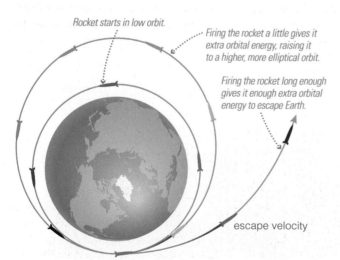

Rocket starts in low orbit.

Firing the rocket a little gives it extra orbital energy, raising it to a higher, more elliptical orbit.

Firing the rocket long enough gives it enough extra orbital energy to escape Earth.

escape velocity

Figure 10.11 | If an object orbiting Earth gains energy, it moves to a higher or more elliptical orbit. With enough extra orbital energy, it may achieve escape velocity. Escape velocity from Earth's surface is about 11 km/s.

gravity above that mass, and the core collapses to form a neutron star, causing a supernova explosion [Section 9.2].

Calculations show that the mass of a neutron star has a limit that lies somewhere between about 2 and 3 solar masses. Above this mass, neutron degeneracy pressure cannot hold off the crush of gravity in a collapsing stellar core. Most supernovae are thought to leave neutron stars behind because the explosion ejects all the matter surrounding the core, keeping its mass below this limit. However, theoretical models show that very massive stars might not succeed in blowing away all their upper layers. If enough matter falls back onto the core, its mass may rise above the neutron star limit. As soon as the core exceeds the neutron star limit, gravity overcomes neutron degeneracy pressure and the core collapses once again. This time, no known force can keep the core from collapsing into oblivion.

Tools of Science: Einstein's Theories of Relativity

Einstein developed his theory of relativity in two parts. The special theory of relativity was published in 1905 and the general theory in 1915. Both are now indispensable tools of modern science.

Special Relativity: Contrary to a common myth, the special theory of relativity does not say that *everything* is relative; rather, it states that *motion* is relative but two things in the universe are absolute:

1. The laws of nature are the same for everyone.

2. The speed of light is the same for everyone.

Statement 1 implies that no one can claim to be standing absolutely still, because the only kinds of motion that matter to the laws of nature are *relative* motions of one object with respect to another. Otherwise, observers moving through space at different speeds would disagree about the laws of physics. Statement 2, that the speed of light is the same for everyone, is much more surprising. Ordinarily, we expect speeds to add and subtract. If you watched someone throw a ball forward from a moving car, you'd see the ball traveling at the speed at which it was thrown *plus* the speed of the car. But if a person shined a light beam from a moving car, you'd see it moving at precisely the speed of light (about 300,000 kilometers per second), no matter how fast the car was going. This strange fact has been experimentally verified countless times.

The fact that the speed of light is always the same leads to several famous consequences. First, it means that no physical object can move faster than light. You can see why by imagining a test ride in the most incredible rocket possible and thinking about your headlight beams. Because everyone measures the speed of light to be the same, both you and someone back on Earth will agree that your headlight beams are traveling at the speed of light: 300,000 kilometers per second. And since you'll also both agree that the headlight beams are shining *ahead* of you, the person on Earth will always conclude that you are going slower than the speed of light. Other consequences of the constant speed of light include Einstein's famous equation $E = mc^2$ and the fact that an object moving by you seems to have slower time, shorter length, and greater mass than it would at rest.

General Relativity. Einstein's general theory extended special relativity by including the effects of gravity, and it led to a radical revision in how we think about space and time. Special relativity showed that, instead of thinking about the three dimensions of space and the one dimension of time as separate, we should think of them as a seamless, four-dimensional entity known as *spacetime*. The general theory of relativity showed that what we perceive as gravity arises from curvature of this four-dimensional spacetime. Mass curves spacetime, and greater spacetime curvature means stronger gravity. We cannot visualize curvature in four dimensions, but we can visualize two-dimensional analogies with rubber sheet diagrams like the one shown in **Figure 1**. According to general relativity, planets orbit the Sun because of the way spacetime is curved by the Sun: Each planet is going as straight as it can, but spacetime is curved in a way that keeps the planets going round and round like marbles in a salad bowl.

Given that we cannot actually perceive all four dimensions of spacetime at once, you may wonder why scientists think spacetime curvature is real. The answer is that Einstein's theory predicts measurable effects from such curvature, and all observations made to date are in accordance with those predictions. Strange as it may seem, we live in a four-dimensional universe in which space and time are intertwined and can never be disentangled.

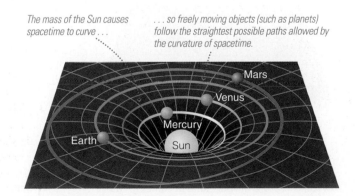

The mass of the Sun causes spacetime to curve . . .

. . . so freely moving objects (such as planets) follow the straightest possible paths allowed by the curvature of spacetime.

Figure 1 | According to the general theory of relativity, planets orbit the Sun for much the same reason that you can make a marble go around in a salad bowl: The planet is going as straight as it can, but the curvature of spacetime causes its path through space to curve.

According to Einstein's theory of relativity, it is highly unlikely that any other as-yet-unknown force could intervene and prevent the collapse. Einstein's theory tells us that energy is equivalent to mass ($E = mc^2$), implying that energy, like mass, must exert some gravitational attraction. The gravity of pure energy is usually very small, but that is not the case for a stellar core collapsing beyond the neutron star limit. Here, the energy associated with the rapidly rising temperature and pressure acts like additional mass, making the crushing power of gravity even stronger. The more the core collapses, the stronger gravity gets. To the best of our understanding, *nothing* can halt the crush of gravity at this point. The core collapses without end, forming a black hole.

The Event Horizon   A collapsing object becomes a black hole when the escape velocity from its surface exceeds the speed of light. At that moment, a boundary called the **event horizon** forms between the inside of the new black hole and the universe outside of it. The event horizon essentially marks the point of no return for objects entering a black hole: It is the boundary around a black hole at which the escape velocity equals the speed of light. Nothing that passes within this boundary can ever escape. The event horizon gets its name from the fact that we have no hope of learning about any events that occur within it.

We usually think of the "size" of a black hole as the radius of its event horizon. This radius is known as the **Schwarzschild radius,** named for the man who derived its formula in 1916. The Schwarzschild radius of a black hole depends only on its mass. A black hole with the mass of the Sun has a Schwarzschild radius of about 3 kilometers—only a little smaller than the radius of a neutron star of the same mass. More massive black holes have larger Schwarzschild radii. For example, a black hole with 10 times the mass of the Sun has a Schwarzschild radius of about 30 kilometers. However, if you actually traveled to a black hole and tried to measure its radius with a ruler, you would run into some difficulty because space and time near a black hole are not like what we are used to here on Earth.

What happens to space and time near a black hole?

The structure of space and time near the event horizon of a black hole is extraordinarily strange. Einstein discovered that space and time are actually bound up together as four-dimensional **spacetime,** and that gravity arises from *curvature of spacetime* (see Tools of Science, p. 169). **Figure 10.12** represents all four dimensions of spacetime with a two-dimensional rubber sheet. In this analogy, the sheet is flat in a region far from any mass (Figure 10.12a). Near a massive object with strong gravity, the sheet becomes curved (Figure 10.12b); the stronger the gravity, the more curved it gets. In this analogy, a black hole is like a bottomless pit in spacetime (Figure 10.12c). Gravity gets stronger and stronger as we get closer and closer to the black hole, so the sheet curves more and more. Keep in mind that the illustration is only an analogy, and black holes are actually spherical, not funnel-shaped. Nevertheless, it captures the key idea that a black hole is like an inescapable hole in the observable universe.

Curved space has different rules of geometry than "flat" space, an idea you can see with a plastic ball and a marker. Draw a straight line going one-quarter of the way around the ball, then make a right-angle turn and **See it for yourself** go one-quarter of the way around the ball in your new direction, and finish a triangle with a straight line back to where you started. Now measure all three angles and add them up. How does the sum of the three angles of a triangle on a sphere compare with the 180° sum you would find for triangles in a flat plane?

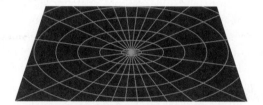

a A two-dimensional representation of "flat" spacetime. Each pair of circles is separated by the same radial distance.

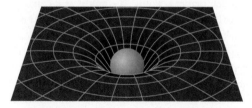

b A mass affects the rubber sheet similarly to the way gravity curves spacetime. The circles become more widely separated—indicating greater curvature—as we look closer to the mass.

event horizon

c The curvature of spacetime becomes greater and greater as we approach a black hole, and a black hole itself is a bottomless pit in spacetime.

Figure 10.12 | We can use two-dimensional rubber sheets to show an analogy to curvature in four-dimensional spacetime.

Singularity and the Limits to Knowledge Our current understanding of physics suggests that space and time get even stranger inside a black hole's event horizon. Because Einstein's theory of relativity says that nothing can stop the crush of gravity in a black hole, it implies that all the matter that forms a black hole must ultimately be crushed to an infinitely tiny and dense point in the black hole's center. We call this point a **singularity.** Unfortunately, this idea of a singularity pushes up against the limits of scientific knowledge today.

The problem is that two very successful theories make very different predictions about the nature of a singularity. Einstein's theory of general relativity, which seems to explain successfully how gravity works throughout the universe, predicts that spacetime should grow infinitely curved as it enters the singularity. Quantum physics, which successfully explains the nature of atoms and the spectra of light, predicts that spacetime should fluctuate chaotically near the singularity. These are very different claims, and no overarching theory that can reconcile them has yet been developed.

A Slowdown in Time Einstein's theory of relativity is thought to be more reliable in the vicinity of the event horizon, and it makes definite, testable predictions about events that happen there. For example, the theory predicts that time should run more slowly as the force of gravity grows stronger, a prediction that has been tested both by experiments with clocks in Earth's gravitational field and through observations of *gravitational redshift,* a redshift that occurs in spectra of the Sun and other stars because the slowdown in time near the strong gravity of their surfaces means an increase in the wavelength of their emitted light. (Notice that this shift to longer wavelength is due only to gravity and not to the Doppler effect.) The effect on time should become quite extreme near a black hole.

You could in principle test this prediction with the aid of two identical clocks whose numerals glow with blue light. Suppose you were orbiting a black hole with a mass of $10M_{Sun}$ and a Schwarzschild radius of 30 kilometers on a circular orbit a few thousand kilometers from the event horizon. This orbit would be perfectly stable—there would be no need to worry about getting "sucked in." You could then perform your experiment by keeping one of the identical clocks aboard the ship and pushing the other one toward the black hole with a small rocket that would fire its engines just enough so that the clock would fall only gradually toward the event horizon (**Figure 10.13**). According to relativity, the clock on the rocket would tick more slowly as it headed toward the black hole, so its light would become increasingly redshifted. When the clock reached a distance of about 10 kilometers above the event horizon, you would see it ticking only half as fast as the clock on your spaceship, and its numerals would appear red instead of blue.

When the rocket ran out of fuel, the clock would start to plunge toward the black hole. From your safe vantage point inside the spaceship, you would see the clock ticking more and more slowly as it fell. However, you would soon need a radio telescope to "see" it, as the light from the clock face would shift from the red part of the visible spectrum to the infrared and on into the radio. The redshift would soon become so great that no telescope could detect the clock's light. As the clock vanished from view, you would see the time on its face freezing to a stop.

Falling into a Black Hole After everything you've just read about black holes, you might wonder what it would be like to fall into one. To satisfy your curiosity, let's imagine that one of your friends on the spaceship climbs into a space suit, grabs the other clock, resets it, and jumps out of the air lock on a trajectory aimed straight for the black hole. Down he falls, clock in hand. He watches the clock, but because he and the clock are traveling together, its time seems to run normally and its numerals stay blue. From his point of view, time

Common Misconceptions

BLACK HOLES DON'T SUCK

What would happen if our Sun suddenly became a black hole? For some reason, the idea that Earth and the other planets would inevitably be "sucked in" by the black hole has become part of our popular culture, but this is not true. Although the sudden disappearance of light and heat from the Sun would be bad news for life, Earth's orbit would not change. Newton's law of gravity tells us that the allowed orbits in a gravitational field are ellipses, hyperbolas, and parabolas. Note that "sucking" is not on the list! Earth would get into trouble only if it came so close to the black hole—within about three times its Schwarzschild radius—that the force of gravity deviated significantly from what Newton's law predicts. Otherwise, it would simply continue on an ordinary orbit.

Space travelers also need not fear running into black holes, because most are so small. Their typical Schwarzschild radii are far smaller than the radius of any star or planet, so a black hole is actually one of the most difficult things in the universe to fall into by accident.

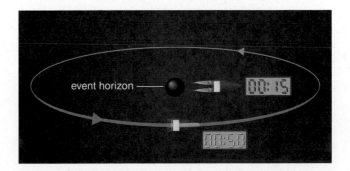

Figure 10.13 | Time runs more slowly on the clock nearer to the black hole, and a gravitational redshift makes its glowing blue numerals appear red from your orbiting spaceship.

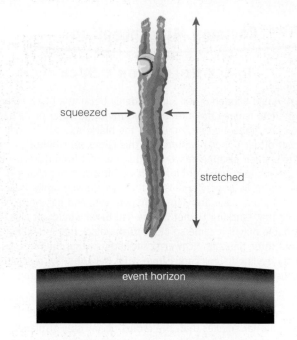

squeezed →   ← stretched

event horizon

Figure 10.14 | Tidal forces would be lethal near a black hole formed by the collapse of a star. The black hole would pull more strongly on the astronaut's feet than on his head, stretching him lengthwise and squeezing him from side to side.

seems to neither speed up nor slow down. In fact, he'd say that *you* were the one with the strange time; he would see your time running increasingly fast and your light becoming increasingly blueshifted. When his clock reads, say, 00:30, he and the clock pass through the event horizon. There is no barrier, no wall, no hard surface. The event horizon is a mathematical boundary, not a physical one. From his point of view, the clock keeps ticking. He is inside the event horizon, the first human being ever to vanish into a black hole.

Back on the spaceship, you watch in horror as your friend plunges to his death. Yet, from your point of view, he will *never* cross the event horizon. You'll see time come to a stop for him and his clock just as he vanishes from view because of the huge gravitational redshift of light. When you return home, you can play a video for the judges at your trial, proving that your friend is still outside the black hole. Strange as it may seem, all of these events are in accordance with Einstein's theory. From your point of view, your friend takes *forever* to cross the event horizon (even though he vanishes from view because of his ever-increasing redshift). From his point of view, it is but a moment's plunge before he passes into oblivion.

The truly sad part of this story is that your friend did not live to experience the crossing of the event horizon. The force of gravity grew so quickly as he approached the black hole that it pulled much harder on his feet than on his head, simultaneously stretching him lengthwise and squeezing him from side to side (**Figure 10.14**). In essence, your friend was stretched in the same way the oceans are stretched by the tides, except that the *tidal force* near the black hole is trillions of times stronger than the tidal force of the Moon on Earth. No human could survive it.

✳ THE PROCESS OF SCIENCE IN ACTION

10.3 Searching for Black Holes

Only a half-century ago, most astronomers who contemplated the idea of black holes thought them too strange to be true. However, science relies on observations, not personal biases, and observational evidence now strongly suggests that black holes really do exist. The search for black holes provides an outstanding example of the process of science in action.

Do black holes really exist?

The fact that black holes emit no light might make it seem as if they should be impossible to detect. However, a black hole's gravity can influence its surroundings in ways that reveal its presence. Astronomers seeking black holes search for telltale signs of an unseen gravitational influence with a large enough mass to suggest that it is a black hole. Objects showing these telltale signs have been found in two different environments. Some reside in binary star systems and are probably black holes born in supernova explosions, while others live at the centers of galaxies and are much too massive to come from a single supernova.

Black Holes in Binary Systems   Strong observational evidence for black holes formed by supernovae comes from studies of X-ray binaries. Recall that the accretion disks around neutron stars in close binary systems can emit strong X-ray radiation, making an X-ray binary. The accretion disk forms because the neutron star's strong gravity pulls in mass from the companion star. Because a black hole has even stronger gravity than a neutron star, a black hole in a close binary system should also be surrounded by a hot, X-ray-emitting accretion disk. In other words, an X-ray binary might contain either a black hole or a neutron star. We can learn which type of corpse resides in an X-ray binary by measuring the object's mass.

One of the most promising black hole candidates is in an X-ray binary called Cygnus X-1 (**Figure 10.15**). This system contains an extremely luminous star with an estimated mass of $18M_{Sun}$. Based on Doppler shifts of its spectral lines, astronomers have concluded that this star orbits a compact, unseen companion with a mass of about $10M_{Sun}$. Although there is some uncertainty in these estimates, the mass of the invisible accreting object clearly exceeds the $3M_{Sun}$ neutron star limit. It is therefore too massive to be a neutron star, so by current knowledge it must be a black hole. A few dozen other X-ray binaries offer similar evidence for black holes formed from the collapse of massive stellar cores.

Think about it Recall that some X-ray binaries that contain neutron stars emit frequent X-ray bursts and are called X-ray bursters. Could an X-ray binary that contains a black hole exhibit the same type of X-ray bursts? Why or why not? (*Hint:* Where do the X-ray bursts occur in an X-ray binary with a neutron star?)

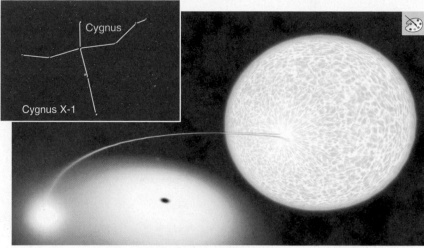

Figure 10.15 | Artist's conception of the Cygnus X-1 system, so named because it is the brightest X-ray source in the constellation Cygnus. The X rays come from the high-temperature gas in the accretion disk surrounding the black hole.

Supermassive Black Holes An even greater body of evidence suggests that *supermassive black holes*—some with masses millions or billions of times that of our Sun—reside at the centers of many galaxies. As we will discuss in Chapter 11, these black holes must have formed differently from the stellar-mass black holes in X-ray binaries, and they are thought to power some of the most luminous objects in the universe. The closest of these supermassive black holes sits at the center of our very own Milky Way galaxy.

The center of the Milky Way contains a source of radio emission called Sagittarius A* (pronounced "Sagittarius A-star"), or Sgr A* for short, which is quite unlike any other radio source in our galaxy. Several hundred stars crowd the region within about 1 light-year of Sgr A*, and careful monitoring of their motions has enabled us to measure the mass of this object (**Figure 10.16**). Applying Newton's version of Kepler's third law to the orbits of the stars passing closest to Sgr A* shows that it has a mass of about 4 million solar masses, all packed into a region of space just a little larger than our solar system. An object that massive within such a small space is almost certainly a black hole. Similar evidence indicates that larger galaxies have black holes of even greater mass at their centers [Section 11.3].

The Bottom Line Confirming that black holes are real with 100% certainty is very difficult because our detection methods remain indirect. However, our current theories successfully explain neutron stars, and the general theory of relativity that leads to the idea of black holes is also well established. Unless something is dramatically wrong in our current theories about the mass limit of neutron stars or some other, unknown type of compact object can have a huge mass, black holes must be real.

We are now able to bring the story of the battle between pressure and gravity, which began in the previous chapter, to its ultimate conclusion. **Figure 10.17** shows how a star's fate depends on its birth mass. Objects with less than $0.08M_{Sun}$ never become stars because degeneracy pressure halts their contraction before fusion can bring them into balance. Stars beginning with less than $8M_{Sun}$ end up as white dwarfs because degeneracy pressure stops their cores from contracting before fusion starts to make iron. Stars with greater masses become either neutron stars or black holes, depending on whether the mass of the core remaining after the supernova explosion exceeds the neutron star limit.

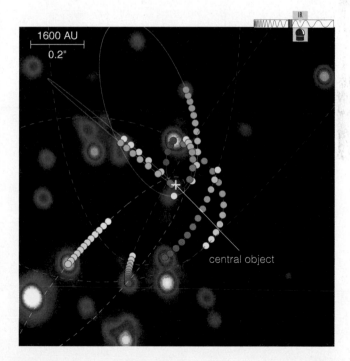

Figure 10.16 | This diagram shows observed stellar positions (colored dots indicate year of observation) and calculated orbits for several stars near the very center of the galaxy. By applying Newton's version of Kepler's third law to these orbits, we infer that the central object has a mass 3 to 4 million times that of our Sun, packed into a space so small that the object is almost certainly a black hole. (The 1600 AU shown on the scale bar is equivalent to about 9 light-days.)

We can understand the entire life cycle of a star in terms of the changing balance between pressure and gravity. This illustration shows how that balance changes over time and why those changes depend on a star's birth mass. *(Stars not to scale.)*

Key
pressure
gravity

(2) Thermal pressure comes into steady balance with gravity when the core becomes hot enough for hydrogen fusion to replace the thermal energy the star radiates from its surface.

(3) The balance tips in favor of gravity after the core runs out of hydrogen. Fusion of hydrogen into helium temporarily stops supplying thermal energy in the core. The core again contracts and heats up. Hydrogen fusion begins in a shell around the core. The outer layers expand and cool, and the star becomes redder.

Hydrogen shell-burning star

Main-sequence star

Pressure balances gravity at every point within a main-sequence star.

> 0.08M_{Sun}

(1) Gravity overcomes pressure inside a protostar, causing the core to contract and heat up. A protostar cannot achieve steady balance between pressure and gravity because nuclear fusion is not replacing the thermal energy it radiates into space.

The balance between pressure and gravity acts as a thermostat to regulate the core temperature:

A drop in core temperature decreases fusion rate, which lowers core pressure, causing the core to contract and heat up.

A rise in core temperature increases fusion rate, which raises core pressure, causing the core to expand and cool down.

Stellar Thermostat: Gravitational Equilibrium

Luminosity continually rises because core contraction causes the temperature and fusion rate in the hydrogen shell to rise.

The balancing point between pressure and gravity depends on a star's mass:

Balance between pressure and gravity in high-mass stars results in a higher core temperature, a higher fusion rate, greater luminosity, and a shorter lifetime.

Protostar

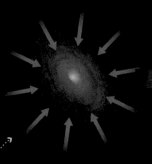

Contraction converts gravitational potential energy into thermal energy.

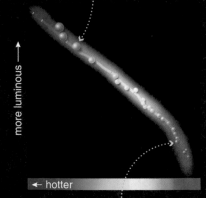

more luminous ⟶

← hotter

Balance between pressure and gravity in low-mass stars results in a cooler core temperature, a slower fusion rate, less luminosity, and a longer lifetime.

Degeneracy pressure balances gravity in objects of less than 0.08M_{Sun} before their cores become hot enough for steady fusion. These objects never become stars and end up as brown dwarfs.

> < 0.08M_{Sun}

④ Balance between thermal pressure and gravity is restored when the core temperature rises enough for helium fusion into carbon, which can once more replace the thermal energy radiated from the core.

⑤ Gravity again gains the upper hand over pressure after the core helium is gone. Just as before, fusion stops replacing the thermal energy leaving the core. The core therefore resumes contracting and heating up, and helium fusion begins in a shell around the carbon core.

⑥ In high mass stars, core contraction continues, leading to multiple shell burning that terminates with iron and a supernova explosion.

⑦ At the end of a star's life, either degeneracy pressure has come into permanent balance with gravity or the star has become a black hole. The nature of the end state depends on the mass of the remaining core.

Multiple shell–burning star

Black hole

Degeneracy pressure cannot balance gravity in a black hole.

Helium shell-burning star

Double shell–burning star

HIGH-MASS

Neutron star ($M < 3M_{Sun}$)

Neutron degeneracy pressure can balance gravity in a stellar corpse with less than about 2–3 M_{Sun}.

Luminosity remains steady because helium core fusion restores balance.

White dwarf ($M < 1.4M_{Sun}$)

LOW-MASS

Electron degeneracy pressure balances gravity in the core of a low-mass star before it gets hot enough to fuse heavier elements. The star ejects its outer layers and ends up as a white dwarf.

Electron degeneracy pressure can balance gravity in a stellar corpse of mass < 1.4 M_{Sun}.

Brown dwarf ($M < 0.08M_{Sun}$)

Degeneracy pressure keeps a brown dwarf stable in size even as it cools steadily with time.

Summary of Key Concepts

10.1 White Dwarfs and Neutron Stars

What are white dwarfs?

A white dwarf is supported against the crush of gravity by **electron degeneracy pressure** and cannot have a mass greater than $1.4M_{Sun}$. If a white dwarf is close enough to another star in a binary system, it can acquire hydrogen from its companion star and trigger a burst of nuclear fusion that produces a **nova**. In extreme cases, a white dwarf may accrete enough matter to exceed the **white dwarf limit** of $1.4M_{Sun}$, at which point it will explode as a **white dwarf supernova.**

What are neutron stars?

A neutron star is supported against the crush of gravity by **neutron degeneracy pressure.** Some neutron stars emit beams of radiation that sweep through space as the neutron star spins. We call such neutron stars **pulsars,** and they provided the first direct evidence for the existence of neutron stars. Additional evidence comes from X-ray-emitting accretion disks around neutron stars in **X-ray binaries.** In some of these systems, helium fusion can ignite on the neutron star's surface, emitting **X-ray bursts.**

10.2 Black Holes

What are black holes?

A black hole is a place where gravity has crushed matter into oblivion, creating an object so massive and compact that nothing can ever escape it, not even light. Such an object can form in a supernova explosion that leaves behind a core with a mass above the neutron star limit of $2M_{Sun}$ to $3M_{Sun}$. The **event horizon** around a black hole marks the boundary between our observable universe and the inside of the black hole; the size of the black hole is characterized by its **Schwarzschild radius.**

What happens to space and time near a black hole?

Einstein's theory of relativity predicts that space and time should act very strangely near a black hole, becoming highly curved near the event horizon. If you could watch an object falling toward a black hole, time would seem to run slowly for the object, and its light would be increasingly redshifted as it approached the event horizon.

✳ THE PROCESS OF SCIENCE IN ACTION

10.3 Searching for Black Holes

Do black holes really exist?

Black holes emit no light, but they reveal their presence through their gravitational effects on nearby matter. Some X-ray binaries include compact objects far too massive to be neutron stars; it is likely that they are black holes. We have also found supermassive black holes in the centers of galaxies, including our own. Motions of stars at the center of our Milky Way show that it contains a black hole with a mass of about 4 million solar masses.

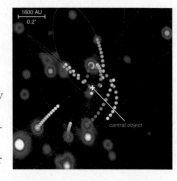

Investigations

Quick Quiz

Choose the best answer to each of the following; answers are in Appendix D. Explain your reasoning with one or more complete sentences.

1. Which of these objects has the smallest radius? (a) a $1.2M_{Sun}$ white dwarf (b) a $0.6M_{Sun}$ white dwarf (c) Jupiter

2. Which of these objects has the largest radius? (a) a $1.2M_{Sun}$ white dwarf (b) a $1.5M_{Sun}$ neutron star (c) the event horizon of a $3.0M_{Sun}$ black hole

3. Which of these objects has the smallest radius? (a) a $1.2M_{Sun}$ white dwarf (b) the event horizon of a $3.0M_{Sun}$ black hole (c) the event horizon of a $10M_{Sun}$ black hole

4. What would happen if the Sun suddenly became a black hole without changing its mass? (a) The black hole would quickly suck in Earth. (b) Earth would gradually spiral into the black hole. (c) Earth's orbit would not change.

5. What happens to a white dwarf that is accreting hydrogen gas from a nearby star? (a) It can become a black hole. (b) A burst of fusion can occur on its surface. (c) Its radius keeps increasing.

6. What happens to a neutron star that is accreting hydrogen gas from a nearby star? (a) A burst of fusion can occur on its surface. (b) It can become a white dwarf. (c) It can undergo a supernova.

7. What happens to a black hole that is accreting hydrogen gas from a nearby star? (a) It can become a neutron star. (b) A burst of fusion can occur on its surface. (c) Its Schwarzschild radius keeps increasing.

8. Why does matter falling toward a neutron star in a binary system form an accretion disk? (a) because some neutron stars spin very rapidly (b) because the infalling matter has some angular momentum (c) because it is emitting X rays

9. Which of these binary systems is most likely to contain a black hole? (a) an X-ray binary containing an O star and another object of equal mass (b) a binary with an X-ray burster (c) an X-ray binary containing a G star and another object of equal mass

10. How would a flashing red light appear as it fell into a black hole? (a) Its flashes would look the same. (b) Its flashes would appear bluer. (c) Its flashes would shift to the infrared part of the spectrum.

11. We measure the mass of a black hole in an X-ray binary from (a) the orbital period of the system and the orbital velocity inferred from its spectral lines. (b) the period of its X-ray pulses. (c) the mass of its companion star.

12. We measure the mass of the black hole at the center of the Milky Way from (a) the orbits of stars in the galactic center. (b) gas clouds in the galactic center. (c) the amount of radiation coming from the galactic center.

Short-Answer/Essay Questions

Explain all answers clearly, with complete sentences and proper essay structure if needed. An asterisk (*) designates a quantitative problem, for which you should show all your work.

13. *Life Stories of Stars.* Write a one- to two-page life story for each of the following scenarios. Each story should be detailed and scientifically correct but also creative. That is, it should be entertaining while at the same time proving that you understand stellar evolution. Be sure to state whether you are a member of a binary system.
    a. You are a white dwarf of $0.8M_{Sun}$.
    b. You are a neutron star of $1.5M_{Sun}$.
    c. You are a black hole of $10M_{Sun}$.
    d. You are a white dwarf in a binary system that is accreting matter from its companion star.

14. *Census of Stellar Corpses.* Which kind of object do you think is most common in our galaxy: white dwarfs, neutron stars, or black holes? Explain your reasoning.

15. *Fate of an X-Ray Binary.* The X-ray bursts that happen on the surface of an accreting neutron star are not powerful enough to accelerate the exploding material to escape velocity. Predict what will happen in an X-ray binary system in which the companion star eventually feeds over 3 solar masses of matter into the neutron star's accretion disk.

16. *Why Black Holes Are Safe.* Explain why the principle of conservation of angular momentum makes it very difficult to fall into a black hole.

17. *Surviving the Plunge.* The tidal forces near a black hole with a mass similar to that of a star would tear a person apart before that person could fall through the event horizon. Black hole researchers have pointed out that a fanciful "black hole life preserver" could help counteract those tidal forces. The life preserver would need to have a mass similar to that of an asteroid and would need to be shaped like a flattened hoop placed around the person's waist. In what direction would the gravitational force from the hoop pull on the person's head? In what direction would it pull on the person's feet? Based on your answers, explain in general terms how the gravitational forces from the "life preserver" would help to counteract the black hole's tidal forces.

*18. *A Black Hole I?* You've just discovered a new X-ray binary, which we will call Hyp-X1 ("Hyp" for hypothetical). The system Hyp-X1 contains a bright, B2 main-sequence star orbiting an unseen companion. The separation of the stars is estimated to be 20 million kilometers, and the orbital period of the visible star is 4 days.
    a. Use Newton's version of Kepler's third law to calculate the sum of the masses of the two stars in the system. (*Hint:* See Tools of Science, p. 96.) Give your answer in both kilograms and solar masses ($M_{Sun} = 2.0 \times 10^{30}$ kg).
    b. Determine the mass of the unseen companion. Is it a neutron star or a black hole? Explain. (*Hint:* A B2 main-sequence star has a mass of about $10M_{Sun}$.)

*19. *A Black Hole II?* You've just discovered another new X-ray binary, which we will call Hyp-X2 ("Hyp" for hypothetical). The system Hyp-X2 contains a bright, G2 main-sequence star orbiting an unseen companion. The separation of the stars is estimated to be 12 million kilometers, and the orbital period of the visible star is 5 days.
    a. Use Newton's version of Kepler's third law to calculate the sum of the masses of the two stars in the system. (*Hint:* See Tools of Science, p. 96.) Give your answer in both kilograms and solar masses ($M_{Sun} = 2.0 \times 10^{30}$ kg).
    b. Determine the mass of the unseen companion. Is it a neutron star or a black hole? Explain. (*Hint:* A G2 main-sequence star has a mass of $1M_{Sun}$.)

*20. *Neutron Star Density.* A typical neutron star has a mass of about $1.5M_{Sun}$ and a radius of 10 kilometers.
    a. Calculate the average density of a neutron star, in *kilograms per cubic centimeter*.
    b. Compare the mass of 1 cm^3 of neutron star material to the mass of Mount Everest ($\approx 5 \times 10^{10}$ kg).

11 Galaxies

This photo shows a galaxy—an island of stars bound together by gravity. Our own Milky Way Galaxy would look similar if we could see it from a few million light-years away. Tens of billions of other galaxies populate the observable universe. Within these galaxies, stars form in dusty clouds of hydrogen gas, fuse new elements out of that hydrogen, and add those new elements to the galaxy's gas clouds when they die. Galaxies therefore function like huge ecosystems in which the cycling of gas through stars gradually builds up the elements necessary to make planets and support life. In this chapter, we'll study how galaxies work, paying special attention to how the history of star formation shapes the appearance of the galaxies we see today.

11.1 Our Galaxy: The Milky Way

We'll begin our study of galaxies with our own Milky Way Galaxy, the only one we can observe in great detail. We'll look at the structure and motion of the galaxy, investigate the galactic processes that maintain the ongoing cycle of stellar life and death, and examine the clues we have collected about how our galaxy formed. Through it all, we'll see that we are not only "star stuff" but also "galaxy stuff"—the product of eons of complex recycling and reprocessing of matter and energy in the Milky Way Galaxy.

What does our galaxy look like?

From Earth, the Milky Way Galaxy looks like a band of light crossing the sky (**Figure 11.1**). The name *Milky Way* comes from the appearance of this band, which looked to ancient people like a ribbon of flowing milk. In fact, even the word *galaxy* comes from the Greek word for milk: *galactos*.

Because we live inside the Milky Way Galaxy, trying to determine its true size and shape is somewhat like trying to draw a picture of your house without ever leaving your bedroom. The task is difficult because much of our galaxy's visible light is hidden from view. Only in the last few decades have we had the technology to observe the galaxy in other wavelengths of light (see Tools of Science, p. 180). Nevertheless, by carefully observing our galaxy and comparing it to others, we have developed a good understanding of its structure and of the processes that shape it.

Structure of the Milky Way   **Figure 11.2** shows the Milky Way Galaxy as it would look from the outside. We say that it is a *spiral galaxy* because of the spectacular **spiral arms** that would be visible in a face-on view. From the side, the spiral arms blend into a thin **disk** that contains more than 100 billion stars. At the center of the disk, we find a bright central **bulge.** Surrounding the disk is a vast, roughly spherical **halo.** The halo is nearly invisible, except for its approximately 200 *globular clusters* of stars [Section 9.3].

The disk is about 100,000 light-years in diameter, but only about 1000 light-years thick. Our Sun is located in the disk about 28,000 light-years from the galactic center—a little more than halfway out from the center to the edge of the disk. Remember that this distance is incredibly vast [Section 1.1]: The few thousand stars visible to the naked eye would fit inside a tiny dot in a picture like that in Figure 11.2.

The galactic disk is filled with interstellar gas and dust—known collectively as the **interstellar medium**—that obscures our view when we try to peer directly through it. The dusty, smoglike nature of the interstellar medium hides most of our galaxy from view when we try to observe it in visible light, and as a result it long fooled astronomers into believing that we lived near our galaxy's center. As we saw in Chapter 9, the dusty gas clouds of the interstellar medium provide the raw material from which all the stars in our galaxy have formed.

Orbits of Disk Stars   The Milky Way's structure reflects how stars move within it. Stars in the disk orbit in roughly circular paths that all go in the same direction in nearly the same plane. If you could stand outside the Milky Way and watch it for a few billion years, the disk would resemble a huge merry-go-round. Like horses on a merry-go-round, individual stars bob up and down through the disk as they orbit. The general orbit of a star around the galaxy arises from its gravitational attraction toward the galactic center, while the bobbing arises from the localized pull of gravity within the disk

Figure 11.1 | This digitally enhanced photo shows the Milky Way over the Kofa Mountains in Arizona.

a Artist's conception of the Milky Way viewed from the outside.

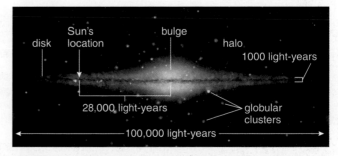

b Edge-on schematic view of the Milky Way.

Figure 11.2 | The Milky Way Galaxy.

Tools of Science: Observing Different Kinds of Light

Much of modern astronomy depends on our ability to detect not just visible light from space but light from across the electromagnetic spectrum (see Tools of Science, p. 80). Two technological advances have given us this ability. The first is the development of specialized instruments capable of detecting forms of light that our eyes cannot see. The second is the development of spacecraft that can carry these instruments high above Earth's atmosphere, which prevents most kinds of light from reaching Earth's surface.

Figure 1 gives an example of how much we can learn by observing wavelengths of light that are invisible to human eyes. It shows seven views of the disk of our Milky Way Galaxy, each made by photographing the Milky Way's disk in every direction from Earth in a particular wavelength band. Four of the images (infrared, X-ray, and gamma-ray images) come from observatories in space. The other three (visible and radio images) are from ground-based telescopes.

Figure 1a is a visible-light view that shows a haze of starlight interrupted by dark blotches running primarily down the center of the disk. These dark blotches occur where dusty clouds of interstellar gas block our view of stars behind them. Figure 1b shows the same view as it appears in

infrared light. Infrared light can pass through the dusty gas clouds, which is why the center of the disk is bright and we can see the Milky Way's bulge in the center of the image. Figure 1c shows long-wavelength infrared light, which comes primarily from dust grains in the interstellar medium. Moving to even longer wavelengths, Figure 1d shows radio emission from atomic hydrogen gas, and Figure 1e shows radio emission from carbon monoxide molecules in molecular clouds. Notice how the regions of bright carbon monoxide emission correspond to the dark patches in Figure 1a.

On the opposite end of the electromagnetic spectrum, at very short wavelengths, we observe the most energetic photons. Figure 1f is an X-ray image showing the distribution of hot gas and X-ray binaries in our galaxy's disk. Figure 1g is a gamma-ray image showing where extremely energetic particles known as *cosmic rays* are colliding with atomic nuclei in the interstellar medium.

You can see from this sequence of images that the ability to observe many different kinds of light allows us to gather much more information about astronomical objects than we would be able to gather from visible light alone. That is why astronomers today observe the universe in many wavelengths of light.

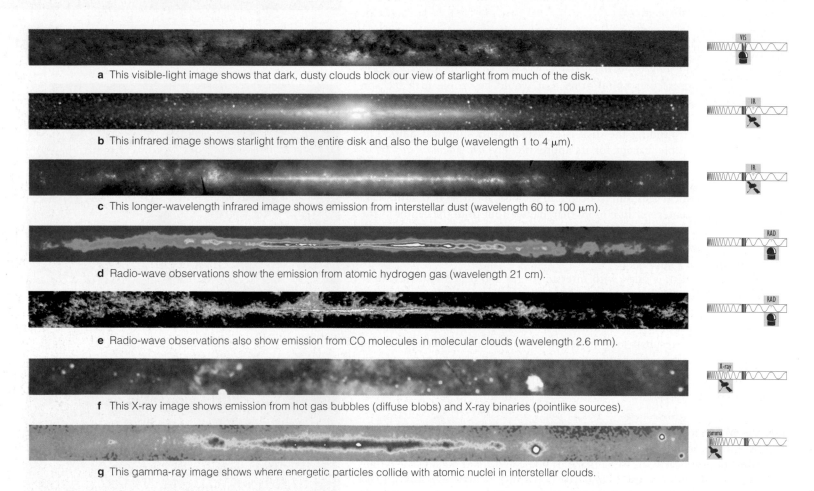

a This visible-light image shows that dark, dusty clouds block our view of starlight from much of the disk.

b This infrared image shows starlight from the entire disk and also the bulge (wavelength 1 to 4 μm).

c This longer-wavelength infrared image shows emission from interstellar dust (wavelength 60 to 100 μm).

d Radio-wave observations show the emission from atomic hydrogen gas (wavelength 21 cm).

e Radio-wave observations also show emission from CO molecules in molecular clouds (wavelength 2.6 mm).

f This X-ray image shows emission from hot gas bubbles (diffuse blobs) and X-ray binaries (pointlike sources).

g This gamma-ray image shows where energetic particles collide with atomic nuclei in interstellar clouds.

Figure 1 | Panoramic views of the Milky Way in different bands of the electromagnetic spectrum. The center of the galaxy is in the center of each strip. The rest of each strip shows all other directions in the Milky Way disk as seen from Earth. (Imagine attaching the left and right ends of each strip to form a circular band that corresponds to the 360° band of the Milky Way in our sky.)

itself (**Figure 11.3**). A star that is far above the disk is pulled back into the disk by gravity. Because the density of interstellar gas is too low to slow the star, it flies through the disk until it is far *below* the disk on the other side. Gravity then pulls it back in the other direction. This ongoing process produces the bobbing of the stars.

The up-and-down motions of the disk stars give the disk its thickness of about 1000 light-years—a great distance by human standards, but only about 1% of the disk's 100,000-light-year diameter. In the vicinity of our Sun, each star's orbit takes more than 200 million years, and each up-and-down bob takes a few tens of millions of years.

The galaxy's rotation is unlike that of a merry-go-round in one important respect: On a merry-go-round, horses near the edge move much faster than those near the center. But in our galaxy's disk, the orbital velocities of stars near the edge and those near the center are about the same. As we will discuss in Chapter 14, the surprisingly high velocities of stars near the edge provide important evidence for the existence of *dark matter*.

Orbits of Halo and Bulge Stars The orbits of stars in the halo and bulge are much less organized than those of disk stars. Individual bulge and halo stars travel around the galactic center on more or less elliptical paths, but the orientations of these paths are relatively random (see Figure 11.3). Neighboring halo stars can circle the galactic center in opposite directions. They swoop from high above the disk to far below it and back again, plunging through the disk at velocities so great that the disk's gravity hardly alters their trajectories. Several fast-moving halo stars are currently passing through the disk not too far from our own solar system.

These swooping orbits explain why the bulge and halo are much rounder and puffier than the disk. Halo and bulge stars soar to heights above the disk far greater than those reached by the disk stars. As we'll soon see, the differences between the orbits of disk stars and halo stars provide an important clue to how our galaxy formed.

 Is there much danger that a halo star swooping through the disk of the galaxy will someday hit the Sun or Earth? Why or why not? (*Hint:* Consider the typical distances between stars on the 1-to-10-billion scale introduced in Chapter 1.)

Galactic Recycling Closer inspection of the Milky Way's disk reveals that stars are continually forming and dying within it through a process we will call the **star–gas–star cycle** (**Figure 11.4**). Several of the steps in this cycle should already be familiar from Chapter 9: Stars are born when gravity causes the collapse of molecular clouds within the interstellar medium. They shine for millions or billions of years with energy produced by nuclear fusion, dying only when they've exhausted their fuel for fusion. As they die, they return much of their material back to the interstellar medium—through a planetary nebula in the case of a low-mass star or through a supernova in the case of a high-mass star.

The remaining stages of the star–gas–star cycle take place in the interstellar medium. Gas ejected by stars, particularly the gas from supernova explosions, often enters the interstellar medium in the form of a hot bubble. **Figure 11.5** shows the bubble produced by a 400-year-old supernova explosion, which contains ionized gas hot enough to emit X rays. This million-degree gas takes many thousands of years to cool, but eventually reaches a temperature of around 10^4 K, which is still quite warm but cool enough for hydrogen atoms to remain neutral rather than being ionized. We refer to this warm gas as **atomic hydrogen gas,** although it contains a significant amount of helium and small amounts of other elements as well. Mapping the distribution of atomic hydrogen gas in the Milky Way using radio observations of its spectral line at a wavelength of 21 centimeters tells us that it is distributed

Figure 11.3 | Characteristic orbits of disk stars (yellow), bulge stars (red), and halo stars (green) around the galactic center. (The yellow path exaggerates the up-and-down motion of the disk star orbits.)

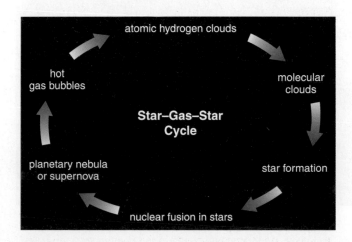

Figure 11.4 | The star–gas–star cycle.

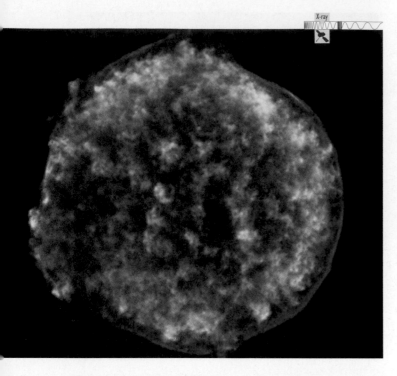

Figure 11.5 | This image shows X-ray emission from hot gas in a young supernova remnant. The most energetic X rays (blue) come from 20-million-degree gas just behind the expanding shock wave. Less energetic X rays (green and red) come from the 10-million-degree debris ejected by the exploded star. The remnant is about 20 light-years across.

Figure 11.6 | Hubble Space Telescope photo of the Orion Nebula, an ionization nebula energized by ultraviolet photons from hot stars.

throughout the galactic disk. Matter remains in the warm atomic hydrogen stage for millions of years, and during this time gravity slowly draws blobs of atomic gas together into tighter clumps, which radiate energy and cool more efficiently as they grow denser. Even so, the interstellar medium remains a nearly perfect vacuum by earthly standards: On average, each cubic centimeter contains only one hydrogen atom.

The temperature of these gas clumps eventually drops well below 100 K, allowing the hydrogen atoms to pair up into hydrogen molecules. The cool, dense clumps then become *molecular clouds* which go on to form stars, thereby completing the star–gas–star cycle. Molecular hydrogen (H_2) is by far the most abundant molecule in molecular clouds, but it is difficult to detect because temperatures are usually too cold for the gas to produce H_2 emission lines. Most of what we know about molecular clouds therefore comes from observing spectral lines of molecules that make up only a tiny fraction of a cloud's mass. Carbon monoxide (CO) is the most abundant of these molecules and produces strong emission lines in the radio portion of the spectrum at the 10–30 K temperatures of molecular clouds.

In addition to making new generations of stars possible, this galactic recycling process gradually changes the chemical composition of the interstellar medium. Recall that the early universe contained only the chemical elements hydrogen and helium; all heavier elements have been produced by stars. The newly created elements mix with other interstellar gas and become incorporated into new generations of stars. That is how our solar system came to have the elements from which our planet was made. Today, thanks to more than 10 billion years of galactic recycling, elements heavier than helium constitute about 2% of the galaxy's gaseous content by mass. The remaining 98% consists of hydrogen (about 70%) and helium (about 28%).

Recall from Chapter 9 that stars in our galaxy's globular clusters are extremely old, while stars in open clusters are relatively young. Based on this fact, which stars do you think contain a higher proportion of heavy elements: stars in globular clusters or stars in open clusters? Explain.

Star-Forming Regions The star–gas–star cycle has operated continuously since the Milky Way's birth, yet new stars are not spread evenly across the galaxy. Some regions seem much more fertile than others. Galactic environments rich in molecular clouds tend to spawn new stars easily, while gas-poor environments do not. However, molecular clouds are dark and hard to see, so we often have to look for other signs of star formation.

Wherever we see hot, massive stars, we know that we have spotted a region of active star formation. Because these stars live fast and die young, they never get a chance to move very far from their birth mates. They therefore signal the presence of star clusters in which many of their lower-mass companions are still forming.

Near these hot stars, we often find colorful, wispy blobs of glowing gas known as **ionization nebulae** (sometimes called *emission nebulae* or *H II regions*). These nebulae glow because electrons in their atoms are raised to high energy levels or ionized when they absorb ultraviolet photons from the hot stars, so they emit light as the electrons fall back to lower energy levels (see Tools of Science, p. 148). The Orion Nebula, about 1500 light-years away in the "sword" of the constellation Orion, is among the most famous (**Figure 11.6**).

Spiral Arms Looking more broadly at our galaxy, we can see that its spiral arms must be full of newly forming stars because they bear all the hallmarks of star formation. They are home both to molecular clouds and to numerous clusters of young, bright blue stars surrounded by ionization

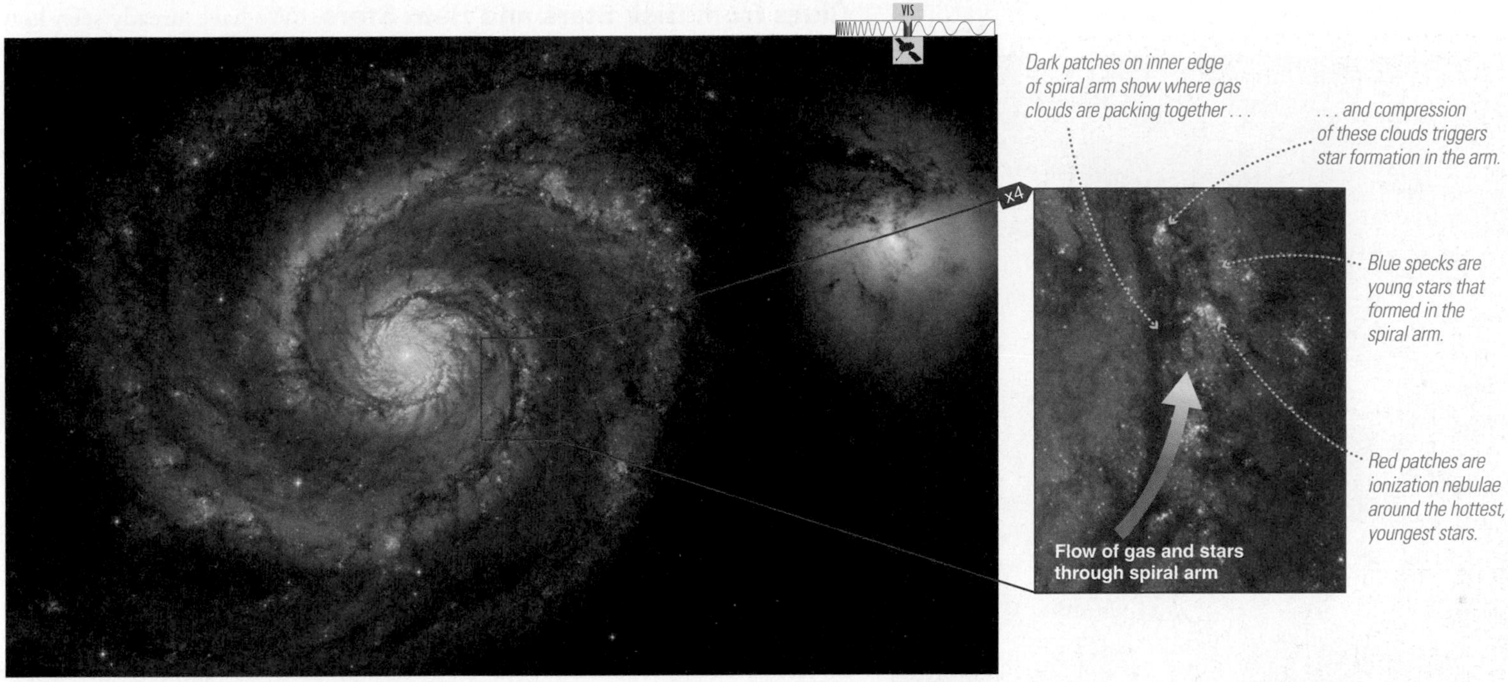

Dark patches on inner edge of spiral arm show where gas clouds are packing together . . .

. . . and compression of these clouds triggers star formation in the arm.

Blue specks are young stars that formed in the spiral arm.

Red patches are ionization nebulae around the hottest, youngest stars.

Flow of gas and stars through spiral arm

Figure 11.7 | This photo from the Hubble Space Telescope shows Galaxy M51's two magnificent spiral arms along with a smaller galaxy that is currently interacting with one of those arms. Notice that the spiral arms are much bluer than the central bulge. Because massive blue stars live for only a few million years, the blueness of the spiral arms tells us that stars must be forming more actively within them than elsewhere in the galaxy. (The large image shows a region roughly 90,000 light-years across.)

nebulae. Photos of other spiral galaxies show these characteristics clearly (**Figure 11.7**). Hot blue stars and ionization nebulae trace out the arms, while the stars between the arms are generally redder and older. We also see enhanced amounts of molecular and atomic gas in the spiral arms, indicating that the spiral arms contain the material necessary to make new stars.

At first glance, spiral arms look as if they ought to move with the stars, like the fins of a giant pinwheel in space. That cannot be the case, however, because stars near the center of the galaxy complete an orbit in much less time than stars far from the center. If the spiral arms simply moved along with the stars, the central parts of the arms would complete several orbits around the galaxy as the outer parts orbited just once. This difference in orbital periods would eventually wind up the spiral arms into a tight coil. Because we generally don't see such tightly wound spiral arms in galaxies, we conclude that spiral arms are more like swirling ripples in a whirlpool than like the fins of a giant pinwheel—a moving pattern rather than a permanent structure.

In fact, spiral arms appear to be enormous waves of star formation that propagate through the disks of spiral galaxies. Theoretical models suggest that disturbances called *spiral density waves* are responsible for the spiral arms. According to these models, spiral arms are places in a galaxy's disk where stars and gas clouds get more densely packed. Pushing the stars closer together has little effect on the stars themselves—they are still much too widely separated to collide with each other. However, large gas clouds do collide, and packing the clouds closer together enhances the force of gravity within them, triggering the formation of new star clusters. Supernova explosions from massive stars in these clusters can compress the surrounding clouds further, triggering even more star formation.

How did the Milky Way form?

All the features of our Milky Way Galaxy that we have discussed, including its structure and the operation of the star–gas–star cycle, provide clues to how it formed. However, some of the most important clues come from a detailed comparison of disk stars with halo stars.

Common Misconceptions

THE SOUND OF SPACE

In many science fiction movies, a thunderous sound accompanies the demolition of a spaceship. If the moviemakers wanted to be more realistic, they would silence the explosion. On Earth, we perceive sound when sound waves—waves of alternately rising and falling pressure—cause trillions of gas atoms to push our eardrums back and forth. Although events like supernova explosions produce sound waves that travel through interstellar gas, the extremely low density of this gas means that only a handful of atoms per second would collide with something the size of a human eardrum. As a result, it would be impossible for a human ear (or a similar-size microphone) to register any sound. Despite the presence of sound waves in space, the sound of space is silence.

A protogalactic cloud contains only hydrogen and helium gas.

Halo stars begin to form as the protogalactic cloud collapses.

Conservation of angular momentum ensures that the remaining gas flattens into a spinning disk.

Billions of years later, the star–gas–star cycle supports ongoing star formation within the disk. The lack of gas in the halo precludes star formation outside the disk.

Clues from Disk Stars and Halo Stars We have already seen how the disorderly orbits of halo stars differ from the orbits of disk stars. Two other differences between halo stars and disk stars give us further clues to their origins. First, we don't find any young stars in the halo, while in the disk we see stars of many different ages. Second, the spectra of halo stars show that they contain fewer heavy elements than disk stars. Because of these striking differences, astronomers divide the Milky Way's stars into two distinct populations.

1. The **disk population** (sometimes called *Population I*) contains both young stars and old stars, all of which have heavy-element proportions of about 2%, like our Sun.

2. The **spheroidal population** (sometimes called *Population II*) consists of stars in the halo and bulge, both of which are roughly spherical in shape. Stars in this population are always old and therefore low in mass, and those in the halo sometimes have heavy-element proportions as low as 0.02%—meaning that heavy elements are about 100 times rarer in these stars than in the Sun.

We can understand why halo stars differ from disk stars by looking at how the Milky Way's gas is distributed. The halo does not contain the cold, dense molecular clouds required for star formation. In fact, the halo contains almost no gas at all, and that small amount of gas is generally quite hot. Because star-forming molecular clouds are found only in the disk, new stars can be born only in the disk and not in the halo.

The relative lack of heavy elements in halo stars indicates that they must have formed early in the galaxy's history—before many supernovae had exploded and added heavy elements to star-forming clouds. We conclude that the halo has lacked the gas needed for star formation for a very long time, and that all the Milky Way's cool gas settled into the disk long ago. The only stars that still survive in the halo are old, low-mass stars. All the stars of greater mass that were once born in the halo died long ago.

Think about it How does the halo of our galaxy resemble the distant future fate of the galactic disk? Explain.

A Model of Galaxy Formation Any model of our galaxy's formation must account for the differences between disk stars and halo stars. The most basic model proposes that our galaxy began as a giant **protogalactic cloud** containing all the hydrogen and helium gas that the galaxy eventually turned into stars. Gravity would have caused such a cloud to contract and fragment, just as in present-day star-forming clouds.

According to this model, stars of the spheroidal population formed first. Early on, the gravity associated with our protogalactic cloud drew in matter from all directions, creating a cloud that was blobby in shape and had little or no measurable rotation. The orbits of stars forming within such a cloud could have had any orientation, accounting for the randomly oriented orbits of stars in the halo.

Later, the remaining gas contracted under the force of gravity and settled into a flattened, spinning disk because of conservation of angular momentum (**Figure 11.8**). This process was much like the process that led to the formation of the solar nebula [Section 4.2], but on a much larger scale. Collisions

Figure 11.8 | This four-picture sequence illustrates a simple schematic model of galaxy formation, showing how a spiral galaxy might develop from a protogalactic cloud of hydrogen and helium gas.

among gas particles tended to average out their random motions, causing them to orbit in the same direction and in the same plane. Stars that formed within this spinning disk were born on orbits moving at the same speed and in the same direction as their neighbors and became members of the disk population of stars.

We can test this basic model by studying the stars of the Milky Way. The clues found to date support the model, but suggest that the full story of galaxy formation may be somewhat more complex. All available evidence confirms that the stars in the Milky Way's halo are indeed old. The main-sequence turnoff points in H-R diagrams of globular clusters show that their stars were born at least 12 billion years ago [Section 9.3]. Individual halo stars (those not in globular clusters) and some of the bulge stars appear to be just as old. Furthermore, the proportions of heavy elements are much lower in halo stars than in the Sun, indicating that they formed before many generations of supernovae had a chance to enrich the Milky Way's interstellar medium with heavy elements.

Additional studies of heavy-element proportions, however, suggest that our galaxy formed from several different gas clouds. If the Milky Way had formed from a single protogalactic cloud, it would have steadily accumulated heavy elements during its inward collapse as stars formed and exploded within it. In that case, the outermost stars in the halo would be the oldest and have the smallest proportion of heavy elements. Stars belonging to different globular clusters in the Milky Way's halo do indeed differ in age and heavy-element content, but these variations do not seem to depend on the stars' distance from the galactic center.

The easiest way to account for these variations is to suppose that the Milky Way's earliest stars formed in relatively small protogalactic clouds, each with a few globular clusters, and that these clouds later collided and combined to create the full protogalactic cloud that became the Milky Way (**Figure 11.9**). Similar processes may still be happening. Two much smaller galaxies are currently crashing through the Milky Way's disk and are being torn apart in the process. A billion years from now, their stars will be indistinguishable from halo stars because they will all be circling the Milky Way on orbits that carry them high above the disk. Recent observations of halo-star motions have confirmed that some halo stars move in organized streams that seem to be the remnants of dwarf galaxies torn apart long ago by the Milky Way's gravity.

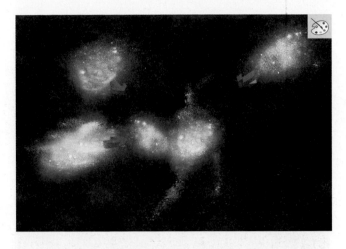

Figure 11.9 | This painting shows a model of how the Milky Way's halo may have formed. The characteristics of stars in the Milky Way's halo suggest that several smaller gas clouds, already containing some stars and globular clusters, may have merged to form the Milky Way's protogalactic cloud. These stars and star clusters remained in the halo while the gas settled into the Milky Way's disk.

11.2 Galaxies Beyond the Milky Way

Take another look at the picture on page 1 of this book. This Hubble Space Telescope photo shows a typical patch of the universe including galaxies of many sizes, colors, and shapes. Some look large, some small. Some are reddish, some whitish. Some appear round, and some appear flat. We would like to understand why these galaxies differ, but this is no easy task. Just as with stars, our observations capture only the briefest snapshot of any galaxy's life, leaving us to piece together the life story of a typical galaxy from the clues we find in pictures of many different galaxies. Let's look more closely at the different types of galaxies and what their differences tell us about their life histories.

What are the major types of galaxies?

Astronomers classify galaxies into three major categories:

- **Spiral galaxies,** such as our own Milky Way, look like flat white disks with yellowish bulges at their centers. The disks are filled with cool gas

Figure 11.10 | The spiral galaxy NGC 4414, whose disk is somewhat tilted to our line of sight. It is about 100,000 light-years in diameter.

Figure 11.11 | NGC 1300, a barred spiral galaxy about 110,000 light-years in diameter.

Figure 11.12 | M87, a giant elliptical galaxy in the Virgo Cluster, is one of the most massive galaxies in the universe. The region shown is more than 120,000 light-years across.

and dust, interspersed with hotter ionized gas, and usually display beautiful spiral arms.

- **Elliptical galaxies** are redder, rounder, and often longer in one direction than in the other, like a football. Compared with spiral galaxies, elliptical galaxies contain very little cool gas and dust, though they often contain very hot, ionized gas.

- **Irregular galaxies** appear neither disklike nor round.

Galaxies have different colors because of the different kinds of stars that populate them: Spiral and irregular galaxies look white because they contain stars of all different colors and ages, while elliptical galaxies look redder because old, reddish stars produce most of their light. Galaxies also come in a wide range of sizes, from *dwarf galaxies* containing as few as 100 million (10^8) stars to *giant galaxies* with more than 1 trillion (10^{12}) stars.

Think about it Take a moment to classify the larger galaxies in the photo on page 1. How many appear spiral? Elliptical? Irregular? Do the colors of galaxies seem related to their shapes?

Spiral Galaxies Like the Milky Way, other spiral galaxies also have a thin *disk* extending outward from a central *bulge* (**Figure 11.10**). The bulge merges smoothly into a nearly invisible *halo* that can extend to a radius of more than 100,000 light-years. All spiral galaxies therefore have both disk and spheroidal populations of stars, although the proportions in each population can vary.

Some spiral galaxies appear to have a straight bar of stars cutting across the center, with spiral arms curling away from the ends of the bar. Such galaxies are known as *barred spiral galaxies* (**Figure 11.11**). Astronomers suspect that the Milky Way is a barred spiral galaxy, because our galaxy's bulge appears to be somewhat elongated.

Other galaxies have disk and spheroidal populations like spiral galaxies but appear to lack spiral arms. These so-called *lenticular galaxies* (*lenticular* means "lens-shaped") are sometimes considered an intermediate class between spirals and ellipticals, because they tend to have less cool gas than normal spirals but more than ellipticals. Galaxies with obvious disks are much rarer among small galaxies. Among large galaxies in the universe, most (75–85%) are spiral or lenticular. Both spiral and lenticular galaxies tend to be found in loose collections of up to a few dozen galaxies, called *groups*. Our Local Group is one example (see Figure 1.1).

Elliptical Galaxies The main difference between elliptical and spiral galaxies is that ellipticals lack a significant disk population. They therefore look much like the bulge and halo of a spiral galaxy that is missing its disk.

Large elliptical galaxies can contain more than a trillion stars (**Figure 11.12**). They are most common in clusters of galaxies, which can contain hundreds and sometimes thousands of galaxies extending over more than 10 million light-years. Elliptical galaxies make up about half the large galaxies in the central regions of clusters, but represent only a small minority (about 15%) of the large galaxies found outside clusters. Much smaller elliptical galaxies, known as *dwarf elliptical galaxies*, are more numerous and are often found near large spiral galaxies. For example, at least 10 dwarf elliptical galaxies, each with fewer than about a billion stars, belong to our Local Group (see Figure 1.1).

Elliptical galaxies usually contain very little dust or cool gas, although some have relatively small and cold gaseous disks rotating at their centers. However, some large elliptical galaxies contain substantial amounts of very hot gas that can emit X rays. The lack of cool gas in elliptical galaxies means that, like the halo of the Milky Way, they generally have little or no ongoing star formation. Elliptical galaxies therefore tend to look red or yellow because they do not have any of the hot, young, blue stars found in the disks of spiral galaxies.

Irregular Galaxies   Some galaxies do not fit either of the two major categories. The irregular class of galaxies is a miscellaneous class, encompassing galaxies that appear to be in disarray (**Figure 11.13**). These blobby star systems are usually white and dusty, like the disks of spirals. Their colors tell us that they contain young, massive stars.

Among nearby galaxies, only a small percentage of those as large as the Milky Way are irregular. Telescopic observations probing deeper into the universe show that distant galaxies are more likely to look irregular than nearby galaxies. Because the light of more distant galaxies has taken longer to reach us, these observations tell us that irregular galaxies were more common when the universe was younger [Section 12.3].

Hubble's Galaxy Classes   Edwin Hubble invented a system for classifying galaxies that organizes the galaxy types into a diagram shaped like a tuning fork (**Figure 11.14**). Elliptical galaxies appear on the "handle" at left, designated by the letter E and a number. The larger the number, the flatter the elliptical galaxy: An E0 galaxy is a sphere, and the numbers increase to the highly elongated type E7. The two "forks" show spiral galaxies, designated by the letter S for ordinary spirals and SB for barred spirals, followed by a lowercase a, b, or c: The bulge size decreases from a to c, while the amount of dusty gas increases. Lenticular galaxies are designated S0, and irregular galaxies are designated Irr.

Astronomers once hoped that the classification of galaxies might yield deep insights, just as the classification of stars did in the early 20th century. The Hubble classification scheme itself was suspected for a time to be an evolutionary sequence in which galaxies flattened and spread out as they aged.

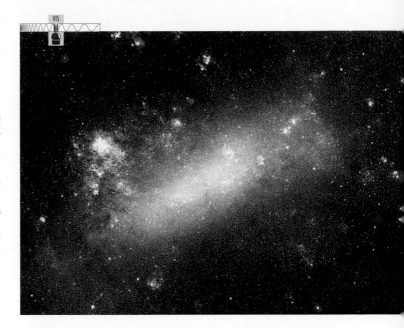

Figure 11.13 | The Large Magellanic Cloud, an irregular galaxy that is a small companion to the Milky Way. It is about 30,000 light-years across.

Figure 11.14 | This "tuning fork" diagram illustrates Hubble's galaxy classes.

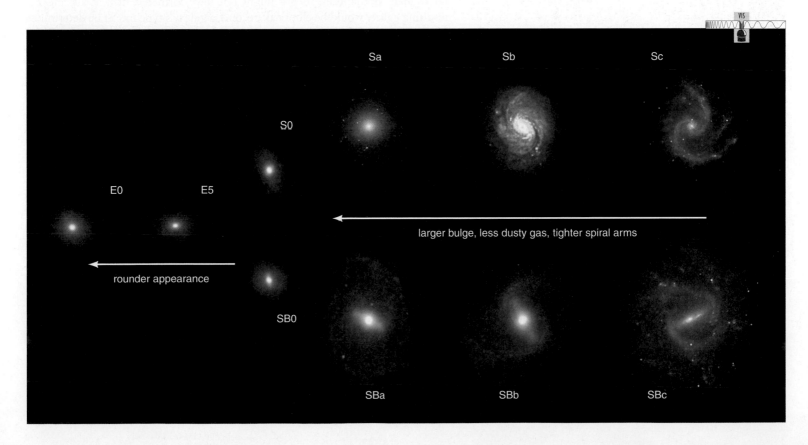

Unfortunately for astronomers, galaxies turn out to be far more complex than stars, and classification schemes like this one have not led to easy insights about their nature.

Why do galaxies differ?

The distinct differences among spiral, elliptical, and irregular galaxies tell us that their life stories must be quite different. We would like to tell the life story of each different type of galaxy from beginning to end as completely as we told the life stories of stars, but too many aspects of the formation and development of galaxies, or **galaxy evolution,** remain mysterious. Because so many important questions remain unanswered, galaxy evolution is one of the most active research areas in astronomy.

Our best models for galaxy formation suggest that all galaxies began their lives in the same basic way as our Milky Way, with gravity pulling matter into a protogalactic cloud that contracted and began to form stars. How, then, did we end up with the different types we see today? Our models suggest two general possibilities: (1) Galaxies may have ended up looking different because they began with slightly different conditions in their protogalactic clouds, or (2) galaxies may have begun their lives similarly but later changed because of interactions with other galaxies. To help differentiate the role of initial conditions in the protogalactic cloud from the role of subsequent interactions, we will explore a single key question: Why do spiral galaxies have gas-rich disks, while elliptical galaxies do not?

Two plausible explanations for the differences between spiral galaxies and elliptical galaxies trace a galaxy's type back to the protogalactic cloud from which it formed. First, a galaxy's type might be determined by the spin of the protogalactic cloud from which it formed. If the original cloud had a significant amount of angular momentum, it would have rotated quickly as it collapsed. The galaxy it produced would therefore have tended to form a disk, and the resulting galaxy would be a spiral galaxy. If the protogalactic cloud had little or no angular momentum, its gas might not have formed a disk at all, and the resulting galaxy would be elliptical. Second, a galaxy's type might be determined by the density of the protogalactic cloud from which it formed. A protogalactic cloud with relatively high gas density would have radiated energy more effectively and cooled more quickly, thereby allowing more rapid star formation. If the star formation proceeded fast enough, all the gas could have been turned into stars before any of it had time to settle into a disk. The resulting galaxy would therefore lack a disk, making it an elliptical galaxy. In contrast, a lower-density cloud would have formed stars more slowly, leaving plenty of gas to form the disk of a spiral galaxy.

These two scenarios describe important parts of the overall story. However, they ignore one key fact: Galaxies rarely evolve in perfect isolation. Think back to our scale-model solar system in Chapter 1. With a scale on which the Sun was the size of a grapefruit, the nearest star was like another grapefruit a few thousand kilometers away. Because the average distances between stars are so huge compared to the sizes of stars, collisions between stars are extremely rare. However, if we rescale the universe so that our *galaxy* is the size of a grapefruit, the Andromeda Galaxy is like another grapefruit only about 3 meters away, and a few smaller galaxies lie considerably closer. The average distances between galaxies are not much larger than the sizes of galaxies, so collisions between galaxies are inevitable.

Collisions between galaxies are spectacular events that unfold over hundreds of millions of years (**Figure 11.15**). During our short lifetimes, we can view only a snapshot of a collision in progress, distorting the shapes of the colliding galaxies. We can learn much more about galactic collisions with the aid of computer simulations that allow us to "watch" collisions that in nature

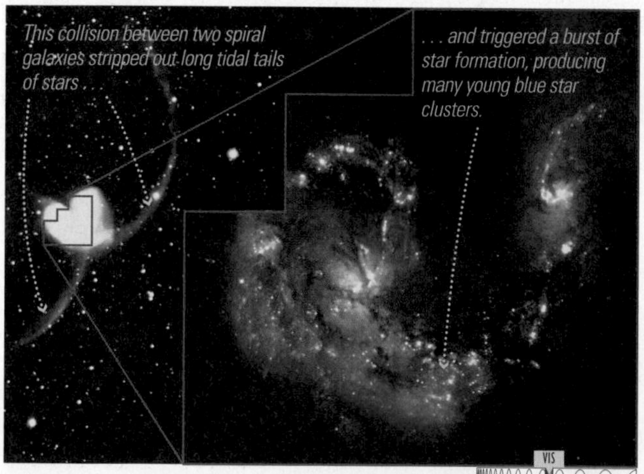

This collision between two spiral galaxies stripped out long tidal tails of stars . . .

. . . and triggered a burst of star formation, producing many young blue star clusters.

VIS

Figure 11.15 | A pair of colliding spiral galaxies known as the Antennae (NGC 4038/4039). The image taken from the ground (left) reveals their vast tidal tails, while the close-up from the Hubble Space Telescope shows the burst of star formation at the center of the collision.

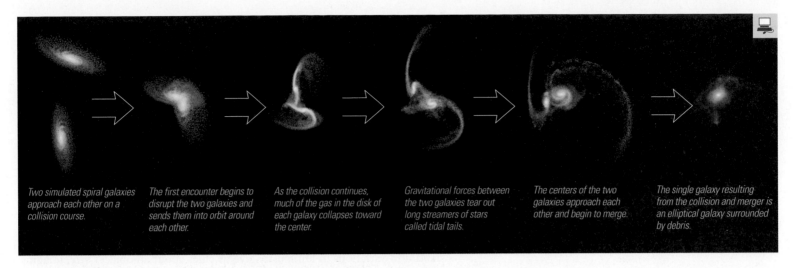

Two simulated spiral galaxies approach each other on a collision course.

The first encounter begins to disrupt the two galaxies and sends them into orbit around each other.

As the collision continues, much of the gas in the disk of each galaxy collapses toward the center.

Gravitational forces between the two galaxies tear out long streamers of stars called tidal tails.

The centers of the two galaxies approach each other and begin to merge.

The single galaxy resulting from the collision and merger is an elliptical galaxy surrounded by debris.

take hundreds of millions of years to unfold. These computer models show that a collision between two spiral galaxies can create an elliptical galaxy (**Figure 11.16**). Tremendous tidal forces between the colliding galaxies tear apart the two disks, randomizing the orbits of their stars. Meanwhile, a large fraction of their gas sinks to the center of the collision and rapidly forms new stars. Supernovae and stellar winds eventually blow away the rest of the gas. When the cataclysm finally settles down, the merger of the two spirals has produced a single elliptical galaxy. Little gas is left for a disk, and the orbits of the stars have random orientations.

Observations of galaxies in clusters support the idea that at least some elliptical galaxies result from collisions and subsequent mergers. Elliptical galaxies dominate the galaxy populations at the cores of dense clusters of galaxies, where collisions should be most frequent. This fact may indicate that any spirals once present became ellipticals through collisions.

Birth conditions and subsequent interactions probably both play important roles in galaxy evolution, but we are not yet certain which is more influential. Nevertheless, when we consider both kinds of mechanisms together, they do seem to account for the basic differences between galaxy types. The formation scenarios explain why the vast majority of galaxies are either spiral or elliptical in shape. The interaction scenarios explain why ellipticals are more common in clusters while spirals are more common outside of clusters. Even the relatively small fraction of galaxies that are irregular may be explained by these ideas: Some irregulars probably are galaxies undergoing a disruptive interaction. However, the overall picture of galaxy evolution remains incomplete. Astronomers hope that a coming generation of larger and more powerful telescopes will provide the remaining pieces of this puzzle.

✳ THE PROCESS OF SCIENCE IN ACTION

11.3 Solving the Mystery of Quasars

So far, our discussion of galaxies has focused largely on what stars do, but some of the most amazing action within galaxies has a different source. A small percentage of the universe's galaxies have extremely bright centers, known as **active galactic nuclei,** that produce light with a spectrum very different from that of starlight. The most luminous active galactic nuclei are known as **quasars,** and in some cases these objects give off more light than 1000 galaxies the size of the Milky Way (**Figure 11.17**). What could possibly be the source of such incredible luminosity? The way in which scientists have answered this question is an excellent example of the process of science in action.

Figure 11.16 | Several stages in a supercomputer simulation of a collision between two spiral galaxies that results in an elliptical galaxy. At least some of the elliptical galaxies in the present-day universe formed in this way. The whole sequence spans about 1.5 billion years.

Figure 11.17 | This Hubble Space Telescope photo shows a quasar named 3C 273, the first quasar to be discovered. Its luminosity is more than 1 trillion times that of the Sun. (The photo shows a region of space 275,000 light-years across at the distance of this quasar.)

What is the energy source for quasars?

The incredible luminosities of quasars became evident soon after they were first discovered in the early 1960s. Standard techniques of distance measurement [Section 12.1] indicated that quasars are generally billions of light-years from Earth, and with these distances the inverse square law for light [Section 8.2] gave luminosities for many quasars that were trillions of times the Sun's luminosity. These results were so astonishing that some scientists feared the distances of quasars had somehow been grossly overestimated. Indeed, for a while there was great debate over the basic nature of quasars, but observations with more powerful telescopes eventually verified their distances and proved that quasars are the bright centers of very distant galaxies. But how could the central region of a galaxy produce so much light?

Observations of variability in quasar luminosities further deepened the mystery, because they indicated that quasars are remarkably compact. To understand how variations in luminosity give us clues about an object's size, imagine that you are a master of the universe and you want to signal one of your fellow masters a billion light-years away. A quasar would make an excellent signal beacon because it is so bright. However, suppose the smallest quasar you can find is 1 light-year across. Each time you flash it on, the photons from the front end of the source reach your fellow master a full year before the photons from the back end. If you flash it on and off more than once a year, your signal will be smeared out. Similarly, if you find a source that is 1 light-day across, you can transmit signals that flash on and off no more than once a day. However, observations show that some quasars are even smaller than this. Occasionally, the luminosity of a quasar is observed to vary within a matter of hours, indicating that the light source must be less than a few light-hours across. In other words, *the incredible luminosities of quasars are being generated in a volume of space not much bigger than our solar system.*

The Black Hole Hypothesis The astonishing luminosities of quasars stimulated a lot of speculation about their energy source, but one hypothesis eventually emerged as the favorite; it proposed that quasars are fueled by matter falling into **supermassive black holes.** The idea that the huge energy outputs of active galactic nuclei can be traced to supermassive black holes is much like the idea we used earlier to explain the emission from X-ray binary star systems [Section 10.1]. The gravitational potential energy of matter falling toward a black hole is converted into kinetic energy, and collisions between infalling particles convert kinetic energy into thermal energy. The resulting heat causes this matter to emit the intense radiation we observe. As in X-ray binaries, we expect that the infalling matter swirls through an accretion disk before it disappears within the event horizon of the black hole.

Explaining the extreme luminosities of quasars was the main motivation for the supermassive black hole model. Matter falling into a black hole can generate awesome amounts of energy. As a chunk of matter falls to the event horizon of a black hole, as much as 10–40% of its mass-energy ($E = mc^2$) can be converted into thermal energy and ultimately to radiation. Accretion by black holes can therefore produce light far more efficiently than nuclear fusion, which converts less than 1% of mass-energy into photons. This mechanism is so efficient that it can account for the enormous luminosity of a quasar if an amount of mass slightly greater than that of the Sun passes through the accretion disk and falls into the black hole each year.

The supermassive black hole model also accounts for the small sizes implied by quasar variability. A black hole accreting a solar mass of matter per year could eventually attain a mass of a billion solar masses if accretion continued at that rate for a billion years. But the event horizon of such a massive black hole, with a radius of about 3 light-hours, would still fit within the orbit of Neptune.

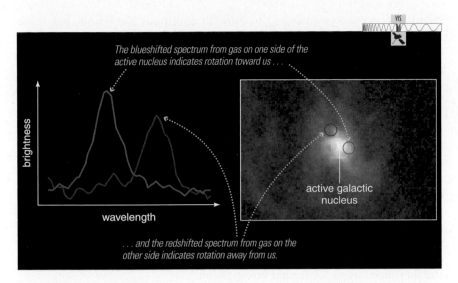

The blueshifted spectrum from gas on one side of the active nucleus indicates rotation toward us . . .

brightness

wavelength

active galactic nucleus

. . . and the redshifted spectrum from gas on the other side indicates rotation away from us.

Figure 11.18 | This Hubble Space Telescope photo shows gas near the center of the galaxy M87, and the graph shows Doppler shifts of spectra from gas 60 light-years from the center on opposite sides (the circled regions in the photo). These Doppler shifts tell us that gas is orbiting around the galactic center, and precise measurements tell us that its orbital speed is about 800 km/s. Using the orbital speed and Newton's laws, we calculate that the central object must have a mass 2 to 3 billion times that of the Sun.

Mounting Evidence Testing the hypothesis that supermassive black holes are the energy sources for quasars and other active galactic nuclei has been tricky. As we discussed in Chapter 10, black holes themselves do not emit any light, so we need to infer their existence from the ways in which they alter their surroundings. In the vicinity of a black hole, matter should be orbiting at high speed around something invisible. We have already seen such evidence for a supermassive black hole at the center of our own Milky Way [Section 10.3], but what about the centers of other galaxies?

The great distances of quasars make the problem of detecting black holes within them particularly difficult, so most of our searches for supermassive black holes have focused on nearer, less luminous active galactic nuclei. One prominent example is the giant elliptical galaxy M87. Spectroscopic observations show blueshifted emission lines on one side of its active galactic nucleus and redshifted emission lines on the other (**Figure 11.18**). This pattern of Doppler shifts is the characteristic signature of orbiting gas: On one side of the orbit, the gas is coming toward us and hence is blueshifted, while on the other side, it is moving away from us and is redshifted. The magnitude of these Doppler shifts shows that gas located up to 60 light-years from the nucleus is orbiting something invisible at a speed of hundreds of kilometers per second. This high-speed orbital motion indicates that the central object in M87 is probably a black hole with a mass 2 to 3 billion times that of our Sun. We may never be 100% certain that this object is a giant black hole. The best we can do is rule out all other possibilities. However, a supermassive black hole is the only thing we know of that could be so massive while remaining invisible.

Even galaxies whose nuclei are not currently active show similar evidence for supermassive black holes, and their mass measurements follow a very interesting pattern: The mass of the black hole at the center of a galaxy appears to be closely related to the properties of the galaxy's bulge. Detailed studies of the orbital speeds of stars and gas clouds in the centers of nearby galaxies show that the mass of the central black hole is typically about 1/500 of the mass of the galaxy's bulge (**Figure 11.19**). Because this relationship holds for galaxies with a wide range of properties, from small spiral galaxies to giant elliptical galaxies, we conclude that the growth of a central black hole must be closely linked with the process of galaxy formation.

Unfortunately, we do not yet understand the link between supermassive black holes and galaxy formation. Some scientists have suggested that the black holes formed first out of gas at the center of a protogalactic cloud and that their energy output regulated the growth of the galaxy around them. Other scientists have suggested that clusters of neutron stars resulting from extremely dense bursts of star formation at the centers of young galaxies

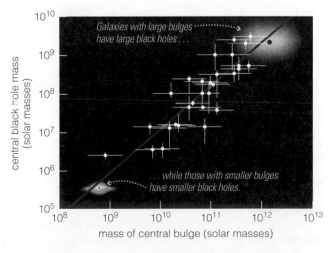

Galaxies with large bulges have large black holes . . .

central black hole mass (solar masses)

. . . while those with smaller bulges have smaller black holes.

mass of central bulge (solar masses)

Figure 11.19 | The relationship between the mass of a galaxy's bulge and the mass of its supermassive black hole.

might have somehow coalesced to form an enormous black hole, but these speculations are still unverified. Once such a black hole forms, it can continue to grow and supply a quasar's energy as it accretes gas from the galaxy around it. However, the fact that quasars are rare indicates that galactic nuclei spend only a small percentage of their time in such an active state.

The Bottom Line   Taken as a whole, the evidence to date strongly favors the supermassive black hole hypothesis for powering quasars. It seems that many galaxies harbor supermassive black holes at their centers, and in some cases those black holes are powering active galactic nuclei nearly as luminous as quasars. Because the behavior of these lower-luminosity active galactic nuclei is very similar to that of quasars, scientists are confident that quasars are also powered in the same way. However, the bulk of our evidence for supermassive black holes comes from relatively nearby galaxies. Quasars themselves are so rare and distant that we have not yet been able to definitively detect a black hole within one and measure its mass. These amazing objects therefore remain prime targets for further investigation.

Summary of Key Concepts

 11.1   ## Our Galaxy: The Milky Way

What does our galaxy look like?

The Milky Way Galaxy consists of a thin **disk** about 100,000 light-years in diameter with a central **bulge** and a spherical region called the **halo** that surrounds the entire disk. The disk contains most of the gas and dust of the **interstellar medium,** in which the **star–gas–star cycle** continually generates new stars and increases the heavy-element content of the galaxy. Active star-forming regions, marked by the presence of hot, massive stars and **ionization nebulae,** are found primarily in **spiral arms.**

How did the Milky Way form?

Our galaxy probably began as a huge blob of gas called a **protogalactic cloud.** Gravity caused the cloud to shrink in size, and conservation of angular momentum caused the gas to form the spinning disk of the galaxy. Stars in the halo formed before the gas finished collapsing into the disk.

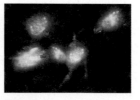

 11.2   ## Galaxies Beyond the Milky Way

What are the major types of galaxies?

Spiral galaxies have prominent disks and spiral arms, and they tend to collect in groups that contain up to several dozen galaxies. **Elliptical galaxies** are rounder and redder than spiral galaxies and contain less cool gas and dust. They are more common in large

clusters of galaxies than elsewhere in the universe. **Irregular galaxies** are neither disklike nor rounded in appearance.

Why do galaxies differ?

Differences between present-day galaxies probably arise both from conditions in their protogalactic clouds and from collisions with other galaxies. Slowly rotating or high-density protogalactic clouds may form elliptical rather than spiral galaxies. Ellipticals may also form through the collision and merger of two spiral galaxies.

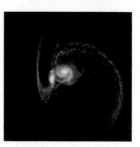

✳ THE PROCESS OF SCIENCE IN ACTION

11.3   ## Solving the Mystery of Quasars

What is the power source for quasars?

Some galaxies have bright centers known as **active galactic nuclei. Quasars** are the most luminous active galactic nuclei, with light outputs than can reach 1000 times the luminosity of the Milky Way, all coming from a region the size of our solar system. The most likely source for this energy is accretion of gas onto a **supermassive black hole.**

Investigations

Quick Quiz

Choose the best answer to each of the following; answers are in Appendix D. Explain your reasoning with one or more complete sentences.

1. Where are most of the Milky Way's globular clusters found? (a) in the disk (b) in the bulge (c) in the halo

2. Why do disk stars bob up and down as they orbit the galaxy? (a) because the gravitational pull of other disk stars pulls them toward the disk (b) because of friction with the interstellar medium (c) because the halo stars keep knocking them back into the disk

3. What is the typical hydrogen content of stars that are forming right now in the vicinity of the Sun? (a) 100% hydrogen (b) 75% hydrogen (c) 70% hydrogen

4. Where would you *least* expect to find an ionization nebula? (a) in the halo (b) in the disk (c) in a spiral arm

5. Where would you *most* expect to find an ionization nebula? (a) in the halo (b) in the bulge (c) in a spiral arm

6. Which kind of star is most likely to be part of the spheroidal population? (a) an O star (b) an A star (c) an M star

7. Which of these galaxies would you most likely find at the center of a large cluster of galaxies? (a) a large spiral galaxy (b) a large elliptical galaxy (c) a small irregular galaxy

8. In which of these galaxies would you be least likely to find an ionization nebula? (a) a large spiral galaxy (b) a large elliptical galaxy (c) a small irregular galaxy

9. About how many galaxies are there in a typical cluster of galaxies? (a) about 10 (b) a few dozen (c) a few hundred

10. A collision between two large spiral galaxies is likely to produce (a) a large elliptical galaxy. (b) a large spiral galaxy. (c) one large spiral galaxy and one large elliptical galaxy.

11. The luminosity of a quasar is generated in a region the size of (a) the Milky Way. (b) a star cluster. (c) the solar system.

12. The primary source of a quasar's energy is (a) chemical energy. (b) nuclear energy. (c) gravitational potential energy.

Short-Answer/Essay Questions

Explain all answers clearly, with complete sentences and proper essay structure if needed. An asterisk (*) designates a quantitative problem, for which you should show all your work.

13. *Research: Discovering the Milky Way.* Humans have been looking at the Milky Way since long before recorded history, but only in the past century did we verify the true shape of the galaxy and our location within it. Learn more about how conceptions of the Milky Way developed through history. What names did different cultures give the band of light they saw? What stories did they tell about it? How have ideas about the galaxy changed in the past few centuries? Write a two- to three-page summary of your findings.

14. *Future of the Milky Way.* Describe in one or two paragraphs how the Milky Way would look from the outside if you could watch it for the next 100 billion years. How would its appearance change during this time?

15. *High-Velocity Star.* The average speed of stars in the solar neighborhood relative to the Sun is about 20 km/s (i.e., the speed at which we see stars moving toward or away from the Sun—*not* their orbital speed around the galaxy). Suppose you discover a star in the solar neighborhood that is moving relative to the Sun at a much higher speed—say, 200 km/s. What kind of orbit does this star probably have around the Milky Way? In what part of the galaxy does it spend most of its time? Explain.

16. *A Nonspinning Galaxy.* How would the development of the Milky Way Galaxy have been different if its original protogalactic cloud had had no angular momentum? Describe how you think our galaxy would look today and explain your reasoning.

17. *Unenriched Stars.* Suppose you discovered a star made purely of hydrogen and helium. How old do you think it would be? Explain.

18. *Enrichment of Star Clusters.* The gravitational pull of an isolated globular cluster is rather weak—a single supernova can blow all the interstellar gas out of a globular cluster. How might this fact be related to observations indicating that stars ceased to form in globular clusters long ago? How might it be related to the fact that globular clusters are deficient in elements heavier than hydrogen and helium? Summarize your answers in one or two paragraphs.

19. *Supernovae in Other Galaxies.* In which type of galaxy would you be most likely to observe a massive star supernova: a giant elliptical galaxy or a large spiral galaxy? Explain.

20. *Hubble's Galaxy Types.* How would you classify the following galaxies using the system illustrated in Figure 11.14? Justify your answers.
    a. Galaxy NGC 4414 (Figure 11.10)
    b. Galaxy NGC 1300 (Figure 11.11)
    c. Galaxy M87 (Figure 11.12)

*21. *Counting Galaxies.* Estimate how many galaxies are shown in the photo on page 1. Explain the method you used to arrive at this estimate. This picture shows about 1/50,000,000 of the sky, so multiply your estimate by 50,000,000 to obtain an estimate of how many galaxies fill the entire sky.

12 Galaxy Distances and Hubble's Law

Learning Goals

12.1 Measuring Cosmic Distances

- How do we measure the distances to galaxies?
- What is Hubble's law?

12.2 The Implications of Hubble's Law

- In what sense is the universe expanding?
- How do distance measurements tell us the age of the universe?

✳ THE PROCESS OF SCIENCE IN ACTION

12.3 Observing Galaxy Evolution

- What do we see when we look back through time?

The photo above is an image of a galaxy 55 million light-years from Earth. We know the distance to this galaxy quite accurately because we have observed a white dwarf supernova within it—the bright dot on the lower left side of the picture. In this chapter, we'll explore how astronomers use observations of white dwarf supernovae and other objects to establish the distances of galaxies out to the edge of the observable universe. We'll then discuss how such observations led to the revelation that we live in an expanding universe and see how measurements of the expansion provide answers to some of the most fundamental questions we can ask: How old is the universe? How big is it? How is it changing with time?

12.1 Measuring Cosmic Distances

Much of what we know about the universe as a whole depends on our ability to measure the distances to galaxies. These measurements are quite challenging, and astronomers are continually working to refine and improve distance measurement techniques. In this section, we'll discuss the key techniques used in modern astronomy and then look at how Edwin Hubble used similar techniques to discover the expansion of the universe.

How do we measure the distances to galaxies?

Our measurements of astronomical distances depend on a chain of methods in which each step allows us to measure greater distances in the universe. Let's follow this chain, which builds link by link from our solar system to the outermost reaches of the observable universe.

Radar Ranging The most basic unit of distance in astronomy is the Sun–Earth distance, or *astronomical unit* (AU) [Section 1.1]. Astronomers today measure the AU with a technique called **radar ranging,** in which radio waves are transmitted from Earth and bounced off Venus. Because radio waves travel at the speed of light, the round-trip travel time for the radar signals tells us Venus's distance from Earth. We can then use Kepler's laws and a little geometry to calculate the length of an AU.

Parallax The next step in the distance chain depends on the effect known as *stellar parallax*, which we first encountered in Chapter 2. Recall that you can observe parallax of your finger by holding it at arm's length and alternately looking at it with first one eye closed and then the other. Your finger appears to shift because each eye sees it from a slightly different point of view. Astronomers see a similar effect when they observe a nearby star from two different points in Earth's orbit around the Sun (**Figure 12.1**). Because we are observing it from two different points of view, the nearby star appears to shift against the background of more distant stars. The maximum shift occurs if we observe the star from opposite sides of Earth's orbit, using observations 6 months apart.

We can calculate a star's distance if we know the precise amount of the star's shift due to parallax over the course of 6 months. This means measuring the star's *parallax angle* (*p*), which is equal to *half* the star's annual back-and-forth shift. Notice that this angle would be smaller if the star in Figure 12.1 were farther away, so more distant stars have smaller parallax angles.

All stars are so far away that they have very small parallax angles, which explains why the ancient Greeks were never able to measure parallax with their naked eyes [Section 2.3]. Even the nearest stars have parallax angles smaller than 1 arcsecond—well below the approximately 1 arcminute shift that a human eye can perceive. By definition, the distance to an object with a parallax angle of 1 arcsecond is 1 parsec (pc), which is equivalent to 3.26 light-years. You'll often hear astronomers state distances in parsecs, kiloparsecs (1000 parsecs), or megaparsecs (1 million parsecs), but in this book we'll stick to giving distances in light-years. As we look to increasingly distant stars, parallax angles quickly become too small to measure even with our best telescopes. Current technology allows us to measure parallax accurately only for stars within a few hundred light-years, a relatively small distance compared with the vast, 100,000-light-year-diameter of our Milky Way Galaxy.

Standard Candles Once we have measured distances to nearby stars through parallax, we can measure distances to other stars in the same way that we might estimate the distance to a street lamp at night. If the street lamp does

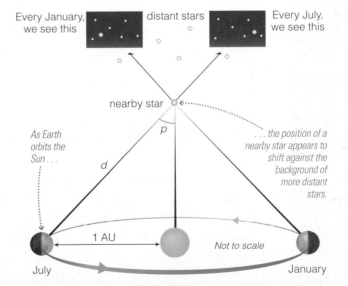

Every January, we see this   distant stars   Every July, we see this

nearby star

As Earth orbits the Sun . . .   *p*   . . . the position of a nearby star appears to shift against the background of more distant stars.

d

1 AU   *Not to scale*

July   January

Figure 12.1 | Parallax makes the apparent position of a nearby star shift back and forth with respect to distant stars over the course of each year. The angle *p*, called the *parallax angle,* represents half the total parallax shift each year. If we measure *p* in arcseconds, the distance *d* to the star is $1/p$ in units of parsecs and $3.26/p$ in units of light-years. The angle in this figure is greatly exaggerated: All stars have parallax angles of less than 1 arcsecond.

not look very bright, then it's probably far away. If it looks very bright, then it's probably quite close.

We can determine the lamp's distance more accurately if we can measure its apparent brightness. For example, suppose we see a distant street lamp and know that all street lamps of its type put out 1000 watts of light. If we then measure the street lamp's apparent brightness, we can calculate its distance by using the inverse square law for light [Section 8.2].

An object such as a street lamp, for which we are likely to know the true luminosity, represents what astronomers call a **standard candle** (see Tools of Science, p. 198). The term *standard candle* is meant to suggest a light source of a known, standard luminosity. Unlike light bulbs, however, astronomical objects do not come marked with wattage. An astronomical object can serve as a standard candle only if we have some way of knowing its true luminosity without first measuring its apparent brightness and distance. Fortunately, many astronomical objects meet this requirement. For example, any star that is a twin of our Sun—a main-sequence star with spectral type G2—should have about the same luminosity as the Sun. If we measure the apparent brightness of a Sun-like star, we can assume it has the same luminosity as the Sun and then use the inverse square law for light to estimate its distance.

Beyond the few hundred light-years for which we can measure distances by parallax, we use standard candles for most cosmic distance measurements. These measurements always have some uncertainty, because no astronomical object is a perfect standard candle. The challenge of measuring astronomical distances comes down to the challenge of finding the objects that make the best standard candles. The more confidently we know an object's true luminosity, the more certain we are of its distance.

Main-Sequence Fitting Although Sun-like stars are reasonably good standard candles, they are of limited use because Sun-like stars are too dim for us to detect beyond relatively short distances. To measure distances beyond 1000 light-years or so, we need brighter standard candles.

Brighter main-sequence stars are an obvious first choice, because all main-sequence stars of a particular spectral type have about the same luminosity [Section 8.2]. However, before we can use any main-sequence star as a standard candle, we must first have some way of knowing its true luminosity. We must therefore follow two steps to use bright main-sequence stars as standard candles:

1. We identify a star cluster that is close enough for us to determine its distance by parallax and plot its H-R diagram. Because we know the distances to the cluster's stars, we can use the inverse square law for light to establish their true luminosities from their apparent brightnesses.

2. We then look at stars in other clusters that are too far away for parallax measurements and measure their apparent brightnesses. If we assume that main-sequence stars in other clusters have the same luminosities as their counterparts in the nearby cluster, we can calculate their distances from the inverse square law for light.

Twentieth-century astronomers laid the groundwork for this technique by calibrating the luminosities on a standard H-R diagram. This calibration relied largely on a single, nearby star cluster—the Hyades Cluster in the constellation Taurus, whose distance is now known from its parallax. We can find the distances to other star clusters by comparing the apparent brightnesses of their main-sequence stars with those in the Hyades Cluster and assuming that all main-sequence stars of the same color have the same luminosity (**Figure 12.2**). This technique of determining distances by comparing main sequences in different star clusters is called **main-sequence fitting**.

The Hyades Cluster does not contain stars of every spectral type. Building a complete H-R diagram therefore required that astronomers use main-sequence

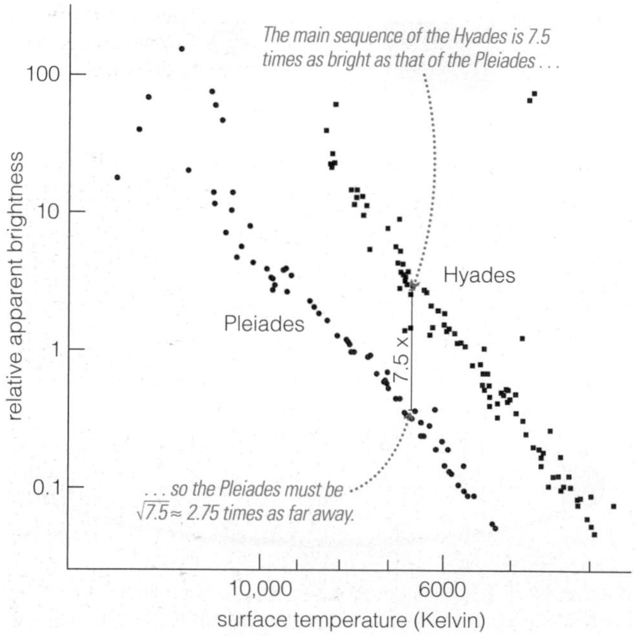

Figure 12.2 | To use the technique of main-sequence fitting, we compare the apparent brightness of the main sequence of a cluster of unknown distance (the Pleiades, in this case) to that of a cluster whose distance we already know (such as the Hyades, whose distance is known from parallax).

fitting to find the distances to many nearby star clusters until every spectral type was represented. Today, with the standard H-R diagram well established (see Figure 8.17), we can use main-sequence fitting to measure distances to any star cluster near enough for us to identify individual main-sequence stars.

Cepheid Variables Main-sequence fitting works well for measuring distances to star clusters throughout the Milky Way, but not for measuring distances to other galaxies. Most main-sequence stars are too faint to be seen in other galaxies, even with our most powerful telescopes. Instead, we need very bright stars to serve as standard candles for distance measurements beyond the Milky Way.

The most useful bright stars for measuring the distances to galaxies are called **Cepheid variable stars,** or **Cepheids** for short. These stars vary in brightness in our sky, alternately dimming and brightening with periods ranging from a few days to a few months. **Figure 12.3** shows brightness variations of a Cepheid with a period of about 50 days. Cepheids vary in luminosity because they have a peculiar problem in matching the amount of energy radiated from the surface with the amount welling up from the core. In a futile quest for steady balance, the upper layers of a Cepheid alternately expand and contract, causing the star's luminosity to rise and fall.

Larger Cepheids take longer to pulsate in and out in size, and because larger ones are also more luminous, we find that Cepheids obey a **period–luminosity relation:** The longer the period, the more luminous the star. **Figure 12.4** shows this relation and how we can use it to determine (within about 10%) a Cepheid's luminosity simply by measuring the time period over which its brightness varies. For example, a Cepheid variable whose brightness peaks every 30 days is effectively screaming out, "Hey, everybody, my luminosity is 10,000 times that of the Sun!" Once we measure a Cepheid's period, we know its luminosity and we can use the inverse square law for light to determine its distance.

Observations of Cepheids have been used to measure distances to galaxies up to 100 million light-years away. This distance may sound very large, but it is still quite small compared with the distances of the vast majority of galaxies in the universe. To go further, we use the distances determined with Cepheids to learn the luminosities of even brighter standard candles.

Distant Standard Candles Astronomers have discovered several techniques for estimating distances beyond those for which we can observe Cepheids. In recent years, the most valuable of these techniques has been the use of white dwarf supernovae as standard candles.

Recall that white dwarf supernovae are thought to be exploding white dwarf stars that have reached the 1.4-solar-mass limit [Section 10.1]. These supernovae should all have nearly the same luminosity, because they all come from stars of the same mass that explode in the same way. Although white dwarf supernovae are rare in any individual galaxy, several have been detected during the past century in galaxies within about 50 million light-years of the Milky Way, including the one in the photo that opens this chapter. Astronomers kept careful records of those events, so today we can determine the true luminosities of these supernovae by using Cepheids to measure the distances to the galaxies in which they occurred. These measurements confirm that the luminosities of all white dwarf supernovae are about the same.

Because white dwarf supernovae are so bright—about 10 billion solar luminosities at their peak—we can detect them even when they occur in galaxies billions of light-years away (**Figure 12.5**). Using white dwarf supernovae as standard candles therefore allows us to measure the distances of galaxies in the far reaches of the observable universe. Although the number of galaxies whose distances we can measure with this technique is relatively small, because white dwarf supernovae occur only once every few hundred years in a typical

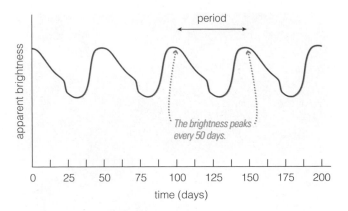

Figure 12.3 | This graph shows how the brightness of a Cepheid varies with time. The period is the time from one peak of brightness to the next.

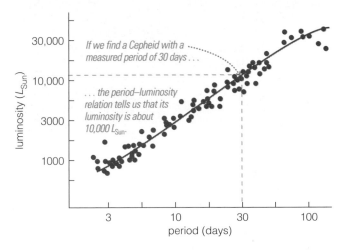

Figure 12.4 | Cepheid period–luminosity relation. The data show that all Cepheids of a particular period have about the same luminosity. Therefore, by measuring a Cepheid's period, we can determine its luminosity and hence its distance.

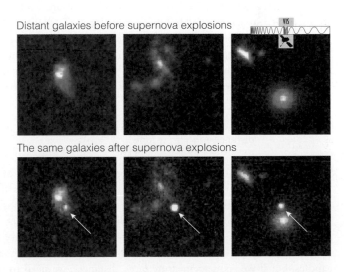

Figure 12.5 | White dwarf supernovae in galaxies approximately 10 billion light-years away. White arrows in the lower images indicate the supernovae, and the upper images show what these galaxies looked like without supernovae.

Figure 12.6 | Edwin Hubble at the Mount Wilson Observatory.

galaxy, these galaxies have enabled us to calibrate yet another technique. It is called *Hubble's law*, and it relies on the expansion of the universe.

What is Hubble's law?

The ability to measure distances to galaxies is the key to much of our modern understanding of the size and age of the universe. We can trace the beginning of this understanding back to discoveries made by Edwin Hubble. Let's explore how Hubble discovered the famous law that bears his name so that we can understand how this law has helped answer fundamental questions about the universe.

Hubble and the Andromeda Galaxy   Although astronomers had seen galaxies through telescopes for at least two centuries, no one at the start of the 1920s knew for certain whether these objects were merely clouds of gas within the Milky Way—and therefore that the Milky Way represented the entire universe—or distant and distinct galaxies. The opinions of astronomers were split on this issue, which became a subject of great debate. The problem was that neither side could prove its case, because the techniques available at the time could not distinguish objects within the Milky Way from those beyond it.

Edwin Hubble put the debate to rest in 1924. Using the new 100-inch telescope atop California's Mount Wilson (**Figure 12.6**)—the largest telescope in the world at that time—he identified Cepheid variables in the Andromeda Galaxy by comparing photographs of the galaxy taken days apart. He then used the Cepheid period–luminosity relation to estimate the galaxy's distance, proving once and for all that the Andromeda Galaxy lay far beyond the outer reaches of stars in the Milky Way.

This single scientific discovery dramatically changed our perspective. Rather than thinking the Milky Way was the entire universe, we suddenly

---

 # Tools of Science: Measuring Distances with Standard Candles

The standard-candle technique for distance measurement is astronomy's most important tool for measuring large distances in the universe. It is based on the inverse square law for light, which relates an object's apparent brightness to its distance and luminosity [Section 8.1]. Rearranging this law with a little algebra gives us a formula for calculating an object's distance from its apparent brightness and luminosity:

$$\text{distance} = \sqrt{\frac{\text{luminosity}}{4\pi \times (\text{apparent brightness})}}$$

If we can measure the apparent brightness of an object of known luminosity, we can use this formula to calculate the object's distance. However, this technique works only for objects whose luminosities we can determine without knowing their distances. We call such objects *standard candles*.

Many different kinds of astronomical objects can be used as standard candles, but if we want to measure large distances, we need to use very luminous objects that can be detected from far away. Cepheid variable stars are particularly useful standard candles because they can have luminosities as great as $30,000L_{Sun}$, making them clearly visible from distances as large as 100 million light-years.

Example: You observe a Cepheid variable star that varies in brightness with a period of 30 days and has an average apparent brightness of 1.0×10^{-18} watt/m². According to the period–luminosity relation for Cepheid variables, the star's luminosity is about $10,000L_{Sun}$. How far away is this star? (*Note:* The Sun's luminosity, $1L_{Sun}$, is 3.8×10^{26} watts.)

Solution: We are given both the star's apparent brightness and its luminosity, so we have all the information we need to calculate its distance using the modified inverse square law for light. However, before we can use this law, we must convert the star's luminosity to units of watts. Multiplying the Sun's luminosity in watts by 10,000 (10^4) gives a luminosity of 3.8×10^{30} watts for the Cepheid. Using this luminosity and the apparent brightness in the formula then gives

$$\text{distance} = \sqrt{\frac{3.8 \times 10^{30} \text{ watts}}{4\pi \times \left(1.0 \times 10^{-18} \frac{\text{watt}}{\text{m}^2}\right)}} \approx 5.5 \times 10^{23} \text{ m}$$

The distance to the star is about 5.5×10^{23} meters, or 58 million light-years (dividing by 9.5×10^{15} meters/light-year).

knew that it is just one among many galaxies in an enormous universe. The stage was set for Hubble to make an even greater discovery.

Distance and Redshift   Astronomers had known since the 1910s that the spectra of most spiral galaxies tend to be *redshifted* (**Figure 12.7**), indicating that they are moving away from us (see Tools of Science, p. 115). However, no one understood the true significance of these motions, because Hubble had not yet proved that the spiral galaxies were separate from the Milky Way.

Following his discovery of Cepheids in Andromeda, Hubble and his coworkers spent the next few years measuring the redshifts of galaxies and estimating their distances. Because even Cepheids were too dim to be seen in most of these galaxies, Hubble needed brighter standard candles for his distance estimates. One of his favorite techniques was to use the brightest object he could see in each galaxy as a standard candle, assuming that they all had about the same luminosity. These bright objects, which turned out to be star clusters, were not very good standard candles, but they were good enough for measuring approximate distances.

In 1929, Hubble announced his conclusion: The more distant a galaxy, the greater its redshift and hence the faster it moves away from us. As we will discuss in the next section, this relationship implies that galaxies all across the universe are moving away from one another. Edwin Hubble had discovered that the universe is expanding.

Hubble's Law   We express the idea that more distant galaxies move away from us faster with a simple formula known as **Hubble's law:**

$$v = H_0 \times d$$

where v stands for a galaxy's velocity away from us (sometimes called a *recession velocity*), d stands for its distance, and H_0 (pronounced "H-naught") is a number called **Hubble's constant.** We usually write Hubble's law in this form to express the idea that galaxies' speeds depend on their distances. However, astronomers more often use the law in reverse—measuring a galaxy's speed from its redshift and then using Hubble's law to estimate its distance.

Today, Hubble's law is our most useful technique for determining distances to galaxies that are very far away. Nevertheless, it's important to keep two practical limitations of the law in mind:

1. Galaxies do not obey Hubble's law perfectly. Hubble's law would give an exact distance only for a galaxy whose speed was determined solely by the expansion of the universe. In reality, nearly all galaxies experience gravitational tugs from other galaxies, and these tugs alter their speeds from the values predicted by Hubble's law.

2. Even when galaxies obey Hubble's law well, the distances we find with it are only as accurate as our best measurement of Hubble's constant.

The first problem is most serious for nearby galaxies. Within the Local Group, for example, Hubble's law does not work at all: The galaxies in the Local Group are gravitationally bound together with the Milky Way and therefore are *not* moving away from us in accord with Hubble's law. However, Hubble's law works fairly well for more distant galaxies. The recession speeds of galaxies at large distances are so great that any motions caused by the gravitational tugs of neighboring galaxies are small in comparison.

The second problem means that, even for distant galaxies, we can know only *relative* distances until we pin down the true value of H_0. For example, Hubble's law tells us that a galaxy moving away from us at 20,000 km/s is twice as far away as one moving at 10,000 km/s, but we can determine the actual distances of the two galaxies only if we know H_0.

The Hubble Space Telescope has helped us obtain an accurate value of H_0. Astronomers used the telescope to discover Cepheid variables in galaxies out to

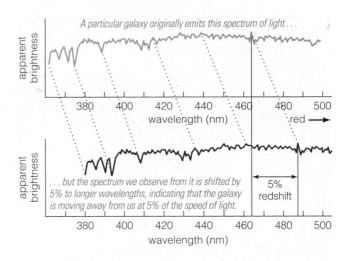

Figure 12.7 | Redshifted galaxy spectrum.

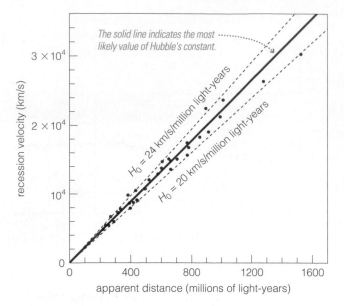

Figure 12.8 | White dwarf supernovae can be used as standard candles to establish Hubble's law out to very large distances. The points on this figure show the apparent distances of white dwarf supernovae and the recession velocities of the galaxies in which they exploded. The fact that these points all fall close to a straight line demonstrates that these supernovae are good standard candles.

about 60 million light-years and then used those distances to determine the luminosities of distant standard candles such as white dwarf supernovae. Plotting galactic distances measured with those distant standard candles against the velocities indicated by their redshifts has pinned down the value of H_0 to somewhere between 20 and 24 *kilometers per second per million light-years* (**Figure 12.8**). In other words, a galaxy's speed away from us is between 20 and 24 kilometers per second for each million light-years of distance away from us. For example, with this range of values for Hubble's constant, Hubble's law predicts that a galaxy located 100 million light-years away would be moving away from us at a speed between 2000 and 2400 kilometers per second.

Distance Chain Summary   **Figure 12.9** summarizes the chain of measurements that allows us to determine ever greater distances. With each link in the distance chain, however, uncertainties become somewhat greater: We know the Earth–Sun distance at the beginning of the chain extremely accurately, but distances to the farthest reaches of the observable universe remain uncertain by about 10%. The six interlocking techniques shown in the figure are as follows:

- *Radar ranging.* We measure the Earth–Sun distance by bouncing radio waves off planets and using some geometry.

- *Parallax.* We measure the distances to nearby stars by observing how their positions change, relative to the background stars, as Earth orbits the Sun. These distances rely on our knowledge of the Earth–Sun distance, determined with radar ranging.

- *Main-sequence fitting.* We know the distance to the Hyades star cluster in the Milky Way Galaxy through parallax. Comparing the apparent brightnesses of its main-sequence stars to those of main-sequence stars in other clusters gives us the distances to other star clusters in our galaxy.

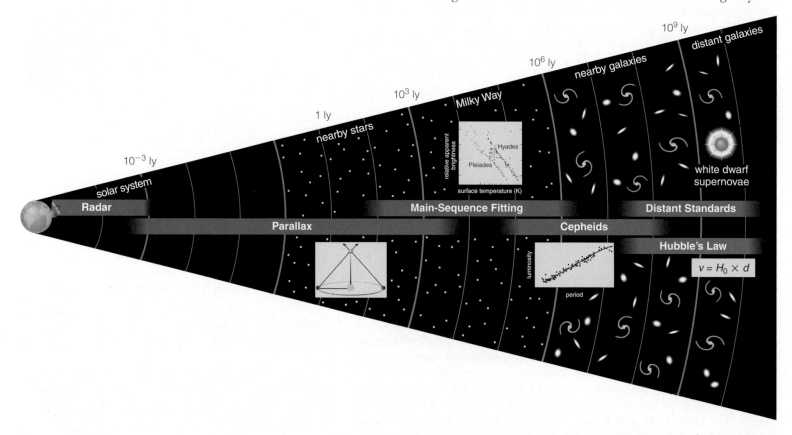

Figure 12.9 | Measurement of cosmic distances relies on a chain of interlocking techniques. The chain begins with radar ranging to determine distances within our solar system, and proceeds through parallax and standard-candle techniques. The use of these techniques allows us to calibrate Hubble's law, which we can then use to estimate distances to galaxies across the observable universe.

- *Cepheid variables.* By studying Cepheids in star clusters with distances measured by main-sequence fitting, we learn the precise period–luminosity relation for Cepheids. When we find a Cepheid in a more distant star cluster or galaxy, we can determine its luminosity by measuring the period between its peaks in brightness and then use this luminosity to determine its distance.

- *Distant standards.* By measuring distances to relatively nearby galaxies with Cepheids, we learn the true luminosities of white dwarf supernovae and other distant standard candles, enabling us to measure great distances throughout the universe.

- *Hubble's law.* Distances measured with white dwarf supernovae and other distant standards allow us to measure Hubble's constant, H_0. Once we know H_0, we can use Hubble's law to determine a galaxy's distance from its redshift.

12.2 The Implications of Hubble's Law

Hubble's law is a powerful tool for understanding the universe. Not only does it tell us that the universe is expanding and give us a way to measure galactic distances, but it also has enormous implications: Tracing the motions of galaxies back through time indicates that all the matter in the observable universe started very close together and that the entire universe we observe today came into being at a single moment, approximately 14 billion years ago. In this section, we'll explore how these implications arise from Hubble's discovery.

In what sense is the universe expanding?

One way to visualize the expansion of the universe is to imagine making a raisin cake in which the distance between adjacent raisins is 1 centimeter. You place the cake in the oven, where it expands as it bakes. After 1 hour, you remove the cake, which has expanded so that the distance between adjacent raisins has increased to 3 centimeters (**Figure 12.10**). The expansion of the cake seems fairly obvious. But what would you see if you lived *in* the cake, as we live in the universe?

Pick any raisin, call it the Local Raisin, and identify it in the pictures of the cake both before and after baking. Figure 12.10 shows one possible choice for the Local Raisin, with three nearby raisins labeled. The accompanying table summarizes what you would see if you lived within the Local Raisin. Notice, for example, that Raisin 1 starts out at a distance of 1 centimeter before baking and ends up at a distance of 3 centimeters after baking, which means it moves a distance of 2 centimeters away from the Local Raisin during the hour of baking. Hence, its speed as seen from the Local Raisin is 2 centimeters per hour. Raisin 2 moves from a distance of 2 centimeters before baking to a distance of 6 centimeters after baking, which means it moves a distance of 4 centimeters away from the Local Raisin during the hour. Hence, its speed is 4 centimeters per hour, or twice as fast as the speed of Raisin 1. The fact that the cake is expanding means that all raisins are moving away from the Local Raisin, with more distant raisins moving away faster.

Hubble's discovery that galaxies are moving in much the same way as the raisins in the cake, with most moving away from us and more distant ones moving away faster, implies that the universe in which we live is expanding much like the raisin cake. If you now imagine the Local Raisin as representing our Local Group of galaxies and the other raisins as representing more distant galaxies or clusters of galaxies, you have a basic picture of the expansion of the

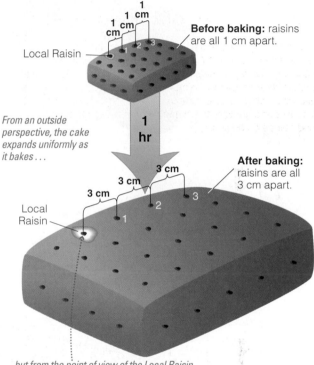

From an outside perspective, the cake expands uniformly as it bakes . . .

. . . but from the point of view of the Local Raisin, all other raisins move farther away during baking, with more distant raisins moving faster.

Distances and Speeds as Seen from the Local Raisin

Raisin Number	Distance Before Baking	Distance After Baking (1 hour later)	Speed
1	1 cm	3 cm	2 cm/hr
2	2 cm	6 cm	4 cm/hr
3	3 cm	9 cm	6 cm/hr

Figure 12.10 | An expanding raisin cake offers an analogy to the expanding universe. Someone living in one of the raisins inside the cake could figure out that the cake was expanding by noticing that all the other raisins were moving away, with more distant raisins moving away faster. In the same way, we know that we live in an expanding universe because all galaxies outside our Local Group are moving away from us, with more distant ones moving faster.

universe. Like the expanding batter between the raisins in the cake, *space itself* is growing between galaxies. More distant galaxies move away from us faster because they are carried along with this expansion like the raisins in the expanding cake.

However, there's an important difference between the raisin cake and our universe: A cake has a center and edges, but we do not think the same is true of the entire universe. To the best of our knowledge, the universe is not expanding *into* anything. As far as we can tell, there is no edge to the distribution of galaxies in the universe. On very large scales, the distribution of galaxies appears to be relatively smooth, meaning that the overall appearance of the universe around you would look more or less the same no matter where you were located. The idea that the matter in the universe is evenly distributed, without a center or an edge, is often called the **Cosmological Principle.** Although we cannot prove it to be true, it is consistent with all our observations of the universe.

So how can the universe be expanding if it's not expanding *into* anything? A better analogy than the raisin cake would be something that can expand but has no center and no edges. We cannot use a three-dimensional object for this purpose, since all such objects have a center and edges. However, the two-dimensional *surface* of a balloon can fit the bill, as can an infinite surface such as a sheet of rubber that extends to infinity in all directions.

Because it's hard to visualize infinity, we'll use the surface of a balloon in our analogy to the expanding universe (**Figure 12.11**). Note that this analogy means we must leave out a dimension: The balloon's two-dimensional *surface* represents all three dimensions of space. The surface of the balloon therefore represents the entire universe, and the spaces inside and outside the balloon have no meaning in this analogy. Aside from the reduced number of dimensions, the analogy works well because the balloon's spherical surface has no center and no edges, just as no city is the center of Earth's surface and no edges exist where you could walk or sail off Earth. Now, remember that although the universe as a whole is expanding, individual galaxies and galaxy clusters do not expand, because gravity holds them together. We can therefore represent galaxies (or clusters of galaxies) with plastic dots attached to the balloon, and we can make our model universe expand by inflating the balloon. The dots move apart as the surface of the balloon expands, but they themselves do not grow in size.

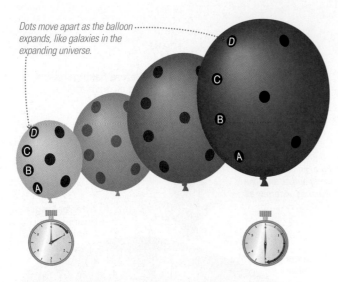

Dots move apart as the balloon expands, like galaxies in the expanding universe.

Figure 12.11 | The expansion of a balloon's surface shows how a finite universe can expand without having a center or edges. As the balloon expands, dots on it move apart in the same way that galaxies move apart in our expanding universe.

> **See it for yourself**
> Make a *one*-dimensional model of the expanding universe with a rubber band and some paper clips. Cut the rubber band so that you can stretch it out along a line, then attach four paper clips along it. Pin down the ends of the rubber band and measure the distance from one paper clip to each of the others. Then unpin one of the ends, stretch the rubber band more, pin it down again, and remeasure the distances. How much have they changed? How do your measurements illustrate Hubble's law?

How do distance measurements tell us the age of the universe?

We are now ready to discuss how Hubble's law leads us to an age estimate for the universe. Imagine that some miniature scientists are living on dot B on the balloon in Figure 12.11. Suppose that, 3 seconds after the balloon begins to expand, they measure the following:

- Dot A is 3 centimeters away and moving at 1 cm/s.

- Dot C is 3 centimeters away and moving at 1 cm/s.

- Dot D is 6 centimeters away and moving at 2 cm/s.

They could summarize these observations as follows: *Every dot is moving away from us with a speed that is 1 cm/s for each 3 centimeters of distance.* Because the expansion of the balloon is uniform, scientists living on any other dot would come to the same conclusion. Each scientist living on the balloon would determine that the following formula relates the distances and velocities of other dots on the balloon:

$$v = \left(\frac{1}{3 \text{ s}}\right) \times d$$

where v and d are the velocity and distance of any dot, respectively.

Think about it Confirm that this formula gives the correct values for the speeds of dots C and D, as seen from dot B, 3 seconds after the balloon begins expanding. How fast would a dot located 9 centimeters from dot B move, according to the scientists on dot B?

If the miniature scientists think of their balloon as a bubble, they might call the number relating distance to velocity—the term $\frac{1}{3 \text{ s}}$ in the preceding formula—the "bubble constant." An especially insightful miniature scientist might flip over the "bubble constant" and find that it is exactly equal to the time elapsed since the balloon started expanding. That is, the "bubble constant" $\frac{1}{3 \text{ s}}$ tells them that the balloon has been expanding for 3 seconds. Perhaps you see where we are heading.

Just as the inverse of the "bubble constant" tells the miniature scientists that their balloon has been expanding for 3 seconds, the inverse of the Hubble constant, or $1/H_0$, tells us something about how long our universe has been expanding. The "bubble constant" for the balloon depends on when it is measured, but it is always equal to 1 divided by the time since the balloon started expanding. Similarly, the Hubble constant changes with time, but it stays roughly equal to 1 divided by the age of the universe. We call it a *constant* because it is the same at all locations in the universe and because its value does not change noticeably on the time scale of human civilization.

Current estimates based on the value of Hubble's constant put the age of the universe between about 12 and 15 billion years. To derive a more precise value for the universe's age, we need to know whether the expansion has been speeding up or slowing down over time, a question we'll examine more closely in Chapter 14. If the gravitational pull of each galaxy on every other galaxy has significantly slowed the expansion rate, then the universe's age is somewhat less than $1/H_0$. If some mysterious force has accelerated the expansion rate, then the universe's age is somewhat more than $1/H_0$. The best available evidence as of 2009 suggests that the universe is about 14 billion years old.

Lookback Time The expansion of the universe leads to a complication in discussing galaxy distances that we have ignored up to this point. To understand the complication, imagine observing a supernova in a distant galaxy. Suppose the supernova occurred 400 million years ago but we are only just now seeing it. The supernova's light must have traveled 400 million light-years to reach Earth, but this simple statement about how far the light has traveled does not translate easily into a distance for the galaxy in which the supernova occurred. The problem is that the universe is expanding, making the distance between Earth and the supernova greater today than it was at the time of the supernova event. If we simply say that the galaxy is 400 million light-years away, it's not clear whether we mean its distance now, its distance at the time of the supernova, or something in between.

Because distances between galaxies are always changing, it is easier to speak about faraway objects in terms of how much time their light takes to reach us—400 million years in the case of the supernova. We call this the **lookback time** to the supernova. A distant object's lookback time is the difference between the current age of the universe and the age of the universe when

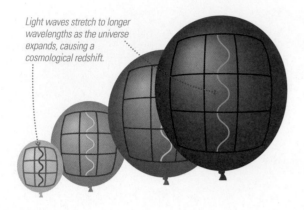

Light waves stretch to longer wavelengths as the universe expands, causing a cosmological redshift.

Figure 12.12 | As the universe expands, photon wavelengths stretch like the wavy lines on an expanding balloon.

Common Misconceptions

BEYOND THE HORIZON

Perhaps you're thinking there must be something beyond the cosmological horizon. After all, we can see farther and farther with each passing second. The new matter we see had to come from somewhere, didn't it? The problem with this reasoning is that the cosmological horizon, unlike a horizon on Earth, is a boundary in time, not in space.

At any one moment, in whatever direction we look, the cosmological horizon lies at the beginning of time and encompasses a certain volume of the universe. At the next moment, the cosmological horizon still lies at the beginning of time, but it encompasses a slightly larger volume. When we peer into the distant universe, we are looking back in both space and time. We cannot look "past the horizon" because we cannot look back to a time before the universe began.

the light left the object. Unlike a statement about distance, a statement about lookback time is unambiguous: If the lookback time is 400 million years, the light traveled through space for a period of 400 million years to reach us.

Cosmological Redshift   An object's lookback time is directly related to its redshift. Recall that redshifts tell us how fast objects are moving away from us. In the context of an expanding universe, redshifts have an additional, more fundamental interpretation. Let's return again to the universe on the balloon. Suppose you draw wavy lines on the balloon's surface to represent light waves. As the balloon inflates, these wavy lines stretch out, and their wavelengths increase (**Figure 12.12**). This stretching closely resembles what happens to photons in an expanding universe. The expansion of the universe stretches out all the photons within it, shifting them to longer, redder wavelengths. We call this effect a **cosmological redshift.**

In a sense, we have a choice when we interpret the redshift of a distant galaxy: We can think of the redshift as being caused either by the Doppler effect, as the galaxy moves away from us, or by a photon-stretching, cosmological redshift. However, as we look to very distant galaxies, the ambiguity in the meaning of *distance* also makes it difficult to specify precisely what we mean by a galaxy's *speed*. It therefore becomes preferable to interpret the redshift as being due to photon stretching in an expanding universe. From this perspective, it is better to think of space itself as expanding, carrying the galaxies along for the ride, than to think of the galaxies as projectiles flying through a static universe. The cosmological redshift of a galaxy tells us how much space has expanded during the time since light from the galaxy left on its journey to us.

The Horizon of the Universe   In our discussion of the expanding universe, we stressed that the universe as a whole does not seem to have an edge. Yet the universe does have a horizon, a place beyond which we cannot see. The **cosmological horizon** that marks the limits of the observable universe is a boundary in time, not in space. It exists because we cannot see back to a time before the universe began. If the universe is 14 billion years old, then no object can have a lookback time greater than 14 billion years. The age of the universe therefore fundamentally limits the size of our observable universe and the distances of all the galaxies that we can observe within it (see Figure 1.3). Now it is time to see what those galaxies have been doing during the last 14 billion years.

✳ THE PROCESS OF SCIENCE IN ACTION

12.3 Observing Galaxy Evolution

The relationship among distance, cosmological redshift, and lookback time is helpful to modern astronomers because it means that powerful telescopes can be used like time machines to observe the life histories of galaxies. We'll conclude this chapter with a brief overview of how the ability to look back through time enables astronomers to test models of galaxy evolution. Through this example, we'll see how technological progress is helping us extend the frontiers of our knowledge of the universe.

What do we see when we look back through time?

The photo on page 1 of this book shows a portion of the Hubble Ultra Deep Field, one of the deepest views of the universe yet captured. To obtain this photo, astronomers used the Hubble Space Telescope to make an 11-day exposure of a single spot in the sky. This long exposure time, combined with Hubble's super-sharp vision, resulted in a photograph that reveals galaxies at a wide range of distances. The farthest, with lookback times exceeding 12 billion years, lie at nearly 90% of the distance to our cosmological horizon. Most of the universe's history is therefore represented in this single image.

A Time Line for Galaxy Evolution We can combine the images in the Hubble Ultra Deep Field with redshift measurements to study the history of galaxies in the universe. Measuring the redshift of each galaxy tells us its distance and lookback time, which allows us to place it on a time line of the universe. Grouping galaxy images by type then shows how spiral, elliptical, and irregular galaxies have changed through time. **Figure 12.13** displays some "family albums" of galaxies assembled from the Hubble Ultra Deep Field using this method. Each picture shows a single galaxy at a single stage in its development. Pictures of the farthest galaxies show galaxies in their childhood, and pictures of the nearest show mature galaxies as they are today.

Think about it In the preceding paragraph, we used the term *today* in a very broad sense. For example, if we look at a relatively nearby galaxy—one located, say, 20 million light-years away—we see it as it was 20 million years ago. In what sense is this "today"? (*Hint:* How does 20 million years compare to the age of the universe?)

Figure 12.13 | Family albums for elliptical, spiral, and irregular galaxies of different ages, plus some very young galaxies shown on the far left. These photos are all zoomed-in images of galaxies from the Hubble Ultra Deep Field, a portion of which is shown in the photo on page 1 of this book. We see more distant galaxies as they were when they were younger, as indicated by the approximate age of the universe along the horizontal axis.

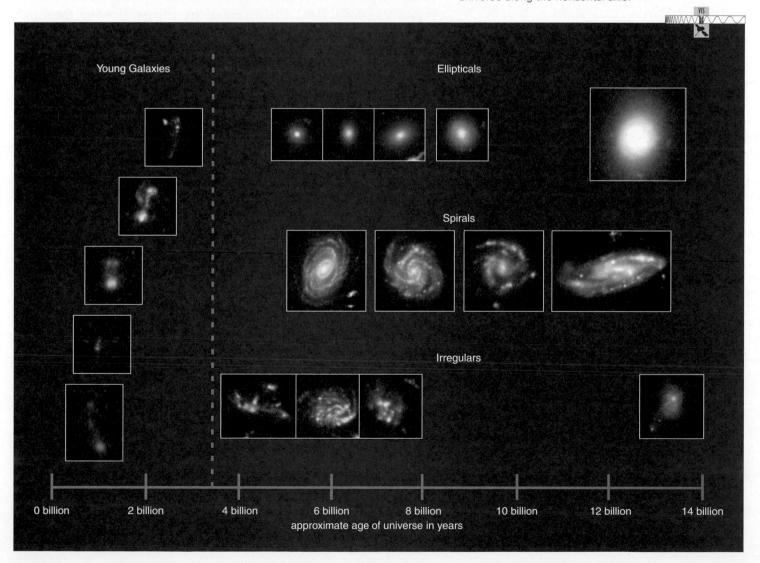

Young Galaxies   Ellipticals

Spirals

Irregulars

0 billion   2 billion   4 billion   6 billion   8 billion   10 billion   12 billion   14 billion

approximate age of universe in years

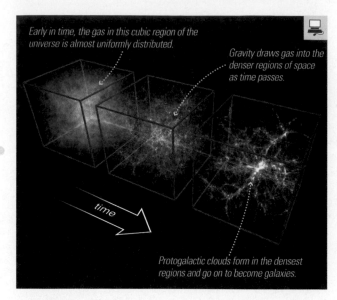

Early in time, the gas in this cubic region of the universe is almost uniformly distributed.

Gravity draws gas into the denser regions of space as time passes.

time

Protogalactic clouds form in the densest regions and go on to become galaxies.

Figure 12.14 | A computer simulation of the formation of protogalactic clouds. The simulated region of space is about 500 million light-years wide and goes on to form numerous galaxies.

One of the most interesting things we learn when we arrange galaxy photos along a time line is that the youngest galaxies are distinctly different from mature galaxies like the Milky Way. For example, look at the "Young Galaxies" in Figure 12.13. They all look disrupted, suggesting that they are in the midst of colliding and merging like the galaxies in Figure 11.15. Disrupted galaxies like these apparently were extremely common before the universe was about 3 billion years old. In contrast, mature galaxies generally look more organized, like the spirals and ellipticals in Figure 12.13.

Models of Galaxy Evolution These findings support models of galaxy evolution in which galaxies build up gradually over time through collisions and mergers with other galaxies and protogalactic clouds. The most successful models of this kind assume the following:

- Hydrogen and helium gas filled all of space more or less uniformly when the universe was very young, in the first million years after its birth.

- The distribution of matter in the universe was not perfectly uniform—certain regions of the universe started out ever so slightly denser than others.

Beginning from these assumptions, for which we have strong observational evidence [Section 13.2], we can model galaxy formation, using well-established laws of physics to trace how the denser regions in the early universe grew into galaxies (**Figure 12.14**). The models show that the regions of enhanced density originally expanded along with the rest of the universe. However, the slightly greater pull of gravity in these regions gradually slowed their expansion. Within about a billion years, the expansion of these denser regions halted and reversed, and the material within them began to contract into *protogalactic clouds* and to form galaxies like the cloud of matter that eventually formed our Milky Way [Section 11.1].

From this point onward, the models predict that galaxy evolution depended critically on how often protogalactic clouds and newly formed galaxies collided with their neighbors. Early in the universe's history, when all the galaxies were much closer together, these collisions should have been frequent. Later in time, after the average distances between galaxies became much greater, collisions became less common, allowing galaxies to settle into the spiral and elliptical shapes typical of the present-day universe.

Next Steps The predictions of galaxy evolution models agree well with what we see in Figure 12.13. Galaxies do indeed look more disrupted early in time and more settled later in the universe's history. However, even the most successful models of galaxy evolution do not yet match our observations in every detail. Further progress in this area will depend on both computer technology and a new generation of telescopes. Current efforts to model galaxy evolution require some of the world's most powerful supercomputers, and as computing power improves, our models should become increasingly accurate.

Our observations of galaxy evolution should also improve in the coming decade. Right now, we can study galaxies extending back to a time when the universe was just 1 or 2 billion years old, but our models suggest that the first stars and galaxies formed even earlier than this. Observing these first stars and galaxies is a challenge not even the Hubble Space Telescope can meet. Detecting such faraway galaxies will require larger telescopes, and the extreme redshifts expected for such galaxies mean the telescopes will need to be particularly sensitive to infrared light. In about 2013, NASA hopes to launch a much larger infrared-sensitive successor to the Hubble Space Telescope. With this telescope (called the *James Webb Space Telescope*), astronomers hope to see galaxies in the earliest stages of formation, allowing the process of science to lead us one step closer to the beginning of time.

Summary of Key Concepts

12.1 Measuring Cosmic Distances

How do we measure the distances to galaxies?

Our measurements of galaxy distances depend on a chain of methods. The chain begins with radar ranging in our own solar system and parallax measurements of distances to nearby stars; then it relies on **standard candles** to measure greater distances.

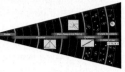

What is Hubble's law?

Hubble's law tells us that more distant galaxies are moving away faster: $v = H_0 \times d$, where H_0 is **Hubble's constant.** It allows us to determine a galaxy's distance from the speed at which it is moving away from us, which we can measure from its Doppler shift.

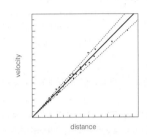

12.2 The Implications of Hubble's Law

In what sense is the universe expanding?

The average distances between galaxies are continually increasing, implying that space itself is expanding with time. Our expanding universe does not appear to have either a center or an edge. In that sense, its expansion is like the stretching of the surface of an inflating balloon: Objects attached to the surface gradually move farther apart, but the surface itself has no center or edges.

How do distance measurements tell us the age of the universe?

Dividing a galaxy's current distance by its current velocity gives us an estimate of how long it took the galaxy to reach that distance. Such measurements allow us to determine **Hubble's constant,** which tells us that it took approximately 14 billion years for the universe's galaxies to reach their current distances, giving us an estimate for the age of the universe. Because distances between galaxies are always changing, it is best to express the distance to a faraway galaxy in terms of its **lookback time**—the time it has taken for the galaxy's light to reach us. The expansion of the universe during that time stretches the light coming from the galaxy, leading to a **cosmological redshift** directly related to the galaxy's lookback time.

✳ THE PROCESS OF SCIENCE IN ACTION

12.3 Observing Galaxy Evolution

What do we see when we look back through time?

Today's telescopes are powerful enough to detect light from galaxies with lookback times almost as large as the age of the universe. Assembling "family albums" of galaxies with different lookback times shows that galaxies were more likely to look disrupted when the universe was very young than at later times. This finding supports models in which galaxies build up gradually through mergers of smaller galaxies and protogalactic clouds. During the next decade, improved telescopes and supercomputers will help us observe and model even younger galaxies.

Investigations

Quick Quiz

Choose the best answer to each of the following; answers are in Appendix D. Explain your reasoning with one or more complete sentences.

1. Which of these galaxies is most likely to be oldest? (a) a galaxy in the Local Group (b) a galaxy observed at a distance of 5 billion light-years (c) a galaxy observed at a distance of 10 billion light-years

2. Which set of star-position measurements is most likely to show the largest parallax shift for nearby stars? (a) measurements made 1 month apart (b) measurements made 6 months apart (c) measurements made 1 year apart

3. If Earth's orbital radius doubled in size, (a) the parallax shifts of nearby stars would double in size. (b) the parallax shifts of nearby stars would be only half their current size. (c) the parallax shifts of nearby stars would stay the same.

4. If all the stars on the main sequence of a star cluster are typically only one-hundredth as bright as their main-sequence counterparts in the Hyades Cluster, then that cluster's distance is (a) 100 times the Hyades' distance. (b) 30 times the Hyades' distance. (c) 10 times the Hyades' distance.

5. Which kind of object is the best standard candle for measuring distances to extremely distant galaxies? (a) a white dwarf (b) a Cepheid variable star (c) a white dwarf supernova

6. If you observe two Cepheid variable stars to have the same period and one is brighter than the other, (a) the brighter one is closer. (b) the brighter one is farther away. (c) there is not enough information to determine which star is closer.

7. If you observe two Cepheid variable stars to have different periods and the brighter one has a longer period, (a) the brighter one is closer. (b) the brighter one is farther away. (c) there is not enough information to determine which star is closer.

8. When the ultraviolet light from hot stars in very distant galaxies finally reaches us, it arrives at Earth in the form of (a) X rays. (b) slightly more energetic ultraviolet light. (c) visible light.

9. Why do virtually all the galaxies in the universe appear to be moving away from our own? (a) because we are located near where the Big Bang happened (b) because we are located near the center of the universe (c) because observers in all galaxies see a similar phenomenon due to the universe's expansion

10. If you observed the redshifts of galaxies at a given distance to be twice as large as they are now, the value you would determine for Hubble's constant would be (a) twice its current value. (b) equal to its current value. (c) half its current value.

11. What would your estimate be for the age of the universe if you measured Hubble's constant to be 11 kilometers per second per million light-years? (a) 7 billion years (b) 14 billion years (c) 28 billion years

12. When we observe a distant galaxy whose photons have traveled for 10 billion years before reaching Earth, we are seeing that galaxy as it was when the universe was about (a) 10 billion years old. (b) 7 billion years old. (c) 4 billion years old.

Short-Answer/Essay Questions

Explain all answers clearly, with complete sentences and proper essay structure if needed. An asterisk (*) designates a quantitative problem, for which you should show all your work.

13. *Distance Measurements.* The techniques astronomers use to measure distances are not so different from the ones you use every day. Describe how using standard-candle measurements is similar to the way you estimate the distance to an oncoming car at night.

14. *Cepheids as Standard Candles.* Suppose you are observing Cepheids in a nearby galaxy. You observe one Cepheid with a period of 8 days between peaks in brightness, and another with a period of 35 days. Estimate the luminosity of each star. Explain how you arrived at your estimate. (*Hint:* See Figure 12.4.)

15. *Galaxies at Great Distances.* The most distant galaxies that astronomers have observed are much easier to see in infrared light than in visible light. Explain why this is the case.

16. *Universe on a Balloon.* In what ways does the surface of a balloon offer a good analogy to the universe? In what ways is this analogy limited?

Explain why a miniature scientist living in a polka dot on the balloon would observe all other dots to be moving away, with more distant dots moving away faster.

17. *Quasar Redshifts.* When quasars were first discovered, they were found to have unusually large redshifts. Astronomers using Hubble's law therefore determined that these quasars must be at unusually large distances, implying unusually great luminosities (see Section 11.3). However, some astronomers did not immediately accept that the large redshifts of quasars necessarily meant that they must be at extremely large distances. What properties would an object need in order to have a large redshift but still be no farther than a billion light-years from the Milky Way? How might you be able to test the hypothesis that redshifts of quasars are not cosmological in origin?

**18. *Supernovae as Standard Candles.* The luminosity of a white dwarf supernova peaks around $10^{10} L_{Sun}$, and it remains above $10^{8} L_{Sun}$ for about 150 days. In comparison, the luminosity of a bright Cepheid variable star is about $10,000 L_{Sun}$. The Hubble Space Telescope is sensitive enough to make accurate measurements of apparent brightness for Cepheid variables at distances up to about 100 million light-years. Estimate the distance of a fading white dwarf supernova of luminosity $10^{8} L_{Sun}$ whose apparent brightness is comparable to that of a bright Cepheid variable star 100 million light-years from Earth. How does the distance of that supernova compare with the size of the observable universe?

**19. *Estimating the Universe's Age.* What would be your estimate of the age of the universe if you measured a value for Hubble's constant of $H_0 = 33$ km/s/Mly? You can assume that the expansion rate has remained unchanged during the history of the universe.

**20. *Distances Between Galaxies.* If you were to divide the present-day universe up into cubes with sides 10 million light-years long, each cube would contain, on average, about one galaxy similar in size to the Milky Way. Now suppose you traveled back in time, to an era when the average distance between galaxies was one-quarter of its current value. How many galaxies similar in size to the Milky Way would you expect to find, on average, in cubes of the same size? In order to simplify the problem, assume that the total number of galaxies of each type has not changed between then and now. Based on your answer, would you expect collisions to be much more frequent at that time or only moderately more frequent?

13 The Early Universe

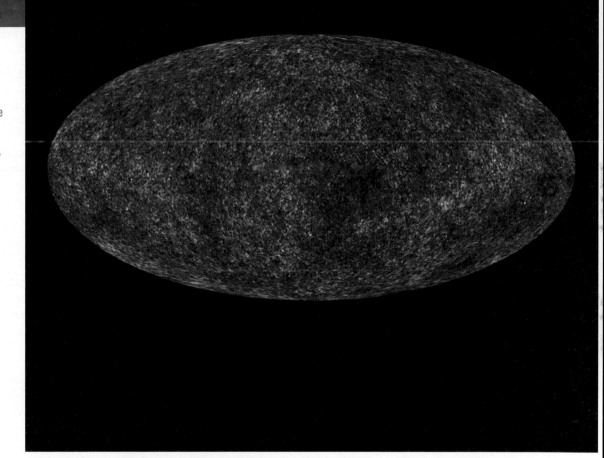

Learning Goals

13.1 The Big Bang
- What were conditions like in the early universe?
- How did the early universe change with time?

13.2 Evidence for the Big Bang Theory
- How do we observe the radiation left over from the Big Bang?
- How do the abundances of elements support the Big Bang theory?

✳ THE PROCESS OF SCIENCE IN ACTION
13.3 Inflation
- Did the universe undergo an early episode of inflation?

This photo shows what we find when we look beyond the most distant galaxies. It shows microwave radiation released at a time when the universe was so young that no stars or galaxies had yet been born. We see the entire sky as it appears in all directions from Earth; the color differences represent slight differences in the temperature of the universe at the time at which the radiation was released. This image is important because it provides strong evidence for the Big Bang theory, which says that the universe began in a very hot and dense state and has been expanding and cooling ever since. In this chapter, we'll explore the Big Bang theory and the evidence that supports it.

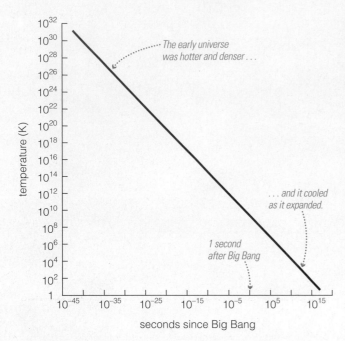

Figure 13.1 | The universe cools as it expands. By using the laws of physics and the current temperature of the universe (about 3 K), we can calculate how hot the universe must have been in the past. This graph shows the results. Notice that both axis scales use powers of 10; therefore, even though most of the graph shows temperatures during the first second after the Big Bang, the far right part of the graph actually extends to the present (14 billion years = 4×10^{17} s).

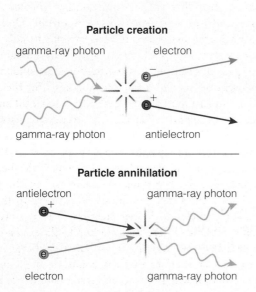

Figure 13.2 | Electron-antielectron creation and annihilation. Reactions like these continually converted photons to particles, and vice versa, in the early universe.

13.1 The Big Bang

Up to this point, we've discussed how the matter produced in the early universe gradually assembled into planets, stars, and galaxies. However, we have not yet answered one big question: Where did *matter itself* come from? To answer this question, we must go beyond even the most distant galaxies. We must go back not only to the origins of matter and energy but to the beginning of time itself.

The scientific theory that predicts what the universe was like in its earliest moments is called the **Big Bang theory.** This theory is based on applying known and tested laws of physics to the idea that all we see today, from Earth to the cosmic horizon, began as an incredibly tiny, hot, and dense collection of matter and radiation. The Big Bang theory successfully describes how the expansion and cooling of this intense mixture of particles and photons could have led to the present universe of stars and galaxies, and it explains many aspects of today's universe with impressive accuracy. Our main goal in this chapter is to understand the evidence supporting the Big Bang theory, but first we must explore what the theory tells us about the early universe.

What were conditions like in the early universe?

Observations demonstrate that the universe is expanding and cooling with time, implying that it must have been hotter and denser in the past. Scientists can calculate exactly how hot and dense the universe must have been when it was more compressed by applying fundamental principles of physics; the calculations are much like those used to predict how the temperature and density of gas in a balloon change when you squeeze it, except that the conditions in the early universe were much more extreme. **Figure 13.1** shows how the temperature of the universe has changed with time, according to these calculations.

Particle Creation and Annihilation The universe was so hot during the first few seconds after the Big Bang that photons could transform themselves into matter, and vice versa, in accordance with Einstein's formula $E = mc^2$. Reactions that create and destroy matter are now relatively rare in the universe, but physicists can reproduce such reactions in laboratories (see Tools of Science, p. 211).

One such reaction is the creation or destruction of an electron-antielectron pair (**Figure 13.2**). When two photons collide with a total energy greater than twice the mass-energy of an electron (the electron's mass times c^2), they can create two brand-new particles: a negatively charged electron and its positively charged twin, the *antielectron* (also known as a *positron*). The electron is a particle of **matter,** and the antielectron is a particle of **antimatter.** The reaction that creates an electron-antielectron pair also runs in reverse. When an electron and an antielectron meet, they **annihilate** each other totally, transforming their mass-energy back into photon energy.

Similar reactions can produce or destroy any particle-antiparticle pair, such as a proton and antiproton or a neutron and antineutron. The early universe therefore was filled with an extremely hot, dense mix of photons, matter, and antimatter, converting furiously back and forth. Despite all these vigorous reactions, describing conditions in the early universe is straightforward, at least in principle. We can use the laws of physics to calculate the proportions of the various forms of radiation and matter at each moment in the universe's early history. The only difficulty is our incomplete understanding of the laws of physics.

To date, physicists have investigated the behavior of matter and energy at temperatures as high as those that existed in the universe just *one ten-billionth* (10^{-10}) of a second after the Big Bang, giving us confidence that we actually understand what was happening at that early time. Our understanding of physics under the more extreme conditions that prevailed even earlier is less certain, but we do have some ideas about what the universe was like when it was a mere 10^{-38} second old, and perhaps a glimmer of what it was like at the age of just 10^{-43} second. These tiny fractions of a second are so small that, for all practical purposes, we are studying the very moment of creation—the Big Bang itself.

Fundamental Forces   To understand the changes that occurred in the early universe, it helps to think in terms of *forces*. Everything that happens in the universe today is governed by four distinct forces: *gravity*, the *electromagnetic force*, the *strong force*, and the *weak force*. We have already encountered examples of some of these forces in action.

Tools of Science: Particle Accelerators

Much of what we know about the universe's earliest moments comes from devices known as *particle accelerators*. Physicists can generate many unusual particles with the aid of these machines, which include Fermilab near Chicago, the Stanford Linear Accelerator in California, the Relativistic Heavy Ion Collider at Brookhaven Labs in New York, and the new Large Hadron Collider (**Figure 1**) on the border between Switzerland and France. Inside a particle accelerator, charged particles are channeled along a narrow tube—typically about as thick as garden hose—by large, powerful magnets. These magnets can accelerate familiar particles such as electrons or protons to very high speeds—often extremely close to the speed of light. When these particles collide with one another or with a stationary target, they release a substantial amount of energy within a very small space. For an instant, the conditions within this small space are similar to those that prevailed a fraction of a second after the Big Bang. The energy released is so great that much of it spontaneously turns into mass, producing a shower of matter and antimatter particles. Recall that energy can turn into mass, just as mass can turn into energy, in accord with Einstein's formula $E = mc^2$.

Scientists identify different types of particles produced in particle accelerators by carefully observing the paths they follow in response to magnetic fields within the accelerator. These paths depend on properties such as the mass and electrical charge of the particles. Whenever scientists observe a particle that has previously unseen characteristics, they catalog it as a new type of particle and give it a name. Dozens of different particles have been discovered in accelerators, but only a few of them, such as *electrons* and *quarks*, appear to be truly fundamental. All the remaining particles can be constructed from the fundamental few according to a scheme known as the *standard model* for the structure of matter. The standard model has proved very successful and has even predicted the existence of new particles that were later discovered in particle accelerators. The standard model has been well tested in particle accelerators at energies equivalent to a temperature of 10^{15} K, giving scientists confidence that we understand how particles behaved a mere 10^{-10} second after the Big Bang. A major goal of the Large Hadron Collider is to learn about conditions even closer to the Big Bang, by accelerating particles to higher energies than any previous accelerator has achieved. One of the Large Hadron Collider's first tasks is to find a particle called the *Higgs boson*, which is predicted by the standard model to exist at energies that the collider can produce. If the Higgs boson is found, it will give us even greater confidence in our understanding of the early universe. If it is not found, then we may have to revisit theories of how the universe worked during its earliest moments.

Figure 1 | Aerial photograph of the Large Hadron Collider. The large circle traces the path of the main particle acceleration ring, which lies underground and has a circumference of 27 kilometers.

Gravity is the most familiar of the four forces, providing the "glue" that holds planets, stars, and galaxies together. The electromagnetic force, which depends on the electrical charge of a particle instead of its mass, is far stronger than gravity. It is therefore the dominant force between particles in atoms and molecules, responsible for all chemical and biological reactions. However, the existence of both positive and negative electrical charges causes the electromagnetic force to lose out to gravity on large scales, even though both forces decline with distance according to an inverse square law. Most large astronomical objects (such as planets and stars) are electrically neutral overall, so they do not significantly interact through the electromagnetic force. Gravity therefore becomes the dominant force for such objects, because more mass always means more gravity.

The strong and weak forces operate only over extremely short distances, making them important within atomic nuclei but not on larger scales. The strong force binds atomic nuclei together [Section 8.1]. The weak force plays a crucial role in nuclear reactions such as fission and fusion.

Although the four forces behave quite differently from one another, current models of fundamental physics predict that they are actually different aspects of a smaller number of more fundamental forces, probably only one or two. These models predict that the four forces would have been merged together at the high temperatures that prevailed in the very early universe (**Figure 13.3**). As an analogy, think about ice, liquid water, and water vapor. These three substances are quite different from one another in appearance and behavior, yet they are all different phases of the single substance H_2O. In a similar way, experiments have shown that the electromagnetic and weak forces lose their separate identities under conditions of very high temperature or energy and merge together into a single **electroweak force.** At even higher temperatures and energies, the electroweak force may merge with the strong force and ultimately with gravity. Theories that predict the merger of the electroweak and strong forces are called **grand unified theories,** or **GUTs** for short. The merger of the strong, weak, and electromagnetic forces is therefore often called the *GUT force.* Many physicists suspect that at even higher energies, the GUT force and gravity merge into a single "super force" that governs the behavior of everything. (You may also hear the names *supersymmetry, superstrings,* and *supergravity* in connection with theories linking all four forces.)

If these ideas are correct, then the universe was once governed solely by the super force in the first instant after the Big Bang. As the universe expanded and cooled, the super force split into gravity and the GUT force, which then split into the strong and electroweak forces. Ultimately, all four forces became distinct. As we'll see shortly, these changes in the fundamental forces probably occurred before the universe was one ten-billionth of a second old.

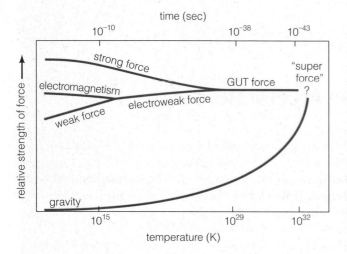

Figure 13.3 | The four forces are distinct at low temperatures but may merge at very high temperatures, such as those that prevailed during the first fraction of a second after the Big Bang.

 Think about it Based on Figure 13.3, about how long after the Big Bang did the strong force first become distinct, as it is in the universe today? At what time did all four forces first have their own individual identities?

How did the early universe change with time?

The Big Bang theory uses the ideas we have described so far to reconstruct the history of the universe. Here we will outline this history as a series of eras, or time periods (**Figure 13.4**). Each era is distinguished from the next by some major change in physical conditions as the universe cools. Notice that the time scale in Figure 13.4 runs by powers of 10, which means that early eras were very brief, even though they appear spread out on the figure. It will take you longer to read this chapter than it took the universe to progress through the first five eras we will discuss, by which point the chemical composition of the universe had already been determined.

The Planck Era

The first era after the Big Bang is called the **Planck era,** named for physicist Max Planck; it represents times before the universe was 10^{-43} second old. We cannot say much about this instant in time, because current theories of physics cannot make firm predictions about the conditions that prevailed then. Nevertheless, we have at least some idea of how the Planck era ended. If you look back at Figure 13.3, you'll see that all four forces are thought to merge into the single, unified super force at temperatures above 10^{32} K—the temperatures that prevailed during the Planck era. In that case, the Planck era would have been a time of ultimate simplicity, when just a single force operated in nature, and it came to an end when the temperature dropped low enough for gravity to become distinct from the other three forces, which were still merged as the GUT force. By analogy to the way ice crystals form as a liquid cools, we say that gravity "froze out" at the end of the Planck era.

The GUT Era

The next era is called the **GUT era,** named for the grand unified theories (GUTs) that predict the merger of the strong, weak, and electromagnetic forces into a single GUT force at temperatures above 10^{29} K (see Figure 13.3). The GUT era is the era during which two forces—gravity and the GUT force—operated in the universe. It came to an end when the GUT force split into the strong and electroweak forces, which happened when the universe was a mere 10^{-38} second old.

Our current understanding of physics allows us to say only slightly more about the GUT era than the Planck era, and none of our ideas about the GUT era have been sufficiently tested to give us great confidence in our notions about what occurred during that time. However, if the grand unified theories are correct, the freezing out of the strong and electroweak forces may have released an enormous amount of energy, causing a sudden and dramatic expansion of the universe called **inflation.** In a mere 10^{-36} second—a trillion trillion trillionth of a second—pieces of the universe the size of an atomic nucleus may have grown to the size of our solar system. Inflation sounds bizarre, but as we will discuss later, it can explain several important features of today's universe.

The Electroweak Era

The splitting of the GUT force marked the beginning of an era during which three distinct forces operated: gravity, the strong force, and the electroweak force. We call this time the **electroweak era,** indicating that the electromagnetic and weak forces were still merged together. Intense radiation continued to fill all of space, as it had since the Planck era, spontaneously producing matter and antimatter particles that almost immediately annihilated each other and turned back into photons.

The universe continued to expand and cool throughout the electroweak era, dropping to a temperature of 10^{15} K when it reached an age of 10^{-10} second. This temperature is still 100 million times hotter than the temperature in the core of the Sun today, but it was low enough for the electromagnetic and weak forces to freeze out from the electroweak force. After this instant, all four forces were forever distinct in the universe.

The Particle Era

As long as the universe was hot enough for the spontaneous creation and annihilation of particles, the total number of particles was roughly in balance with the total number of photons. Once it became too cool for this spontaneous exchange of matter and energy to continue, photons became the

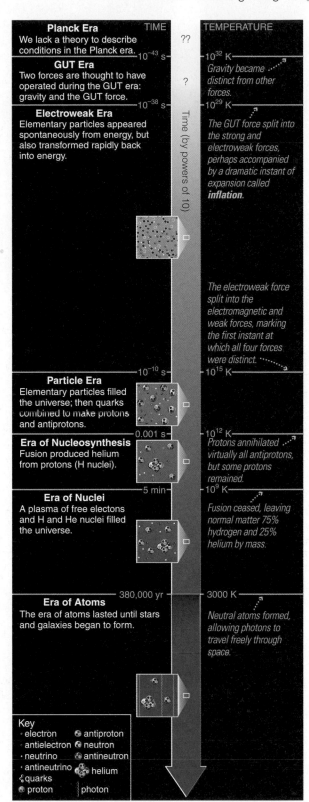

Figure 13.4 | A time line for the eras of the early universe. The only era not shown is the era of galaxies, which began with the birth of stars and galaxies when the universe was a few hundred million years old.

dominant form of energy in the universe. We refer to the time between the end of the electroweak era and the moment when spontaneous particle production ceased as the **particle era,** to emphasize the importance of subatomic particles during this period.

During the early parts of the particle era (and earlier eras), photons produced all sorts of exotic particles that we no longer find freely existing in the universe today, including *quarks*—the building blocks of protons and neutrons. As the universe cooled, all the quarks eventually combined into protons and neutrons, which then shared the universe with other particles such as electrons and neutrinos.

The particle era ended when the universe reached an age of 1 millisecond (0.001 second), at which point it was no longer hot enough to produce protons and antiprotons spontaneously from pure energy. If the universe had contained equal numbers of protons and antiprotons at that time, they all would have annihilated each other, creating photons and leaving essentially no matter in the universe. Since the universe contains a significant amount of matter today, we conclude that protons must have slightly outnumbered antiprotons at the end of the particle era.

We can estimate the size of the imbalance between matter and antimatter by comparing the present numbers of protons and photons in the universe. The two numbers should have been similar in the very early universe, but today photons outnumber protons by about a billion to one. This ratio indicates that for every billion antiprotons in the early universe, there must have been about a billion and one protons. As a result, for each 1 billion protons and antiprotons that annihilated each other at the end of the particle era, a single proton was left over. This slight excess of matter over antimatter makes up everything we can see in the present-day universe.

The Era of Nucleosynthesis The eras we have discussed so far all occurred within the first 0.001 second of the universe's existence—less time than it takes you to blink an eye. At this point, the protons and neutrons left over after the annihilation of antimatter began to fuse into heavier nuclei. However, the temperature of the universe remained so high that most nuclei were blasted apart by gamma rays as fast as they formed. This dance of fusion and demolition marked the **era of nucleosynthesis,** which ended when the universe was about 5 minutes old. By this time, the density in the expanding universe had dropped so much that fusion no longer occurred, even though the temperature was still about a billion Kelvin (10^9 K)—much hotter than the temperature of the Sun's core.

When fusion ceased at the end of the era of nucleosynthesis, the chemical content of the universe had become (by mass) about 75% hydrogen and 25% helium, along with trace amounts of deuterium (hydrogen with a neutron) and lithium (the next heaviest element after hydrogen and helium). Except for the small proportion of matter (2%) that stars later forged into heavier elements, the chemical composition of the universe remains the same today.

The Era of Nuclei After fusion ceased, the universe consisted of a very hot plasma of hydrogen nuclei, helium nuclei, and free electrons. The fully ionized nuclei moved independently of electrons during this period (rather than being bound with electrons in neutral atoms), so we call it the **era of nuclei.** Throughout this era, photons bounced rapidly from one electron to the next, just as they do deep inside the Sun today [Section 8.1], never traveling far between collisions. Any time a nucleus managed to capture an electron to form a complete atom, one of the photons quickly ionized it.

The era of nuclei lasted until the universe was about 380,000 years old, by which point the universe had cooled to a temperature of about 3000 K—roughly half the temperature of the Sun's surface today. Hydrogen and helium nuclei finally captured electrons for good, forming stable, neutral atoms for the first time. With electrons now bound into atoms, the universe became transparent, as if a thick fog had suddenly lifted. Photons, formerly trapped among the electrons, began to stream freely through the universe. We still see these photons today as the **cosmic microwave background** (pictured in the photo that opens this chapter).

The Era of Atoms   We've already discussed the rest of the universe's history in earlier chapters. The end of the era of nuclei marked the beginning of the **era of atoms,** when the universe consisted of a mixture of neutral atoms and plasma (ions and electrons), along with a large number of photons. Because the density of matter in the universe differed slightly from place to place, gravity slowly drew atoms and plasma into the higher-density regions, which assembled into protogalactic clouds [Section 11.1, Section 12.3]. Stars then formed in the clouds, transforming them into galaxies.

The Era of Galaxies   The first full-fledged galaxies had formed by the time the universe was about 1 billion years old, beginning what we call the **era of galaxies,** which continues to this day. Generation after generation of star formation in galaxies steadily builds elements heavier than helium and incorporates them into new star systems. Some of these star systems develop planets, and on at least one of these planets life burst into being a few billion years ago. Now here we are, thinking about it all.

Figure 13.5 (on pages 216–217) summarizes the major ideas we have discussed about the history of the universe as described by the Big Bang theory.

13.2 Evidence for the Big Bang Theory

What makes us think that the Big Bang theory really describes events that occurred nearly 14 billion years ago? Like any scientific theory, the Big Bang theory is a model of nature designed to explain a set of observations. The model was inspired by Edwin Hubble's discovery of the universe's expansion: If the universe has been expanding for billions of years, then simple physical reasoning suggests that conditions ought to have been much denser and hotter in the past. However, the model was not accepted as a valid scientific theory until its major predictions were verified through additional observations and experiments. The Big Bang theory has gained wide scientific acceptance for two key reasons:

- It predicts that the radiation released at the end of the era of nuclei should still be present today in the form of a *cosmic microwave background*. Sure enough, we have detected this background and find that its characteristics precisely match those predicted by the theory.

- It predicts the precise fraction of the original hydrogen in the universe that should have fused into helium during the era of nucleosynthesis. Observations of the actual helium content of the universe closely match the predicted value.

Let's take a closer look at this evidence, starting with the cosmic microwave background.

How do we observe the radiation left over from the Big Bang?

The discovery of the cosmic microwave background was announced in 1965. Arno Penzias and Robert Wilson, two physicists working at Bell Laboratories in New Jersey, were calibrating a sensitive microwave antenna designed for

The Big Bang theory is a scientific model that explains how the present-day universe developed from an extremely hot and dense beginning. This schematic diagram shows how conditions in the early universe changed as the universe expanded and cooled with time.

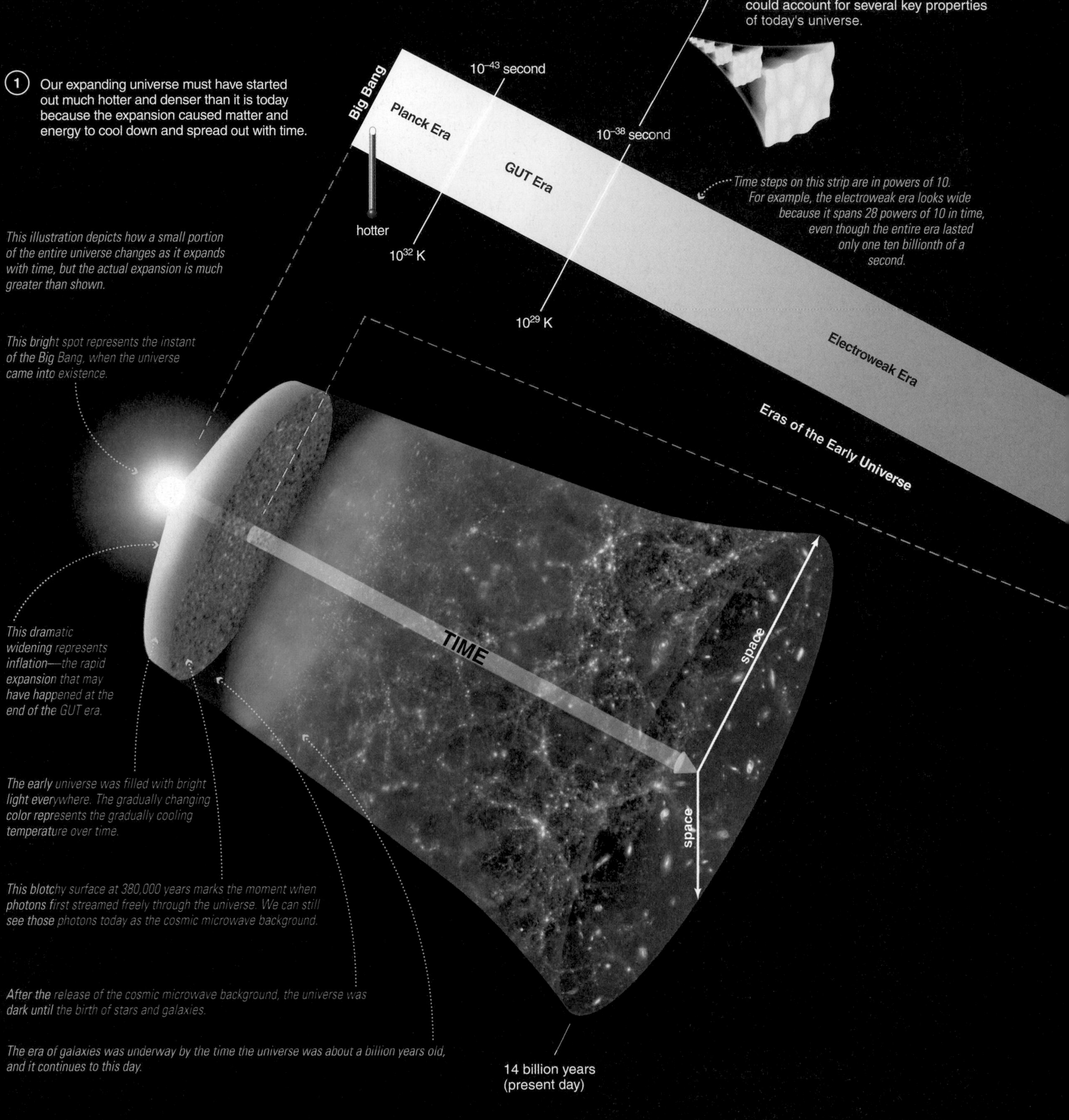

1 Our expanding universe must have started out much hotter and denser than it is today because the expansion caused matter and energy to cool down and spread out with time.

2 As the universe cooled down, it may have undergone a brief period of very rapid expansion known as inflation that could account for several key properties of today's universe.

Big Bang

Planck Era

10^{-43} second

GUT Era

10^{-38} second

hotter

10^{32} K

10^{29} K

Time steps on this strip are in powers of 10. For example, the electroweak era looks wide because it spans 28 powers of 10 in time, even though the entire era lasted only one ten billionth of a second.

This illustration depicts how a small portion of the entire universe changes as it expands with time, but the actual expansion is much greater than shown.

This bright spot represents the instant of the Big Bang, when the universe came into existence.

Electroweak Era

Eras of the Early Universe

This dramatic widening represents inflation—the rapid expansion that may have happened at the end of the GUT era.

TIME

space

space

The early universe was filled with bright light everywhere. The gradually changing color represents the gradually cooling temperature over time.

This blotchy surface at 380,000 years marks the moment when photons first streamed freely through the universe. We can still see those photons today as the cosmic microwave background.

After the release of the cosmic microwave background, the universe was dark until the birth of stars and galaxies.

The era of galaxies was underway by the time the universe was about a billion years old, and it continues to this day.

14 billion years (present day)

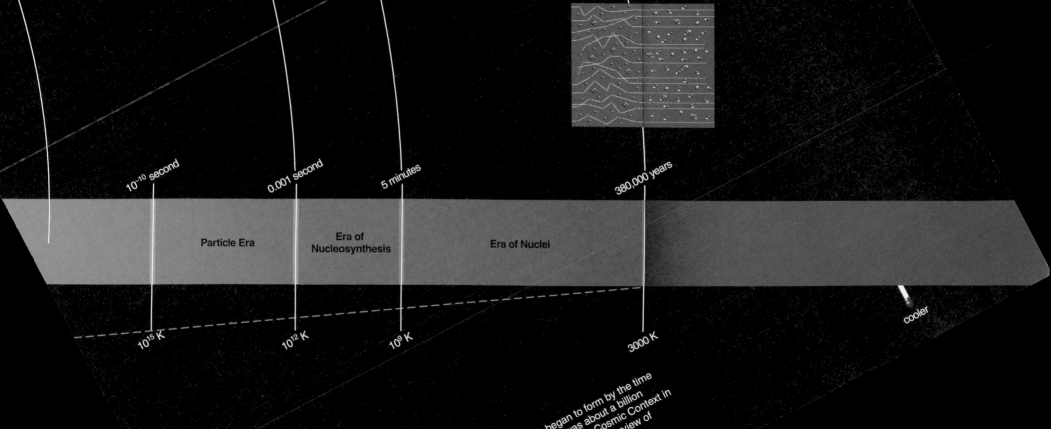

10⁻¹⁰ second

0.001 second

5 minutes

380,000 years

Particle Era

Era of
Nucleosynthesis

Era of Nuclei

10^{15} K

10^{12} K

10^{9} K

3000 K

cooler

began to form by the time
was about a billion
Cosmic Context in
view of

Figure 13.6 | Arno Penzias and Robert Wilson, discoverers of the cosmic microwave background, with the Bell Labs microwave antenna.

Photons bounced around among the free electrons early in time . . .

time →

380,000 years

. . . but they moved freely through the universe after atoms captured the electrons.

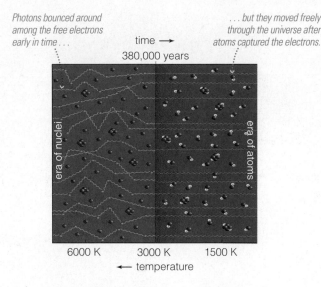

era of nuclei

era of atoms

6000 K   3000 K   1500 K

← temperature

Figure 13.7 | This diagram shows the transition that occurred when the universe was 380,000 years old, when photons (yellow squiggles) could first travel freely through the universe. We detect these photons today as the cosmic microwave background.

satellite communications (**Figure 13.6**). Much to their chagrin, they kept finding unexpected "noise" in every measurement they made. The noise was the same no matter where they pointed their antenna, indicating that it came from all directions in the sky and ruling out the possibility that it came from any particular astronomical object or from any place on Earth.

Meanwhile, physicists at nearby Princeton University were busy calculating the expected characteristics of the radiation left over from the heat of the Big Bang. They concluded that, if the Big Bang had really occurred, this radiation should be permeating the entire universe and should be detectable with a microwave antenna. The Princeton group soon met with Penzias and Wilson to compare notes, and both teams realized that the "noise" from the Bell Labs antenna was the predicted cosmic microwave background—the first strong evidence that the Big Bang really did happen.

Origin of the Cosmic Microwave Background The cosmic microwave background consists of microwave photons that have traveled through space since the end of the era of nuclei, when most of the electrons in the universe joined with nuclei to make neutral atoms. With very few free electrons left to block them, most of the photons from that time have traveled unobstructed through the universe ever since (**Figure 13.7**). When we observe the cosmic microwave background, we essentially are seeing back to the end of the era of nuclei, when the universe was only 380,000 years old, and studying light from the most distant observable region of the universe.

Note that the cosmic microwave background represents a limit on how far back we can see in time. The universe was filled with an impenetrable fog of light before this radiation was released, so we cannot see light from earlier times. However, the cosmic microwave background provides clues about those earlier eras, because conditions in the universe at the time the background radiation was released were determined by events that occurred even earlier in time.

Characteristics of the Cosmic Microwave Background The Big Bang theory predicts that the cosmic microwave background should have an essentially perfect thermal radiation spectrum (see Tools of Science, p. 80), because it came from the heat of the universe itself. Moreover, the theory predicts the approximate wavelength at which this thermal radiation spectrum should peak. When the radiation of the cosmic microwave background broke free, the temperature of the universe was about 3000 K, similar to the surface temperature of a red giant star. The spectrum of the cosmic microwave background therefore originally peaked in visible light, just like the thermal radiation from a red star, with wavelengths of a few hundred nanometers. However, the universe has expanded by a factor of about 1000 since that time, stretching the wavelengths of these photons by the same amount [Section 12.2]. Their wavelengths have therefore shifted to about a millimeter, squarely in the microwave portion of the spectrum and corresponding to a temperature of a few degrees above absolute zero.

In the early 1990s, a NASA satellite called the *Cosmic Background Explorer (COBE)* was launched to test these ideas about the cosmic microwave background. The results were a stunning success for the Big Bang theory. As shown in **Figure 13.8**, the cosmic microwave background does indeed have a perfect thermal radiation spectrum, with a peak corresponding to a temperature of 2.73 K.

 Think about it Suppose the cosmic microwave background did not really come from the heat of the universe itself, but instead came from many individual stars and galaxies. Explain why, in that case, we would not expect it to have a perfect thermal radiation spectrum. How does the spectrum of the cosmic microwave background lend support to the Big Bang theory?

COBE and a successor satellite, the *Wilkinson Microwave Anisotropy Probe (WMAP)*, have also mapped the temperature of the cosmic microwave background in all directions (**Figure 13.9**). It turns out to be extraordinarily uniform throughout the universe—just as the Big Bang theory predicts it should be—with temperature variations of only a few parts in 100,000 across the sky. Observations of these variations have been very important to studies of galaxy evolution because they arise from the variation in density that needed to be present in the early universe in order to form galaxies and other larger structures later on [Section 14.2]. Measuring the patterns in these variations therefore tells us both about what must have happened at early times to create these variations and about the starting conditions that we should use in models of galaxy evolution.

How do the abundances of elements support the Big Bang theory?

The discovery of the cosmic microwave background in 1965 quickly solved another long-standing astronomical problem: the origin of cosmic helium. The chemical composition of the universe is about one-quarter helium by mass, no matter where we look. Helium fusion in stars cannot account for this large abundance of helium, because even small galaxies that have experienced very little star formation still have helium fractions of 25%. Larger galaxies like the Milky Way have slightly more helium, up to about 28% by mass, and this extra 3% is consistent with the amount expected from stars. We therefore conclude that the majority of the helium in the universe must already have been present in the protogalactic clouds that preceded the formation of galaxies. In other words, the universe itself must once have been hot enough to fuse hydrogen into helium. The current microwave background temperature of 2.73 K tells us precisely how hot the universe was in the distant past and exactly how much helium it should have made. The result—25% helium—is another impressive success of the Big Bang theory.

Helium Formation in the Early Universe We can see why 25% of protons and neutrons became helium by thinking about what protons and neutrons were doing during the 5-minute-long era of nucleosynthesis. Early in this era, the universe's temperature was so high (about 10^{11} K) that nuclear reactions could convert protons into neutrons, and vice versa. These reactions kept the numbers of protons and neutrons nearly equal. But as the temperature fell, neutron–proton conversion reactions began to favor protons.

Neutrons are slightly more massive than protons, so reactions that convert protons to neutrons require energy to proceed (in accordance with $E = mc^2$); these reactions therefore proceed more slowly as the temperature falls. In

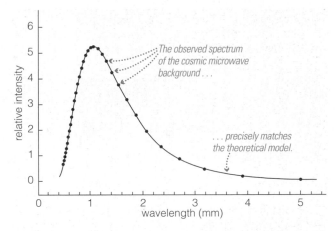

Figure 13.8 | This graph shows the spectrum of the cosmic microwave background recorded by NASA's *COBE* satellite. A theoretically calculated thermal radiation spectrum (smooth curve) for a temperature of 2.73 K perfectly fits the data (dots). This excellent fit is important evidence in favor of the Big Bang theory.

The observed spectrum of the cosmic microwave background . . .

. . . precisely matches the theoretical model.

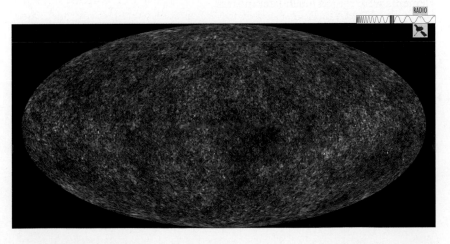

Figure 13.9 | This all-sky map shows temperature differences in the cosmic microwave background measured by *WMAP*. The background temperature is about 2.73 K everywhere, but the brighter regions of this picture are about 0.0001 K hotter than the darker regions—indicating that the early universe was very slightly lumpy at the end of the era of nuclei. We are essentially seeing what the universe was like at the surface marked "380,000 years" in Figure 13.5. Gravity later drew matter toward the centers of these lumps, forming galaxies and other structures that we see in the universe today.

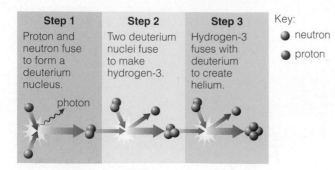

Figure 13.10 | During the 5-minute-long era of nucleosynthesis, virtually all the neutrons in the universe fused with protons to form helium. This figure illustrates one way in which helium formed.

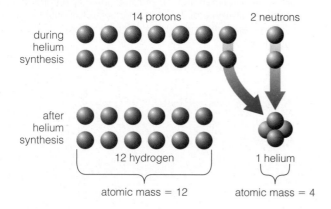

Figure 13.11 | Calculations show that protons outnumbered neutrons 7 to 1, which is the same as 14 to 2, during the era of nucleosynthesis. The result was 12 hydrogen nuclei (individual protons) for each helium nucleus. As shown, this means a hydrogen-to-helium mass ratio of 12 to 4, which is the same as 75% to 25%.

contrast, reactions that convert neutrons into protons release energy and so are unhindered by cooler temperatures. Just seconds into the era of nucleosynthesis, the temperature had fallen enough (to about 10^{10} K) that protons began to outnumber neutrons because the conversion reactions ran in only one direction. Neutrons changed into protons, but the protons didn't change back. The universe was still hot and dense enough for nuclear fusion, so protons and neutrons combined to form *deuterium*—a rare form of hydrogen nuclei that contains a neutron in addition to a proton—and deuterium nuclei fused to form helium (**Figure 13.10**).

At first, the helium nuclei were almost immediately blasted apart by one of the many gamma rays that filled the early universe. But fewer gamma rays remained as the universe cooled, and helium nuclei were able to survive by the time the universe was about 1 minute old. Calculations show that the proton-to-neutron ratio at this time should have been about 7 to 1. Moreover, almost all the available neutrons should have become incorporated into nuclei of helium. **Figure 13.11** shows that, based on the 7-to-1 ratio of protons to neutrons, the universe should have had a composition of 75% hydrogen and 25% helium by mass at the end of the era of nucleosynthesis.

The Big Bang theory therefore makes a very concrete prediction about the chemical composition of the universe: It should be 75% hydrogen and 25% helium by mass, in good agreement with the observed proportions of hydrogen and helium.

Think about it Briefly explain why it should not be surprising that some galaxies contain a little more than 25% helium, but why it would be very surprising if some galaxies contained less.

Abundances of Other Light Elements Why didn't the Big Bang produce heavier elements? By the time stable helium nuclei formed, when the universe was about a minute old, the temperature and density of the rapidly expanding universe had already dropped too low to sustain a process like carbon production (three helium nuclei fusing into carbon [Section 9.2]). Reactions between protons, deuterium nuclei, and helium were still possible, but most of these reactions led nowhere because the resulting nuclei were unstable and quickly split apart.

A few reactions involving hydrogen-3 (also known as tritium) or helium-3 *can* create long-lasting nuclei. However, the contributions of these reactions to the overall composition of the universe were minor because hydrogen-3 and helium-3 were so rare. Models of element production in the early universe show that, before the cooling of the universe halted fusion entirely, such reactions generated only trace amounts of lithium, the next lightest element after helium. Aside from hydrogen, helium, and this trace of lithium, all other elements were forged much later in stars.

✳ THE PROCESS OF SCIENCE IN ACTION

13.3 Inflation

The Big Bang theory has gained wide acceptance because of the strong evidence from the cosmic microwave background and the abundance of helium in the universe. However, the theory in its simplest form leaves several major features of our universe unexplained. The three most pressing questions are the following:

- *Where do structures like galaxies come from?* Recall that successful models of galaxy formation start from the assumption that gravity could collect matter together in regions of the early universe that had slightly enhanced density [Section 12.3]. We know that such regions of enhanced density were present in the universe at an age of 380,000 years from our observations of variations in the cosmic microwave background, but we have not yet explained how these density variations came to exist.

- *Why is the large-scale universe nearly uniform?* Although the slight variations in the cosmic microwave background show that the universe is not *perfectly* uniform on large scales, the fact that it is smooth to within a few parts in 100,000 is remarkable.

- *Why is the geometry of the universe flat?* Einstein's general theory of relativity tells us that the overall geometry of the universe can be curved, like the surface of a balloon or a saddle. However, observational efforts to measure the large-scale geometry of the universe have not yet detected any curvature. As far as we can tell, the large-scale geometry of the universe is flat.

The process of science demands that we seek natural explanations for these features of the universe. We will therefore conclude this chapter by examining the hypothesis of *inflation*, which may provide the necessary explanations, and how it has been tested by recent observations.

Did the universe undergo an early episode of inflation?

The idea of inflation first emerged in 1981, when physicist Alan Guth was considering how the separation of the strong force from the GUT force at the end of the GUT era might have affected the expansion of the universe. Some theories of high-energy physics predict that this separation of forces would have released enormous amounts of energy, and Guth realized that this energy could have caused the universe to expand dramatically, perhaps by a factor of 10^{30} in a time of less than 10^{-36} second. This period of rapid expansion is what scientists refer to as inflation, and it can explain the three key mysteries.

Structure: Giant Quantum Fluctuations To understand how inflation can explain the origin of structure, we need to recognize a special feature of energy fields. Laboratory-tested principles of quantum mechanics, especially the uncertainty principle (see Tools of Science, p. 148), tell us that the energy fields at any point in space are always fluctuating on very small scales, so even a complete vacuum is filled with tiny quantum "ripples" that can be characterized by a wavelength that corresponds roughly to their size. In principle, quantum ripples in the very early universe could have been the seeds for density enhancements that later grew into galaxies. However, the wavelengths of the original ripples were far too small to explain density enhancements like those that left imprints on the cosmic microwave background.

Inflation would have dramatically increased the wavelengths of these quantum fluctuations. The rapid growth of the universe during the period of inflation would have stretched tiny ripples from a size smaller than that of an atomic nucleus to the size of our solar system (**Figure 13.12**), making them large enough to become the density enhancements from which galaxies and larger structures later formed. In that case, the structure of today's universe may be all a result of tiny quantum fluctuations that rippled through the universe when it was less than a trillion trillion trillionth of a second old.

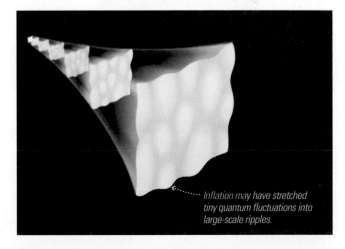

Inflation may have stretched tiny quantum fluctuations into large-scale ripples.

Figure 13.12 | During inflation, quantum ripples in spacetime would have stretched by a factor of perhaps 10^{30}. The peaks of these ripples then would have become the density enhancements that produced all the structure we see in the universe today.

Uniformity: Equalizing Temperatures and Densities The remarkable uniformity of the universe that the cosmic microwave background reveals might at first seem quite natural, but on further inspection it becomes difficult to explain. Imagine observing the cosmic microwave background in a certain part of the sky. You are seeing microwaves that have traveled through the universe since the end of the era of nuclei, which means you are seeing a region of the universe as it was about 14 billion years ago, when the universe was only 380,000 years old. Now imagine turning around and looking at the cosmic microwave background coming from the opposite direction. You are also seeing this region as it was 14 billion years ago, and it looks virtually identical in temperature and density. The surprising part is this: The two regions are billions of light-years apart—on opposite sides of our observable universe—but we are seeing them as they were when they were only 380,000 years old and couldn't possibly have exchanged light or any other information. A signal traveling at the speed of light from one to the other would barely have started its journey. So how did they come to have the same temperature and density?

The idea of inflation answers this question by saying that even though the two regions cannot have had any contact *since* the time of inflation, they were in contact prior to that time. Before the onset of inflation, when the universe was 10^{-38} second old, the two regions were less than 10^{-38} light-second away from each other. Radiation traveling at the speed of light would therefore have had time to bounce between the two regions, and this exchange of energy equalized their temperatures and densities. Inflation then pushed these equalized regions to much greater distances, far out of contact with each other. Like criminals getting their stories straight before being locked in separate jail cells, the two regions (and all other parts of the observable universe) came to the same temperature and density before inflation spread them far apart.

Because inflation caused different regions of the universe to separate so vastly in such a short period of time, many people wonder whether it violates Einstein's theories, which say that nothing can move faster than the speed of light. It does not, because nothing actually *moves* through space as a result of inflation or the ongoing expansion of the universe. Instead, the expansion of the universe is the expansion of *space itself*. Objects may be separating from one another at a speed faster than the speed of light, but no matter or radiation is able to travel between them during that time. In essence, inflation opens up a huge gap in space between objects that were once close together. The objects get very far apart, but nothing ever travels between them at a speed that exceeds the speed of light.

Geometry: Flattening the Universe The third question asks why the overall geometry of the universe is "flat." To understand this question, we must consider the overall geometry of the universe in a little more detail.

Recall that Einstein's general theory of relativity tells us that the presence of matter can curve spacetime [Section 10.2]. We cannot visualize this curvature in all four dimensions of spacetime, but we can detect its presence by its effects on how light travels through the universe. Although the curvature of the universe can vary from place to place, the universe as a whole must have some overall shape. Almost any shape is possible, but all the possibilities fall into just three general categories (**Figure 13.13**). Using analogies to objects that we can see in three dimensions, scientists refer to these three categories of shape as *flat*, *spherical*, and *saddle shaped*.

According to general relativity, the overall geometry of the universe depends on the average density of matter and energy within it, and the geometry can be flat only if the density of matter plus energy is precisely equal to a

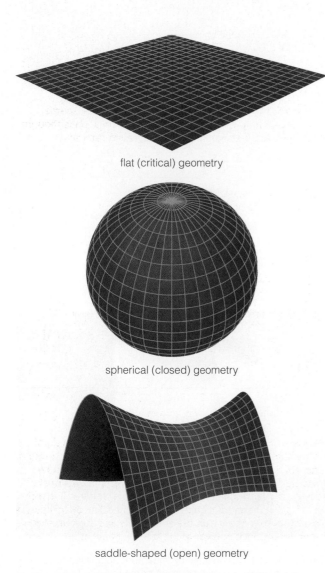

flat (critical) geometry

spherical (closed) geometry

saddle-shaped (open) geometry

Figure 13.13 | The three possible categories of overall geometry for the universe. Keep in mind that the real universe has these shapes in more dimensions than we can see.

value known as the **critical density.** If the universe's average density is less than the critical density, then the overall geometry is saddle shaped. If its average density is greater than the critical density, then the overall geometry is spherical.

Inflation can explain why the overall geometry of the universe is so close to being flat. In terms of Einstein's theory, the effect of inflation on space-time curvature is similar to the flattening of a balloon's surface when you blow it up (**Figure 13.14**). The flattening of space caused by inflation would have been so enormous that any curvature the universe might have had previously would be noticeable only on scales much larger than the observable universe. Inflation therefore makes the observable universe appear flat, which means that the average density of matter plus energy must be very close to the critical density.

Evidence for Inflation We've seen that inflation answers some remaining mysteries about the universe, but did it really happen? We cannot directly observe the universe at the very early time when inflation is thought to have occurred. Nevertheless, we can test the idea of inflation by exploring whether its predictions are consistent with our observations of the universe at later times. Scientists are only beginning to make observations that test inflation, but the findings to date are consistent with the idea that inflation made the universe uniform and flat while planting the seeds of structure formation.

The strongest tests of inflation to date come from detailed studies of the cosmic microwave background, especially the map made by the *WMAP* satellite (see Figure 13.9). Remember that this map shows tiny temperature differences corresponding to density variations in the universe at the end of the era of nuclei, when the universe was about 380,000 years old. The inflation hypothesis predicts that these density variations were actually created much earlier, when tiny quantum ripples expanded to much longer wavelengths. Careful measurements of the temperature variations in the microwave background support this prediction. **Figure 13.15** shows an analysis of the temperature variations observed by *WMAP* in the cosmic microwave background. The graph shows how the typical temperature differences between patches of sky depend on their angular size on the celestial sphere. The dots represent data from the observations, and the red curve shows the inflation-based model that best fits the observations; notice the close agreement between the model and the data. This model also makes specific and accurate predictions about other characteristics of our universe, such as its overall geometry, composition, and age, lending further credence to the idea of inflation.

The bottom line is that, all things considered, inflation accounts for features of our universe that are otherwise unaccounted for in the Big Bang theory. Many astronomers and physicists therefore suspect that some process akin to inflation really did occur in the early universe, though many of the details remain unclear. If these details can be worked out successfully, we face an amazing prospect—a breakthrough in our understanding of physical laws that govern the very smallest particles, achieved by studying the universe on the largest observable scales.

A very large curved surface seems flat to something small that lives on it.

Figure 13.14 | As a balloon expands, its surface seems increasingly flat to an ant crawling along it. Inflation is thought to have made the universe seem flat in a similar way.

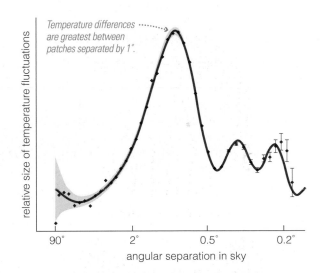

Temperature differences are greatest between patches separated by 1°.

relative size of temperature fluctuations

angular separation in sky

Figure 13.15 | This graph shows how detailed analysis of temperature differences in the cosmic microwave background supports the idea of inflation. The data points indicate how the typical temperature differences between patches of sky depend on their angular size on the celestial sphere. The red curve shows the prediction of a model that relies on inflation. Close agreement between the data points and the model is therefore evidence in favor of the idea of inflation.

Summary of Key Concepts

13.1 The Big Bang

What were conditions like in the early universe?

The early universe was filled with radiation and elementary particles. It was so hot and dense that the energy of radiation could turn into particles of **matter** and **antimatter,** which then collided and turned back into radiation.

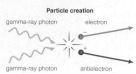

How did the early universe change with time?

According to the Big Bang Theory, the universe's history is marked by a number of eras, each with unique physical conditions. We know little about the **Planck era,** when the four forces may have all behaved as one. Gravity became distinct at the start of the **GUT era,** which may have ended with a period of rapid expansion called **inflation.** Electromagnetism and the weak force became distinct at the end of the **electroweak era.** Matter particles annihilated all the antimatter particles at the end of the **particle era.** Fusion of protons and neutrons into helium ceased at the end of the **era of nucleosynthesis.** Hydrogen nuclei captured all the free electrons, forming hydrogen atoms, at the end of the **era of nuclei.** Galaxies began to form at the end of the **era of atoms.** The **era of galaxies** continues to this day.

13.2 Evidence for the Big Bang Theory

How do we observe the radiation left over from the Big Bang?

Microwave telescopes allow us to observe the **cosmic microwave background**— radiation left over from the Big Bang. Its spectrum matches the characteristics expected of the radiation released at the end of the era of nuclei, confirming a key prediction of the Big Bang theory.

How do the abundances of elements support the Big Bang theory?

The Big Bang theory predicts the ratio of protons to neutrons during the era of nucleosynthesis and, from this, predicts that the chemical composition of the universe should be about 75% hydrogen and 25% helium by mass. This prediction matches observations of the element abundances in the universe, providing another piece of evidence to support the Big Bang theory.

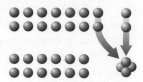

✴ THE PROCESS OF SCIENCE IN ACTION

13.3 Inflation

Did the universe undergo an early episode of inflation?

Scientists hypothesize that the universe underwent an early episode of inflation in order to explain (1) the density enhancements that led to galaxy formation, (2) the smoothness of that cosmic microwave background, and (3) the flat geometry of the observable universe. We can test the idea of inflation because it makes specific predictions about patterns we observe in the cosmic microwave background, and observations made by microwave telescopes so far match those predictions.

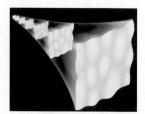

Investigations

Quick Quiz

Choose the best answer to each of the following; answers are in Appendix D. Explain your reasoning with one or more complete sentences.

1. What is the current temperature of the universe? (a) absolute zero (b) a few degrees K (c) a few thousand degrees K
2. What is the charge of an antielectron? (a) positive (b) negative (c) neutral
3. What happens when a proton collides with an antiproton? (a) They repel each other. (b) They fuse together. (c) They convert into two photons.
4. Which of the following does *not* provide strong evidence for the Big Bang theory? (a) observations of the cosmic microwave background (b) observations of the amount of hydrogen in the universe (c) observations of the ratio of helium to hydrogen in the universe
5. What kinds of new particles are produced in particle accelerators? (a) only matter (b) only antimatter (c) both matter and antimatter
6. What kinds of particles were present in the universe during the particle era? (a) mostly matter (b) mostly antimatter (c) approximately equal amounts of matter and antimatter
7. What kinds of particles were present in the universe during the era of nucleosynthesis? (a) mostly matter (b) mostly antimatter (c) approximately equal amounts of matter and antimatter
8. What kinds of atomic nuclei were present in the universe during the era of nuclei? (a) all kinds of nuclei (b) mostly hydrogen and helium nuclei (c) only hydrogen nuclei

9. If you had been present in the universe at the beginning of the era of atoms, what color light would you have seen? (a) white light (b) red light (c) no visible light, because only microwave light was present

10. Which of the following does inflation help to explain? (a) the uniformity of the cosmic microwave background (b) the amount of helium in the universe (c) the temperature of the cosmic microwave background

11. Adding the idea of inflation to the Big Bang theory accounts for (a) the origin of hydrogen. (b) the origin of galaxies. (c) the origin of atomic nuclei.

12. Which of these pieces of evidence supports the idea that inflation really happened? (a) observations showing that the universe is expanding (b) measurements of the abundance of helium in the universe (c) observations of the cosmic microwave background that indicate a flat geometry for the universe

Short-Answer/Essay Questions

Explain all answers clearly, with complete sentences and proper essay structure if needed. An asterisk (*) designates a quantitative problem, for which you should show all your work.

13. *Life Story of a Proton.* Tell the life story of a proton from its formation shortly after the Big Bang to its presence in the nucleus of an oxygen atom you have just inhaled. Your story should be creative and imaginative, but it should also demonstrate your scientific understanding of as many stages in the proton's life as possible. Your story should be three to five pages long.

14. *Creative History of the Universe.* The story of creation as envisioned by the Big Bang theory is quite dramatic, but it is usually told in a fairly straightforward, scientific way. Write a more dramatic version of the story, in the form of a short story, play, or poem. Be as creative as you wish, but be sure to remain scientifically accurate.

15. *Re-Creating the Big Bang.* Particle accelerators on Earth can push particles to extremely high speeds. When these particles collide, the amount of energy associated with the colliding particles is much greater than the mass-energy these particles have when at rest. As a result, these collisions can produce many other particles out of pure energy. Explain in your own words how the conditions that occur in these accelerators are similar to the conditions that prevailed shortly after the Big Bang. Also point out some of the differences between what happens in particle accelerators and what happened in the early universe.

16. *Wavelength of the Microwave Background.* Suppose you could have observed the spectrum of the cosmic microwave background radiation (as in Figure 13.8) 1 billion years after the Big Bang. How would the wavelength of radiation at the peak of that spectrum compare with the peak wavelength of today's background spectrum? Explain your reasoning.

17. *"Observing" the Early Universe.* The only way we have of studying the era of nucleosynthesis is through the abundances of the nuclei that it left behind. Explain why we will never be able to observe that era through direct detection of photons emitted at that time.

18. *Element Production in the Big Bang.* Nucleosynthesis in the early universe was unable to produce more than trace amounts of elements heavier than helium. Using the information in Figure 9.18, which shows the mass per nuclear particle for many different elements, explain why producing elements like lithium (3 protons), boron (4 protons), and beryllium (5 protons) was so difficult.

19. *Evidence for the Big Bang.* Alternatives to the Big Bang theory need to propose alternative explanations for many of the observations scientists have made of the universe. Make a list of at least six observed features of the universe that are satisfactorily explained by the Big Bang theory when it is combined with the idea of inflation.

*20. *Energy from Antimatter.* The total annual U.S. power consumption is about 2×10^{20} joules. Suppose you could supply that energy by combining pure matter with pure antimatter. Estimate the total mass of matter-antimatter fuel you would need in order to supply the United States with energy for 1 year. How does that mass compare with the amount of matter in your car's gas tank? (A gallon of gas has a mass of about 4 kilograms.) (*Hint:* Consider the meaning of the formula $E = mc^2$ and use the fact that 1 joule $= 1 \text{ kg} \times \text{m}^2/\text{s}^2$.)

*21. *Uniformity of the Cosmic Microwave Background.* The temperature of the cosmic microwave background differs by only a few parts in 100,000 across the sky. Compare this level of uniformity to that of the surface of a table in the following way. Consider a square table that is 1 meter on each side. How big would the largest bumps on that table be if its surface were smooth to 1 part in 100,000? Could you see bumps of that size on the table's surface with your naked eyes?

14 Dark Matter and Dark Energy

This photograph is a composite image of a cluster of galaxies currently undergoing an enormous collision. Many yellow and white galaxies are visible, but they represent only a small fraction of the cluster's total mass. X-ray observations, shown in red, reveal hot gas with mass several times as great as the mass in stars. But most of the mass of the cluster seems to consist of *dark matter* that is completely invisible—blue regions show where it is located. In this chapter, we'll discuss the evidence for dark matter and explore its role in galaxy formation. We'll also present evidence for something even more mysterious—a form of *dark energy* that is overpowering the gravity of dark matter on large scales and causing the expansion of the universe to accelerate.

14.1 Evidence for Dark Matter

What is the universe made of? Ask an astronomer this seemingly simple question, and you might see a professional scientist blush with embarrassment. Based on all the available evidence today, the answer to this simple question is "We do not know."

It might seem incredible that we still do not know the composition of most of the universe, and you might wonder why we should be so clueless. After all, astronomers can measure the chemical composition of distant stars and galaxies from their spectra, so we know that stars and gas clouds are made almost entirely of hydrogen and helium, with small amounts of heavier elements mixed in. But notice the key words "chemical composition." When we say these words, we are talking about the composition of material built from atoms of elements such as hydrogen, helium, carbon, and iron.

While it is true that all familiar objects—including people, planets, and stars—are built from atoms, the same may not be true of the universe as a whole. In fact, we now have good reason to think that the vast majority of the matter in the universe is *not* composed of atoms. We call this stuff **dark matter** because it emits no detectable light—it reveals its presence only through its gravitational influence on things we *can* observe.

What is the evidence for dark matter?

We infer the existence of dark matter by observing the motions of objects such as stars and gas clouds in galaxies and clusters of galaxies. From these motions, we can calculate how much matter must be present. These calculations reveal much more matter than we can see, leading us to conclude that most of the matter must be dark matter. Let's explore this evidence in more detail, beginning with the evidence for dark matter in our own Milky Way Galaxy.

Dark Matter in the Milky Way   Recall that Newton's law of gravity determines the speeds (or velocities) at which objects orbit one another, allowing us to measure the mass of an orbiting system. In the case of the Milky Way Galaxy, we can use the Sun's orbital velocity and its distance from the galactic center to determine the mass lying *within* the Sun's orbit (**Figure 14.1**). Similarly, we can use the orbital motion of any other star or gas cloud to measure the mass of the galaxy within that star or gas cloud's orbit (see Tools of Science, p. 229). Because interstellar dust obscures our view of stars throughout most of our galaxy's disk, most measurements of motion in the Milky Way are based on observations of atomic hydrogen gas clouds, which emit radio waves that penetrate interstellar dust [Section 11.1].

We can summarize the results of these orbital velocity measurements with a diagram called a **rotation curve,** which plots the rotational velocity of stars or gas clouds against their distance from the center of the galaxy. As a simple example of the concept, let's construct a rotation curve for a merry-go-round. Every object on a merry-go-round goes around the center with the same rotational period, but objects farther from the center move in larger circles, so they must move at faster speeds. The rotation curve for a merry-go-round is therefore a straight line that rises steadily with distance (**Figure 14.2a**).

Next, consider the rotation curve for our solar system; it drops off with distance from the Sun, because inner planets orbit at faster speeds than outer planets (**Figure 14.2b**). This drop-off in speed with distance occurs because virtually all the mass of the solar system is concentrated in the Sun. The gravitational force holding a planet in its orbit decreases with distance from the Sun, and a smaller force means a lower orbital speed. The rotation curve of any astronomical system whose mass is concentrated toward the center must drop similarly.

Figure 14.1 | Our Sun orbits the Milky Way at a distance 28,000 light-years from the center and at a speed of 220 km/s. From these orbital measurements, we determine that the mass of the portion of the Milky Way lying within the Sun's orbit is about 2×10^{41} kg, equivalent to 100 billion times the mass of the Sun.

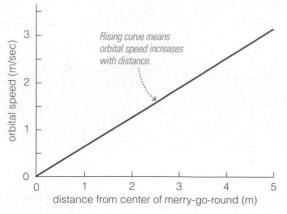

Rising curve means orbital speed increases with distance.

a A rotation curve for a merry-go-round.

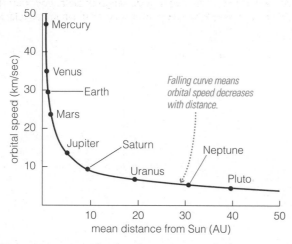

Falling curve means orbital speed decreases with distance.

b The rotation curve for the planets in our solar system.

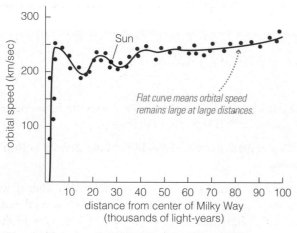

Flat curve means orbital speed remains large at large distances.

c The rotation curve for the Milky Way Galaxy. Dots represent stars or gas clouds whose orbital speeds have been measured.

Figure 14.2 | Rotation curves show how the orbital speed of a system depends on distance from its center. The solar system's rotation curve declines with radius because its mass is concentrated at the center. The Milky Way's rotation curve is flat, indicating that its mass extends well beyond the Sun's orbit.

Figure 14.2c shows the rotation curve for the Milky Way Galaxy. Each individual dot represents the distance from the galactic center and the orbital speed of a particular star or gas cloud. Notice that the orbital velocities remain approximately constant beyond the inner few thousand light-years, making the rotation curve look flat. This behavior contrasts sharply with the steeply declining rotation curve of the solar system. Therefore, most of the mass of the Milky Way must *not* be concentrated at its center, as it is in the solar system. Instead, the orbits of progressively more distant gas clouds must encircle more and more mass. The Sun's orbit encompasses about 100 billion solar masses, but a circle twice as large surrounds twice as much mass, and a larger circle surrounds even more mass.

The flatness of the Milky Way's rotation curve therefore implies that most of our galaxy's mass lies well beyond our Sun, tens of thousands of light-years from the galactic center. A more detailed analysis suggests that most of this mass is located in the spherical halo that surrounds the disk of our galaxy, and that the total amount of this mass might be *10 times* the total mass of all the stars in the disk. Because we have detected very little radiation coming from this enormous amount of mass, it qualifies as dark matter. If we are interpreting the evidence correctly, the luminous part of the Milky Way's disk is like the tip of an iceberg, marking only the center of a much larger, unseen clump of mass (**Figure 14.3**).

Think about it Suppose we made a rotation curve for the moons orbiting Jupiter. Would it rise, fall, or stay flat with increasing distance from Jupiter?

Dark Matter in Other Galaxies Other galaxies also seem to contain vast quantities of dark matter. We can determine the amount of dark matter in a galaxy by comparing the galaxy's mass to its luminosity. The procedure is simple in principle. First, we use the galaxy's luminosity to estimate the amount of mass that the galaxy contains in the form of stars. Next, we determine the galaxy's total mass by applying the law of gravity to observations of

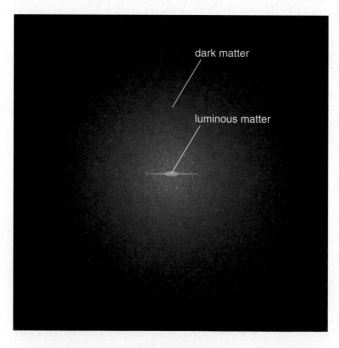

Figure 14.3 | The dark matter associated with a spiral galaxy like the Milky Way occupies a much larger volume than the galaxy's luminous matter. The radius of this dark-matter halo may be 10 times as large as that of the galaxy's halo of stars.

the orbital velocities of stars and gas clouds. If this total mass is larger than the mass that we can attribute to stars, then we infer that the excess mass must be dark matter.

Measuring the galaxy's total mass requires measuring orbital speeds of stars or gas clouds as far from the galaxy's center as possible. Atomic hydrogen gas clouds can be found in spiral galaxies at greater distances from the center than stars, so most of our data come from radio observations of these clouds. We use Doppler shifts to determine how fast a cloud is moving toward us or away from us, then combine observations for clouds at varying orbital distances to construct the galaxy's rotation curve.

The rotation curves of most spiral galaxies turn out to be remarkably flat as far out as we can see (**Figure 14.4**). Just as in the Milky Way, these flat rotation curves imply that a great deal of matter lies far out in the halos of these spiral galaxies. A detailed analysis tells us that other spiral galaxies also have at least 10 times as much mass in dark matter as they do in stars. In other words, the composition of typical spiral galaxies is 90% or more dark matter and 10% or less matter in stars.

Elliptical galaxies contain very little atomic hydrogen gas and hence do not produce detectable 21-centimeter radiation, so we generally measure the masses of elliptical galaxies by observing the motions of their stars. When we compare the Doppler shifts of spectral lines from different regions of an elliptical galaxy, we find that the speeds of stars remain fairly constant as we look farther from the galaxy's center. Just as in spirals, most of the matter in elliptical galaxies must lie beyond the radii where we no longer observe many stars, and hence must be dark matter. The evidence for dark matter becomes even more convincing for cases in which we can measure the speeds of globular star clusters orbiting at large distances from the center of an elliptical galaxy. These

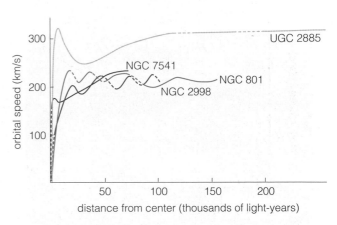

Figure 14.4 | Actual rotation curves of four spiral galaxies. They are all nearly flat over a wide range of distances from the center, indicating that dark matter is common in spiral galaxies.

Tools of Science: The Orbital Velocity Law

Newton's laws of motion and gravity are astronomers' primary tools for measuring mass. We have already seen how Newton's version of Kepler's third law can tell us the mass of an orbiting system if we know its period and average orbital distance (see Tools of Science, p. 96). With a little algebra, we can transform this law into a related law that allows us to calculate the amount of mass contained within any object's orbit, if we know the radius and velocity of that orbit. For an object on a circular orbit, this *orbital velocity law* is

$$M_r = \frac{r \times v^2}{G},$$

where v is the object's orbital velocity, r is its orbital radius, M_r is the amount of mass contained within its orbit, and $G = 6.67 \times 10^{-11} \frac{m^3}{kg \times s^2}$ is the gravitational constant. Notice that for a given radius, a larger orbital velocity indicates a larger amount of mass, reflecting the fact that a stronger gravitational pull is necessary to hold a faster-moving object in orbit.

This law is one of our best tools for measuring the amount of dark matter in the universe. When we apply it to the orbits of stars and gas clouds on the outskirts of spiral galaxies, we find that these galaxies contain at least 10 times as much matter as we can see in the form of stars. When we apply it to the orbits of galaxies within clusters of galaxies, we

find that the total amount of matter in these clusters is as much as 50 times the amount of matter in the form of stars.

Example: Calculate the mass of the Milky Way Galaxy within the Sun's orbit.

Solution: The Sun orbits the center of the Milky Way Galaxy with velocity $v = 220$ km/s $= 2.2 \times 10^5$ m/s. The Sun's orbital radius (distance) is 28,000 light-years, equivalent to $(28,000 \text{ ly}) \times (9.46 \times 10^{15} \text{ m/ly}) = 2.6 \times 10^{20}$ m. Using these values in the orbital velocity law, we find

$$M_r = \frac{r \times v^2}{G} = \frac{(2.6 \times 10^{20} \text{ m}) \times (2.2 \times 10^5 \frac{m}{s})^2}{6.67 \times 10^{-11} \frac{m^3}{kg \times s^2}}$$

$$= 1.9 \times 10^{41} \text{ kg}$$

The mass of the Milky Way Galaxy within the Sun's orbit is about 2×10^{41} kg, which we can convert to solar masses by dividing by the Sun's mass of about 2×10^{30} kg; it is equivalent to about 10^{11} (or 100 billion) solar masses. Note that most of this mass is not dark matter; evidence for dark matter comes from using the same equation for stars or gas clouds orbiting much farther from the center of the galaxy than our Sun.

Figure 14.5 | A distant cluster of galaxies in both visible light and X-ray light. The visible-light photo shows the individual galaxies. The blue-violet overlay represents different levels of X-ray emission from extremely hot gas in the cluster. Evidence for dark matter comes both from the observed motions of the visible galaxies and from the temperature of the hot gas. (The region shown is about 8 million light-years across.)

Figure 14.6 | This Hubble Space Telescope photo shows a galaxy cluster acting as a gravitational lens. The yellow elliptical galaxies are cluster members. The many small blue ovals (such as those indicated by the arrows) are multiple images of a single galaxy that lies almost directly behind the cluster's center. (The picture shows a region about 1.4 million light-years across.)

measurements suggest that elliptical galaxies, like spirals, also contain far more matter than we can see in the form of stars.

Dark Matter in Clusters of Galaxies   Observations of galaxy clusters suggest that the total proportion of dark matter is even greater than we find in galaxies. The evidence for dark matter in clusters comes from three different ways of measuring cluster masses: from the speeds of galaxies orbiting the center of the cluster, from the temperature of hot gas between the cluster galaxies, and from how the clusters bend light as *gravitational lenses.*

The technique of measuring a cluster's mass from the orbital speeds of its galaxies is similar to the techniques we've discussed for individual galaxies. We measure the velocities of galaxies orbiting within the cluster from the Doppler shifts of their spectral lines. We then combine this information with the distances of those galaxies to calculate how much mass the cluster contains.

Think about it   What would happen to a cluster of galaxies if you could remove all the dark matter without changing the velocities of the galaxies?

A second technique relies on the fact that clusters of galaxies generally contain vast amounts of very hot, X-ray-emitting gas (**Figure 14.5**). We can measure the temperature of this gas from its X-ray spectrum, and this temperature tells us about dark matter because it depends on the total mass of the cluster. The gas in most clusters is nearly in a state of *gravitational equilibrium*—that is, the outward gas pressure balances gravity's inward pull [Section 9.1]. In this state of balance, the average kinetic energies of the gas particles are determined primarily by the strength of gravity and hence by the amount of mass within the cluster. Because the temperature of a gas reflects the average kinetic energies of its particles, it tells us the average speeds of the gas particles, which we can then use to determine the cluster's total mass. The mass measurements obtained with this technique agree with those found by studying the orbital motions of the cluster's galaxies.

The third technique relies on Einstein's general theory of relativity (see Tools of Science, p. 169), which tells us that masses distort spacetime—the "fabric" of the universe. Massive objects can therefore act as **gravitational lenses** that bend light beams passing by them. Because the light-bending angle of a gravitational lens depends on the mass of the object doing the bending, we can measure the masses of objects by observing how much they distort light paths.

Figure 14.6 shows a striking example of how a cluster of galaxies can act as a gravitational lens. Many of the yellow elliptical galaxies concentrated toward the center of the picture belong to the cluster, but at least one of the galaxies pictured does not. Notice the multiple blue ovals that appear at several positions around the central clump of yellow galaxies. These ovals are all images of the same galaxy. This galaxy lies almost directly behind the center of the cluster, at a much greater distance. We see multiple images of this single galaxy because photons do not follow straight paths as they travel from the galaxy to Earth. Instead, the cluster's gravity bends the photon paths, allowing light from the galaxy to arrive at Earth from a few slightly different directions (**Figure 14.7**). Each alternative path produces a separate, distorted image of the blue galaxy.

Multiple images of a gravitationally lensed galaxy are rare. They occur only when a distant galaxy lies directly behind the lensing cluster. However, single distorted images of gravitationally lensed galaxies are quite common. **Figure 14.8** shows a typical example. This picture shows numerous normal-looking galaxies and several arc-shaped galaxies. The oddly curved galaxies are not members of the cluster, nor are they really curved. Rather, each is a normal galaxy lying far beyond the cluster, and their distorted images are a result of the bending of light by the cluster's gravity.

Careful analyses of these images enable us to measure how massive these clusters must be to generate the observed distortions. It is reassuring that cluster masses derived in this way generally agree with those derived from galaxy velocities and X-ray temperatures. The three different methods all indicate that the mass of dark matter in clusters of galaxies is as much as 50 times the mass in stars and at least 5 times the mass in hot gas.

Alternatives to Dark Matter   Astronomers have made a strong case for the existence of dark matter, but is it possible that there's a completely different explanation for the observations we've discussed? All the evidence for dark matter relies on our understanding of gravity. For individual galaxies, the case for dark matter rests primarily on applying Newton's laws of motion and gravity to observations of the orbital speeds of stars and gas clouds. We've used the same laws to make the case for dark matter in clusters, along with additional evidence based on gravitational lensing predicted by Einstein's general theory of relativity. It therefore seems that one of the following must be true:

1. Dark matter really exists, and we are observing the effects of its gravitational attraction.

2. There is something wrong with our understanding of gravity that is causing us to mistakenly infer the existence of dark matter.

We cannot yet rule out the second possibility, but most astronomers consider it very unlikely. Newton's laws of motion and gravity are among the most trustworthy tools in science. We have used them time and again to measure masses of celestial objects from their orbital properties. We found the masses of Earth and the Sun by applying Newton's version of Kepler's third law to objects that orbit them. We used this same law to calculate the masses of stars in binary star systems, revealing the general relationships between the masses of stars and their outward appearances. Newton's laws have also told us the masses of things we can't see directly, such as the masses of orbiting neutron stars in X-ray binaries and of black holes in active galactic nuclei. Einstein's general theory of relativity likewise stands on solid ground, having been repeatedly tested and verified to high precision in many observations and experiments. We therefore have good reason to trust our current understanding of gravity. While we should always keep an open mind about the possibility of future changes in our understanding, we will proceed with the rest of the chapter under the assumption that dark matter really does exist.

 Think about it   Should the fact that we have three different ways of measuring cluster masses give us greater confidence that we really do understand gravity and that dark matter really does exist? Why or why not?

What might dark matter be made of?

What is all this dark stuff in galaxies and clusters of galaxies? There are two basic possibilities: (1) It could be made of ordinary matter that is built from protons, neutrons, and electrons, but in forms too dark for us to detect with current technology; or (2) it could be made of exotic particles that we have yet to discover. Current evidence indicates that while some of the dark matter might be ordinary, most of it must be exotic. To understand this evidence, let's start with the first possibility.

Astronomers consider matter to be "dark" as long as it is too dim to be visible in the halo of our galaxy or beyond. Your body is dark matter, because you would be far too dim for our telescopes to detect if you were somehow flung into the halo of our galaxy. Similarly, Earth and other planets are dark

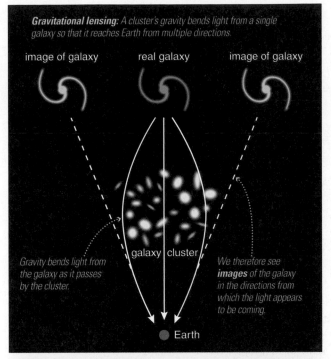

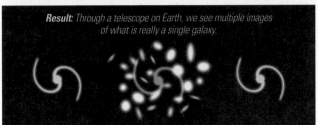

Figure 14.7 | A cluster's powerful gravity bends light paths from background galaxies to Earth. If light arrives from several different directions, we see multiple images of the same galaxy.

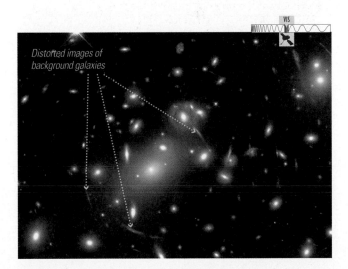

Figure 14.8 | Hubble Space Telescope photo of the cluster Abell 2218. The thin, elongated shapes around the main clump of galaxies are images of background galaxies distorted by the cluster's gravity. By measuring these distortions, astronomers can determine the total amount of mass in the cluster. (The region pictured is about 1.4 million light-years from side to side.)

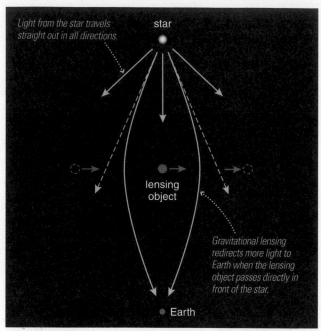

Light from the star travels straight out in all directions.

star

lensing object

Gravitational lensing redirects more light to Earth when the lensing object passes directly in front of the star.

Earth

Result: The lensed star appears brighter when the lensing object is in front.

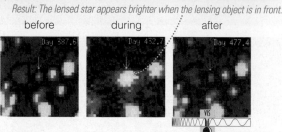

before · during · after

Figure 14.9 | When an object similar in mass to a star passes in front of another, more distant star, gravitational lensing temporarily makes the more distant star appear brighter. Searches for such events show that our galaxy's halo does indeed contain dim objects with masses similar to those of small stars, but that these objects do not constitute the majority of the galaxy's dark matter.

matter because we could not see them at great distances either. The "failed stars" known as *brown dwarfs* [Section 9.1] and even faint red main-sequence stars of spectral type M [Section 8.2] also qualify as dark matter, because they are too dim for current telescopes to see in the halo.

Searches for dark matter in the form of objects like planets, brown dwarfs, or dim stars have shown that such objects do exist in our galaxy's halo, but account for only a small fraction of the halo's mass. One innovative technique for detecting such objects (sometimes known as MACHOs, for *massive compact halo objects*) takes advantage of gravitational lensing on a much smaller scale than the examples we studied for clusters of galaxies. Every once in a while, when one of these objects drifts across our line of sight to a more distant star, its gravity can focus more of the distant star's light directly toward Earth. The distant star appears much brighter than usual for several days or weeks as the lensing object passes in front of it (**Figure 14.9**). We cannot see the lensing object itself, but the duration of the lensing event reveals its mass. Counting these gravitational lensing events has shown that there simply cannot be enough compact objects made of ordinary matter to account for the bulk of our galaxy's dark matter. Similar measurements rule out the possibility that the dark matter consists of large numbers of black holes formed by the deaths of massive stars.

Additional evidence indicating that most of the dark matter cannot be made of familiar particles comes from studies of the early universe. Recall that the universe was hot enough during the first 5 minutes after the Big Bang to fuse protons and neutrons into deuterium and deuterium nuclei into helium [Section 13.2]. Most of the neutrons ended up in helium nuclei, but a small fraction remained in deuterium nuclei. Models of nuclear fusion tell us that the amount of deuterium left over depends on how many protons and neutrons were made by the Big Bang. Measurements of the amount of deuterium in the universe therefore allow us to infer the overall number of proton and neutrons in the observable universe. These measurements show that matter made from protons and neutrons represents only about one-sixth of the total amount of matter in the universe.

Such observations have forced astronomers to seriously consider the possibility that most of the matter in the universe is made of unfamiliar particles. Let's begin to explore this possibility by taking another look at a peculiar type of particle that we first encountered in connection with nuclear fusion: neutrinos. Large numbers of these particles were made in the Big Bang, and they are dark by nature because they have no electrical charge and cannot emit electromagnetic radiation of any kind. Moreover, they are never bound together with charged particles in the way that neutrons are bound in atomic nuclei, so their presence cannot be revealed by associated light-emitting particles. Particles like neutrinos interact with other forms of matter through only two of the four forces: gravity and the weak force. For this reason, they are said to be *weakly interacting particles*.

The dark matter in galaxies cannot be made of neutrinos, because these very-low-mass particles travel through the universe at enormous speeds and can easily escape a galaxy's gravitational pull. But what if other weakly interacting particles exist that are similar to neutrinos but considerably more massive? They, too, would have been made in the Big Bang and would evade direct detection, but they would move more slowly, so their mutual gravity could hold together a large collection of them. Such hypothetical particles are called **weakly interacting massive particles,** or **WIMPs** for short. Note that they are subatomic particles, so the "massive" in their name is relative—they are massive only in comparison with lightweight particles like neutrinos. Such particles could make up most of the mass of a galaxy or cluster of galaxies, but they would be completely invisible in all wavelengths of light.

Supporting evidence for weakly interacting massive particles comes from careful analysis of the cosmic microwave background. Look back at Figure 13.15, which shows the model that best explains observed temperature patterns in this background radiation. In this model, one-sixth of the matter in the universe consists of ordinary matter made from protons and neutrons and five-sixths consists of WIMPs, in agreement with the conclusions we have drawn from deuterium observations and the proportions of dark matter and hot gas in clusters of galaxies. If the proportion of ordinary matter were greater, the model would no longer fit the microwave data. Astronomers therefore consider it likely that weakly interacting massive particles make up the vast majority of dark matter and hence the majority of all matter in the universe.

 Think about it What do you think of the idea that much of the universe is made of as-yet-undiscovered particles? Can you think of other instances in the history of science in which the existence of something was predicted before it was discovered?

14.2 Gravity versus Expansion

Whatever dark matter turns out to be, its presence in the universe appears to have been crucial to our existence today. According to the Big Bang theory, the expansion of the universe initially carried each part of the universe away from every other part of the universe. In order for stars and planets to form, gravity somehow had to halt the expansion of clumps of matter large enough to make the protogalactic clouds that eventually formed stars and galaxies [Section 11.1]. Because dark matter seems to provide most of the gravity that holds galaxies together, we strongly suspect that the gravitational attraction of dark matter is what stopped protogalactic clouds from expanding and pulled them together in the first place. Without dark matter, expansion would have dispersed all the matter in the universe before any galaxies could form.

In fact, the gravity of dark matter appears to have halted the expansion of regions much larger than individual galaxies, producing structure in the universe on very large scales. In this section, we'll explore how such structures formed and then consider an even bigger question: Could the gravity of dark matter be powerful enough to halt the expansion of the universe as a whole?

What is the role of dark matter in structure formation?

During the first few million years after the Big Bang, the universe expanded everywhere. However, the density of the universe was not completely uniform; recall that the temperature patterns we observe in the cosmic microwave background (see Figure 13.9) show that the density of matter varied slightly from place to place. Over time, gravity drew matter toward higher-density regions and away from lower-density regions, which made the density differences greater. Gradually, the stronger gravity in regions of enhanced density stopped the expansion of these regions, which then contracted to form protogalactic clouds, even as the universe as a whole continued (and still continues) to expand.

Models of structure formation predict what should have happened once gravity collected matter into a protogalactic cloud. Ordinary matter consisted of hydrogen and helium gas produced by the Big Bang. Because this matter is made of normal atoms that respond to the electromagnetic force, collisions among gas particles could convert some of their orbital energy into radiative energy that escaped the cloud in the form of photons. This loss of orbital energy allowed these particles to sink toward the center of the cloud, where they formed a rotating disk. In contrast, dark matter made of weakly interacting particles could not produce photons and rarely interacted and exchanged energy with other particles. These dark matter particles therefore remained

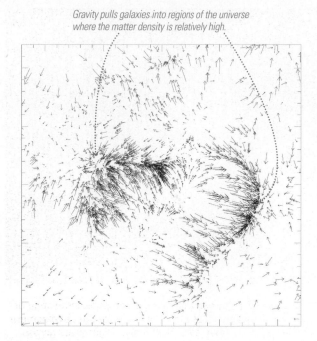

Gravity pulls galaxies into regions of the universe where the matter density is relatively high.

Figure 14.10 | This diagram represents the motions of galaxies attributable to effects of gravity. Each arrow represents the amount by which a galaxy's actual velocity (inferred from a combination of observations and modeling) differs from the velocity we'd expect it to have from Hubble's law alone. The Milky Way is at the center of the picture, which shows an area about 600 million light-years from side to side. Notice how the galaxies tend to flow into regions where the density of galaxies is already high. These vast, high-density regions are probably superclusters in the process of formation.

stuck in orbits far out in the galactic halo. The fact that most dark matter seems to be located at large distances from the centers of galaxies provides additional support for the idea that dark matter is made primarily of weakly interacting particles.

These models also account for the formation of galaxy clusters in a process that echoed the formation of galaxies. Early on, all the galaxies that would eventually constitute a cluster were flying apart with the expansion of the universe, but the overall gravity of the cluster—most of which is associated with dark matter—eventually reversed the trajectories of these galaxies. The galaxies ultimately fell back inward to form the cluster. On even larger scales, clusters themselves seem to be tugging on one another, hinting that even larger structures, called **superclusters,** may still be in the early stages of formation (**Figure 14.10**).

Large-Scale Structures Maps of the distribution of galaxies across large regions of space reveal **large-scale structures** much vaster than clusters or even superclusters of galaxies. Making these maps requires an enormous amount of data. A long-exposure photo showing galaxy positions is not enough, because it does not tell us the galaxy distances. We must also measure the redshift of each individual galaxy so that we can estimate its distance by applying Hubble's law.

Figure 14.11 shows the distribution of galaxies in three slices of the universe, each extending farther out in distance. Our Milky Way Galaxy is located at the vertex at the far left, and each dot represents another galaxy. The slice at the left comes from one of the first surveys of large-scale structures, performed

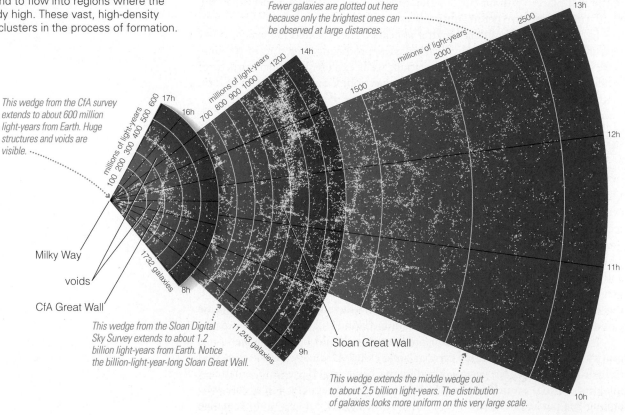

Fewer galaxies are plotted out here because only the brightest ones can be observed at large distances.

This wedge from the CfA survey extends to about 600 million light-years from Earth. Huge structures and voids are visible.

Milky Way

voids

CfA Great Wall

This wedge from the Sloan Digital Sky Survey extends to about 1.2 billion light-years from Earth. Notice the billion-light-year-long Sloan Great Wall.

Sloan Great Wall

This wedge extends the middle wedge out to about 2.5 billion light-years. The distribution of galaxies looks more uniform on this very large scale.

1732 galaxies

11,243 galaxies

millions of light-years

Figure 14.11 | Each of these three wedges shows a slice of the universe extending outward from our own Milky Way Galaxy. The dots represent galaxies, shown at their measured distances from Earth. We see that galaxies are not scattered randomly but instead trace out long chains and sheets surrounded by huge voids containing very few galaxies.

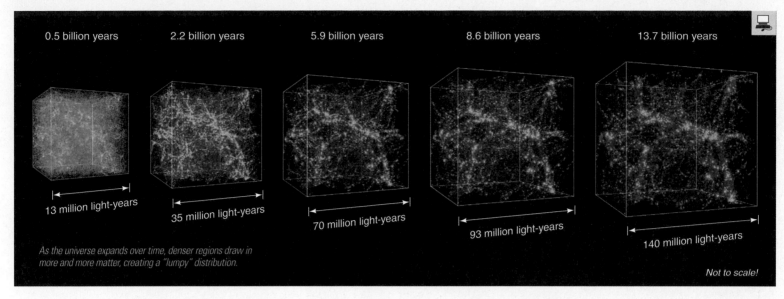

0.5 billion years   2.2 billion years   5.9 billion years   8.6 billion years   13.7 billion years

13 million light-years

35 million light-years

70 million light-years

93 million light-years

140 million light-years

As the universe expands over time, denser regions draw in more and more matter, creating a "lumpy" distribution.

Not to scale!

Figure 14.12 | Frames from a supercomputer simulation of structure formation. These five boxes depict the development of a cubical region that is now 140 million light-years across. The labels above the boxes give the age of the universe, and the labels below give the size of the box as it expands with time. Notice that the distribution of matter is only slightly lumpy when the universe is young (left frame). Structures grow more pronounced with time as the densest lumps draw in more and more matter.

in the 1980s. This map, which required years of effort by many astronomers, dramatically reveals the complex structure of our corner of the universe. It shows that galaxies are not scattered randomly through space, but are instead arranged in huge chains and sheets that span many millions of light-years. Clusters of galaxies are located at the intersections of these chains. Between these chains and sheets of galaxies lie giant empty regions called **voids.** The other two slices show data from the more recent Sloan Digital Sky Survey, which takes advantage of technology that gives us redshift measurements—and hence estimated distances—for hundreds of galaxies during a single night of telescopic observation. The Sloan Survey has measured redshifts for nearly a million galaxies spread across about one-fourth of the sky. Some of the structures in these pictures are amazingly large: The so-called *Sloan Great Wall*, clearly visible in the center slice, extends more than 1 billion light-years from end to end.

The universe may still be growing structures on these very large scales. However, there seems to be a limit to the size of the largest structures. If you look closely at the rightmost slice in Figure 14.11, you'll notice that the overall distribution of galaxies appears nearly uniform on scales larger than about a billion light-years. In other words, on very large scales the universe looks much the same everywhere, in agreement with what we expect from the *Cosmological Principle* [Section 12.2].

The Origin of Large Structures

Why is gravity collecting matter on such enormous scales? As with galaxies, we suspect that these larger structures formed from regions of enhanced density in the early universe. Galaxies, clusters, superclusters, and the Sloan Great Wall probably all started as different-sized regions of slightly higher density. The voids in the distribution of galaxies probably started as regions of slightly lower density.

If this picture of structure formation is correct, then the structures we see in today's universe mirror the original distribution of dark matter very early in time. Supercomputer models of structure formation in the universe can now simulate the growth of galaxies, clusters, and larger structures from tiny density enhancements as the universe evolves (**Figure 14.12**). Models of extremely large regions reveal how dark matter should be distributed throughout the entire observable universe. The results look remarkably similar to the slices of the universe in Figure 14.11, giving us confidence in our ideas about structure formation. **Figure 14.13** summarizes our understanding of how galaxies and clusters of galaxies developed from slight density enhancements in the early universe.

All galaxies, including our Milky Way, developed as gravity pulled together matter in regions of the universe that started out slightly denser than surrounding regions. The central illustration depicts how galaxies formed over time, starting from the Big Bang in the upper left and proceeding to the present day in the lower right, as space gradually expanded according to Hubble's law.

Big Bang

380,000 years

1 billion years

TIME

1 Dramatic inflation early in time is thought to have produced large-scale ripples in the density of the universe. All the structure we see today formed as gravity drew additional matter into the peaks of these ripples.

Inflation may have stretched tiny quantum fluctuations into large-scale ripples.

2 Observations of the cosmic microwave background show us what the regions of enhanced density were like about 380,000 years after the Big Bang.

RADIO

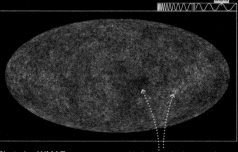

Photo by *WMAP*

Variations in the cosmic microwave background show that regions of the universe differed in density by only a few parts in 100,000.

3 Large-scale surveys of the universe show that gravity has gradually shaped early regions of enhanced density into a web-like structure, with galaxies arranged in huge chains and sheets.

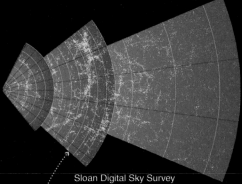

Sloan Digital Sky Survey

The web-like patterns of structure observed in large-scale galaxy surveys agree with those seen in large-scale computer simulations of structure formation.

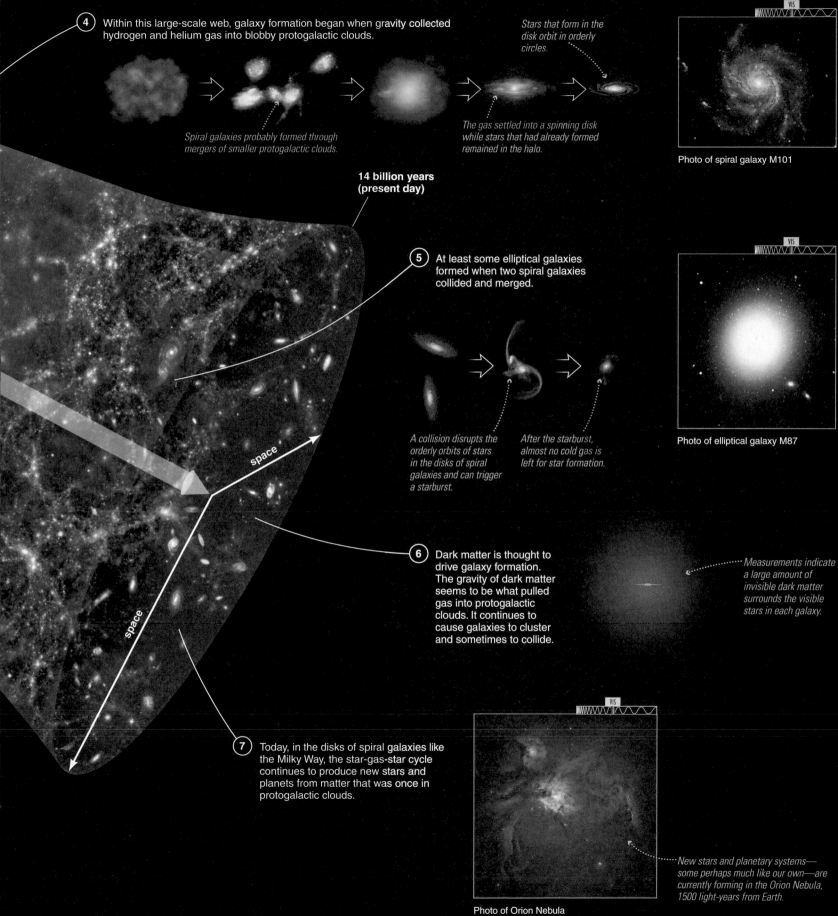

4 Within this large-scale web, galaxy formation began when gravity collected hydrogen and helium gas into blobby protogalactic clouds.

Spiral galaxies probably formed through mergers of smaller protogalactic clouds.

Stars that form in the disk orbit in orderly circles.

The gas settled into a spinning disk while stars that had already formed remained in the halo.

Photo of spiral galaxy M101

14 billion years (present day)

5 At least some elliptical galaxies formed when two spiral galaxies collided and merged.

A collision disrupts the orderly orbits of stars in the disks of spiral galaxies and can trigger a starburst.

After the starburst, almost no cold gas is left for star formation.

Photo of elliptical galaxy M87

space

space

6 Dark matter is thought to drive galaxy formation. The gravity of dark matter seems to be what pulled gas into protogalactic clouds. It continues to cause galaxies to cluster and sometimes to collide.

Measurements indicate a large amount of invisible dark matter surrounds the visible stars in each galaxy.

7 Today, in the disks of spiral galaxies like the Milky Way, the star-gas-star cycle continues to produce new stars and planets from matter that was once in protogalactic clouds.

New stars and planetary systems— some perhaps much like our own—are currently forming in the Orion Nebula, 1500 light-years from Earth.

Photo of Orion Nebula

Will the universe continue expanding forever?

Our discussion of how structure has formed on large scales in the universe leads to one of the ultimate questions in astronomy: What is the fate of the universe itself? We have seen that the gravity of dark matter has already halted and reversed the expansion of regions as large as clusters of galaxies and also affects structures as large as superclusters. Could the gravity of dark matter be strong enough to someday halt the expansion of the universe and cause it to collapse in a cataclysmic "big crunch"?

Just over a decade ago, astronomers thought they could answer this question simply by measuring the amount of matter in the universe. The greater the matter density, the greater the overall strength of gravity and the higher the likelihood that gravity will someday halt the expansion. Precise calculations show that, in the absence of any other forces, gravity can win out over expansion if the current matter density of the universe exceeds a seemingly minuscule 10^{-29} gram per cubic centimeter—roughly equivalent to a few hydrogen atoms in a volume the size of a closet. This density of matter, which marks the dividing line between eternal expansion and eventual collapse if there are no other forces acting on the universe, is exactly the same as the *critical density* of matter plus energy needed to make the overall geometry of the universe flat [Section 13.3].

Observations indicate that the total amount of matter, including dark matter, falls short of this critical density. Adding up the luminosities of galaxies shows that all the mass contained in stars amounts to only about 0.5% of the critical density. That means the overall matter density could exceed the critical density only if dark matter contributed about 200 times as much mass as the luminous matter. While our observations of galaxy clusters indicate that they contain about 50 times as much matter in the form of dark matter as stars, this is still only about a quarter of the amount of dark matter needed to halt the expansion. If the proportion of dark matter in the universe at large is similar to that in clusters, the universe seems destined to expand forever.

For gravity to reverse the expansion, even more dark matter would have to lie beyond the boundaries of clusters. But evidence indicates that it does not. If large-scale structures contained a higher proportion of dark matter than clusters, the influence of that extra dark matter should show up in the velocities of galaxies near those large-scale structures: Larger amounts of dark matter would cause greater deviations from Hubble's law. We have not observed such deviations, so we conclude that the gravity of dark matter is too weak to keep the universe from expanding forever.

In fact, recent measurements of the universe's expansion show that gravity is not slowing it down at all. Instead, as we'll discuss in the next section, these observations indicate that the expansion of the universe is accelerating, suggesting that a force even more powerful than the gravity of dark matter is acting to push the universe's galaxies apart. This finding seems to strengthen the case for an ever-expanding universe: If the universe is already short of the amount of matter needed to halt the expansion and there is also a force that accelerates the expansion, the universe appears destined to expand forever with ever-increasing speed. Of course, it's important to remember that forever is a very long time. The universe may hold other surprises that we haven't yet discovered and that might force us to reconsider what could happen between now and the end of time.

THE PROCESS OF SCIENCE IN ACTION

14.3 Evidence for Dark Energy

The discovery that the expansion of the universe is accelerating came as a great surprise. Before this discovery, Einstein's theory of general relativity allowed for the possibility that a large-scale repulsive force might accelerate the universe's expansion. However, generating such a powerful force requires an extremely large amount of energy. We still do not know the source of the mysterious energy that appears to be responsible for the observed acceleration, so we refer to it as **dark energy,** a term meant to reflect our ignorance about it, just as the term *dark matter* reflects our ignorance about the unseen form of mass whose gravity holds together galaxies and clusters of galaxies. Nevertheless, the evidence for dark energy is growing stronger with each passing year, and it is this chapter's example of the process of science in action, illustrating how careful observations sometimes require scientists to change their thinking about major features of our universe.

What is the evidence for dark energy?

The evidence for dark energy is based on observations of distant white dwarf supernovae that have enabled us to probe the fate of the universe in an entirely new way. Recall that white dwarf supernovae are excellent standard candles that allow us to measure the distances of very distant galaxies [Section 12.1]. We can therefore use observations of white dwarf supernovae to determine how the expansion rate of the universe has changed over long periods of time. In the 1990s, astronomers began seeking out and measuring white dwarf supernovae, hoping to learn the rate at which gravity has been slowing the universe's expansion over time. Much to their surprise, they discovered that the expansion is not slowing down at all, but speeding up.

Four Expansion Patterns   To understand the evidence for accelerating expansion, we must first become more familiar with the various ways in which gravity could affect the expansion of the universe. Astronomers divide these possibilities into four broad categories. Each represents a particular pattern of change in the future expansion rate (**Figure 14.14**). We will call these four possible patterns *recollapsing, critical, coasting,* and *accelerating.* The first three possibilities assume that gravity is the only force that affects the expansion rate

Figure 14.14 | Four models for the expansion of the universe. Each diagram shows how the size of a circular slice of the universe changes with time in a particular model. The slices are the same size at the present time, marked by the red line, but the models make different predictions about the sizes of the slices in the past and future. The first three cases assume that there is no dark energy, so the fate of the universe depends only on how its actual density compares to the critical density. The last case assumes that a repulsive force, perhaps produced by dark energy, is accelerating the expansion over time. (The diagram assumes continuous acceleration, but it is also possible that the universe initially slowed before the acceleration began.)

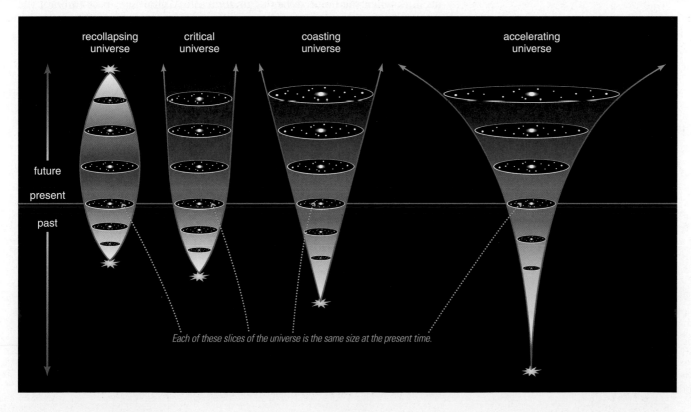

recollapsing universe · critical universe · coasting universe · accelerating universe

future · present · past

Each of these slices of the universe is the same size at the present time.

of the universe, while the fourth adds a repulsive force (due to dark energy) that opposes gravity:

- A **recollapsing universe.** If there is no dark energy and the matter density of the universe is *larger* than the critical density, then the collective gravity of all its matter will eventually halt the universe's expansion and reverse it. Galaxies will come crashing back together, and the entire universe will end in a fiery "Big Crunch." We call this a recollapsing universe because the final state, with all matter collapsed together, would look much like the state in which the universe began in the Big Bang.

- A **critical universe.** If there is no dark energy and the matter density of the universe *equals* the critical density, then the collective gravity of all the matter is exactly the amount needed to balance the expansion. The universe will never collapse but will expand more and more slowly as time progresses. We call this a critical universe because its matter density equals the critical density.

- A **coasting universe.** If there is no dark energy and the matter density of the universe is *smaller* than the critical density, then the collective gravity of all the matter cannot halt the expansion. The universe will keep expanding, with little change in its rate of expansion, coasting forever.

- An **accelerating universe.** If dark energy exerts a repulsive force that causes the expansion of the universe to *accelerate* with time, then the expansion rate will grow with time. Galaxies will recede from one another with increasing speed, and the universe will become cold and dark more quickly than would a coasting universe.

Evidence for Acceleration **Figure 14.15** illustrates how the average distance between galaxies should change with time for each possibility. The lines for the accelerating, coasting, and critical universes always continue upward as time increases, because in these cases the universe is always expanding. The steeper the slope, the faster the expansion. For the recollapsing universe, the line begins on an upward slope but eventually turns around and declines when the universe begins to contract. All the lines pass through the same point and have the same slope at the moment labeled "now," because the

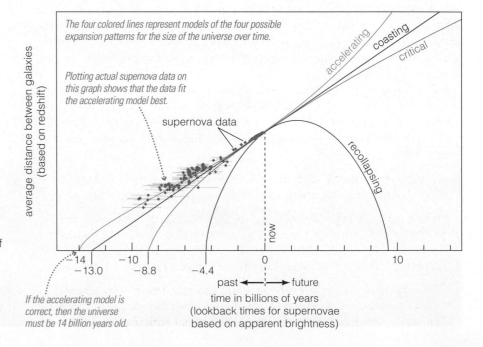

Figure 14.15 | Data from white dwarf supernovae are shown along with four possible models for the expansion of the universe. Each curve shows how the average distance between galaxies changes with time for a particular model. A rising curve means that the universe is expanding, and a falling curve means that the universe is contracting. Notice that the supernova data fit the accelerating universe better than the other models.

current separation between galaxies and the current expansion rate in each case must agree with observations of the present-day universe.

Notice that the age we infer for the universe from its expansion rate differs in each case. A recollapsing universe requires the least amount of time to arrive at the current separation between galaxies—the example in Figure 14.15 goes from zero separation to the current separation in less than 5 billion years. The cases for which gravity is less important require more time to achieve the current separation between galaxies. The ages that we infer from the examples in Figure 14.15 are 8.8 billion years for a critical universe, 13 billion years for a coasting universe, and around 14 billion years for an accelerating universe.

> **See it for yourself** Try tossing a ball in the air, observing how it rises and falls. Then make a graph to illustrate your observations, with time on the horizontal axis and height on the vertical axis. Which universe model does your graph most resemble? What is the reason for that resemblance? How would your graph look different if Earth's gravity were not as strong? Would the time for the ball to rise and fall be longer or shorter?

The data points in Figure 14.15 represent actual observations of white dwarf supernovae. These supernovae occur only rarely, but when we happen to observe one, we can use its apparent brightness and the known luminosity of white dwarf supernovae to determine its distance. This distance essentially tells us the look-back time to the supernova, and we can use its redshift to determine how much the universe has expanded since the time of the supernova. In Figure 14.15, these data have been converted into a form in which each supernova is used to indicate the approximate size of the observable universe (or, equivalently, the average distance between galaxies) at the time it occurred. Although there is some scatter in these data, they appear to fit the curve for an accelerating universe better than any of the other models, and they do not fit either a critical or a recollapsing universe. In other words, the observations to date seem to favor an accelerating universe.

Flatness and Dark Energy Additional support for the existence of dark energy comes from comparing observations of the universe's overall geometry to measurements of its total matter content. Recall that observations of the temperature patterns in the cosmic microwave background show that the overall geometry is indistinguishable from being completely flat [Section 13.3]. According to Einstein's general theory of relativity, a flat geometry means that the total density of matter plus energy must be equal to the critical density. However, as we have already discussed, the total matter density of the universe is not large enough to make the geometry flat on its own; it amounts to only about one-quarter of the critical density. The remaining three-quarters of the critical density must therefore be in the form of energy. **Figure 14.16** summarizes the evidence for dark matter and dark energy.

The amount of dark energy required to explain the supernova observations in Figure 14.15 can also explain the observations of temperature patterns in the cosmic microwave background. According to the model that best explains the observed temperature patterns (see Figure 13.15), today's universe has the following features:

- The density of matter in the form of protons and neutrons is 4.6% of the critical density, in agreement with observations of deuterium in the universe.

- The total matter density is 28% of the critical density. Subtracting the 4.6% for ordinary matter, we conclude that weakly interacting massive particles make up about 23% of the critical density, in agreement with what we infer from measurements of the masses of clusters of galaxies.

- The combination of a flat geometry and a matter density lower than the critical density implies the existence of a repulsive *dark energy* that currently accelerates the expansion, in agreement with observations of

Scientists suspect that most of the matter in the universe is *dark matter* we cannot see, and that the expansion of the universe is accelerating because of a *dark energy* we cannot directly detect. Both dark matter and dark energy have been proposed to exist because they bring our models of the universe into better agreement with observations, in accordance with the process of science. This figure presents some of the evidence supporting the existence of dark matter and dark energy.

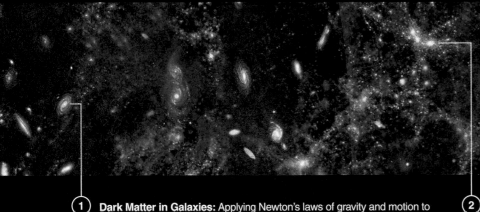

(1) Dark Matter in Galaxies: Applying Newton's laws of gravity and motion to the orbital speeds of stars and gas clouds suggests that galaxies contain much more matter than we observe in the form of stars and glowing gas.

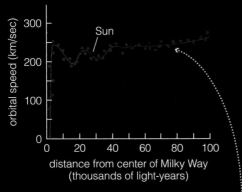

Orbital speeds of stars and gas clouds remain high even quite far from our galaxy's center . . .

dark matter

luminous matter

. . . indicating that the visible portion of our galaxy lies at the center of a much larger volume of dark matter.

HALLMARK OF SCIENCE **A scientific model must seek explanations for observed phenomena that rely solely on natural causes.** Orbital motions within galaxies demand a natural explanation, which is why scientists proposed the existence of dark matter.

(2) Dark Matter in Clusters: Further evidence for dark matter comes from studying galaxy clusters. Observations of galaxy motions, hot gas, and gravitational lensing all suggest that galaxy clusters contain far more matter than we can directly observe in the form of stars and gas.

This cluster of galaxies acts as a gravitational lens to bend light from a single galaxy behind it into the multiple blue shapes in this photo. The amount of bending allows astronomers to calculate the total amount of matter in the cluster.

HALLMARK OF SCIENCE **Science progresses through creation and testing of models of nature that explain the observations as simply as possible.** Dark matter accounts for our observations of galaxy clusters more simply than alternative hypotheses.

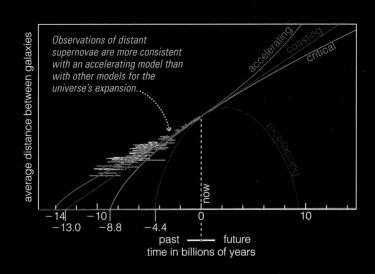

③ Structure Formation: If dark matter really is the dominant source of gravity in the universe, then its gravitational force must have been what assembled galaxies and galaxy clusters in the first place. We can test this prediction using supercomputers to model the formation of large-scale structures both with and without dark matter. Models with dark matter provide a better match to what we observe in the real universe.

④ Universal Expansion and Dark Energy: The expansion of a universe consisting primarily of dark matter would slow down over time (because of gravity), but observations suggest that the expansion is actually speeding up. Scientists hypothesize that a mysterious *dark energy* is causing the expansion to accelerate. Models that include both dark matter and dark energy agree more closely with observations than models containing dark matter alone.

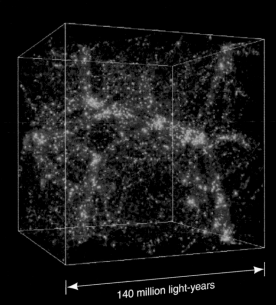

Observations of distant supernovae are more consistent with an accelerating model than with other models for the universe's expansion.

average distance between galaxies

accelerating

coasting

critical

recollapsing

now

−14
−13.0
−10
−8.8
−4.4
0
10

past ⊢———⊣ future

time in billions of years

140 million light-years

Supercomputer models in which dark matter is the dominant source of gravity show galaxies organized into strings and sheets similar in size and shape to those we observe in the real universe.

HALLMARK OF SCIENCE **A scientific model makes testable predictions about natural phenomena. If predictions do not agree with observations, the model must be revised or abandoned.** Observations of the universe's expansion have forced us to further modify our models of the universe to include dark energy along with dark matter.

distant supernovae. Because the total mass-energy of the universe is the critical density and matter accounts for only 28% of it, dark energy must account for the remaining 72% of the mass-energy of the universe.

- The universe's age is about 13.7 billion years, in agreement with what we infer from Hubble's constant and the ages of the oldest stars.

The excellent agreement between this model and our current observations of the large-scale universe gives scientists confidence that dark matter and dark energy are real. Nevertheless, the fact that we have not identified either of these mysterious "dark" substances means that this story of the process of science is still nowhere near its conclusion.

Summary of Key Concepts

 14.1 **Evidence for Dark Matter**

What is the evidence for dark matter?

Applying Newton's laws of gravitation and motion to the orbits of stars and gas clouds in galaxies leads to the conclusion that the total mass of a galaxy is far larger than the mass of its stars. Because no detectable visible light is coming from this matter, we call it **dark matter.** Similar observations of galaxy clusters indicate that the total mass of dark matter is about 50 times the mass in stars.

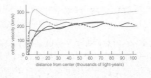

What might dark matter be made of?

The universe does not appear to contain enough ordinary matter made from protons and neutrons to account for all the dark matter we observe in the universe. We therefore suspect that most of the dark matter probably consists of **weakly interacting massive particles** of a kind not yet observed here on Earth.

 14.2 **Gravity versus Expansion**

What is the role of dark matter in structure formation?

Galaxies can form in regions of the universe where the gravity of dark matter is strong enough to stop the expansion and pull gas into protogalactic clouds. The gravity of dark matter also appears to be responsible for arranging galaxies into **large-scale structures** such as the gigantic chains and sheets that surround great voids.

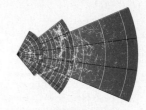

Will the universe continue expanding forever?

If the gravity of dark matter were the only force affecting the expansion, it still would not be strong enough to prevent the universe from expanding forever. The overall matter density of the universe appears to be only about 25% of the critical density of matter that the universe would need for gravity to eventually halt the expansion.

✳ **THE PROCESS OF SCIENCE IN ACTION**

 14.3 **Evidence for Dark Energy**

What is the evidence for dark energy?

Observations of distant supernovae indicate that the expansion of the universe is speeding up. No one knows the nature of the mysterious force (due to **dark energy**) that could be causing this acceleration, but it is a key area of modern research that illustrates how careful observations sometimes require us to change our thinking about the universe.

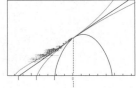

Investigations

Quick Quiz

Choose the best answer to each of the following; answers are in Appendix D. Explain your reasoning with one or more complete sentences.

1. The flat part of the Milky Way Galaxy's rotation curve tells us that stars in the outskirts of the galaxy (a) orbit the galactic center just as fast as stars closer to the center. (b) rotate rapidly on their axes. (c) travel in straight, flat lines rather than elliptical orbits.

2. Dark matter is inferred to exist because (a) we see lots of dark patches in the sky. (b) it explains how the expansion of the universe can be accelerating. (c) we can observe its gravitational influence on visible matter.

3. Strong evidence for the existence of dark matter comes from observations of (a) our solar system. (b) the center of the Milky Way. (c) clusters of galaxies.

4. A photograph of a cluster of galaxies shows distorted images of galaxies that lie behind it at greater distances. This is an example of what astronomers call (a) dark energy. (b) spiral density waves. (c) a gravitational lens.

5. Based on observational evidence, is it possible that dark matter doesn't really exist? (a) No, the evidence for it is too strong for us to think that our understanding could be in error. (b) Yes, but only if there is something wrong with our current understanding of how gravity works on large scales. (c) Yes, but only if all the observations themselves are in error.

6. Based on current evidence, which of the following is considered a likely candidate for the majority of the dark matter in galaxies? (a) subatomic particles that we have not yet detected in particle physics experiments (b) clusters of relatively dim, red stars (c) supermassive black holes

7. Which of the following provides evidence that the majority of the dark matter consists of weakly interacting massive particles? (a) measurements of the amount of deuterium made in the early universe (b) measurements of the rotation curves of galaxies (c) measurements of the universe's current expansion rate

8. Which region of the early universe was most likely to become a galaxy? (a) a region whose matter density was lower than average (b) a region whose matter density was higher than average (c) a region whose matter density was average

9. Dark energy has been hypothesized to exist in order to explain (a) observations suggesting that the expansion of the universe is accelerating. (b) the high orbital speeds of stars far from the center of our galaxy. (c) explosions that seem to create giant voids between galaxies.

10. The major evidence for the idea that the expansion of the universe is accelerating comes from observations of (a) white dwarf supernovae. (b) the orbital speeds of stars within galaxies. (c) the motions of galaxies within galaxy clusters.

11. Which of these possible types of universe would *not* expand forever? (a) a critical universe (b) an accelerating universe (c) a recollapsing universe

12. We determine the age of the universe from Hubble's constant and the overall expansion pattern. For which type of expansion pattern is this age the greatest? (a) a critical universe (b) a coasting universe (c) a recollapsing universe

Short-Answer/Essay Questions

Explain all answers clearly, with complete sentences and proper essay structure if needed. An asterisk (*) designates a quantitative problem, for which you should show all your work.

13. *What Is Dark Matter?* Describe at least three possible constituents of dark matter. Explain how we would expect each to interact with light and how we might go about detecting its existence.

14. *Dark Matter and Life.* Explain at least two reasons one might argue that dark matter is (or was) essential for life to exist on Earth.

15. *Rotation Curves.* Draw and label a rotation curve for each of the following three hypothetical situations. Make sure the radius axis has approximate distances labeled.
    a. All the mass of the galaxy is concentrated in the center of the galaxy.
    b. The galaxy has a constant mass density inside 20,000 light-years, and zero density outside that.
    c. The galaxy has a constant mass density inside 20,000 light-years, and its enclosed mass increases proportionally to the radius outside that.

16. *Alternative Gravity.* How would gravity have to be different in order to explain the rotation curves of galaxies without the need for dark matter? Would gravity need to be stronger or weaker than expected at very large distances? Explain.

17. *The Future Universe.* Based on current evidence concerning the growth of structure in the universe, briefly describe what you would expect the universe to look like on large scales about 10 billion years from now.

18. *Dark Energy and Supernova Brightness.* When astronomers began measuring the brightnesses and redshifts of distant white dwarf supernovae, they expected to find that expansion of the universe was slowing down. Instead, they found that it was speeding up. Were the distant supernovae brighter or fainter than expected? Explain why. (*Hint:* In Figure 14.15, the position of a supernova point on the vertical axis depends on its redshift. Its position on the horizontal axis depends on its brightness—supernovae seen farther back in time are not as bright as those seen closer in time.)

*19. *Mass from Rotation Curves.* Study the rotation curve for the spiral galaxy NGC 7541, which is shown in Figure 14.4.
    a. Use the orbital velocity law to determine the mass (in solar masses) of NGC 7541 enclosed within a radius of 30,000 light-years from its center. (*Hint:* 1 light-year = 9.461×10^{15} m.)
    b. Use the orbital velocity law to determine the mass of NGC 7541 enclosed within a radius of 60,000 light-years from its center.
    c. Based on your answers to parts (a) and (b), what can you conclude about the distribution of mass in this galaxy?

15 Life in the Universe

Learning Goals

15.1 The Search for Life in the Solar System

- What are the necessities of life on Earth?
- Could there be life elsewhere in our solar system?

15.2 The Search for Life Among the Stars

- Where might we find habitable planets?
- Is there intelligent life beyond Earth?

❋ THE PROCESS OF SCIENCE IN ACTION
15.3 The Evolution of Life on Earth

- What is the evidence for evolution?

We've discussed many fundamental questions about the universe in this book, but we have not yet discussed one of the most profound questions of all: Are we alone? The universe is filled with worlds beyond imagination—more than 100 billion star systems in our Milky Way Galaxy and some 100 billion galaxies in the observable universe—but we do not yet know whether any other world has ever been home to life. In this chapter, we'll discuss the search for life in the universe, which includes the *search for extraterrestrial intelligence*, or SETI, which is conducted with radio telescopes like those in the photo above.

15.1 The Search for Life in the Solar System

The idea of life on other worlds is not new. Not long after he discovered the laws of planetary motion, Kepler wrote a story imagining the Moon to be inhabited. William Herschel, co-discoverer of Uranus, suspected life on virtually all the planets. Most famously, in the late 19th century Percival Lowell claimed to see networks of canals on Mars, which he argued were the mark of an advanced civilization. Lowell's ideas became the basis for H. G. Wells's novel *The War of the Worlds* and led to widespread public belief in Martians.

We now have enough spacecraft images to say confidently that no other world in our solar system has ever been home to a civilization. Nevertheless, we have good reason to suspect that some of these worlds have conditions that could allow more primitive life to survive, and hundreds of scientists around the world are actively engaged in research that should help us learn whether life exists elsewhere in our solar system.

What are the necessities of life on Earth?

The first step in the search for life in our solar system is deciding what it is that we are looking for. Defining life turns out to be surprisingly difficult, and science fiction writers have imagined all sorts of exotic life forms, such as crystal life or silicon-based life. Our initial searches, however, have generally focused on looking for life that is at least somewhat Earth-like and depends on living conditions similar to those required by life forms on Earth.

The Nature of Life on Earth   If we hope to learn the requirements of Earth-like life, we must determine the characteristics that all life forms on Earth share. Scientists studying life on Earth have found numerous shared characteristics. For example, all known life uses DNA as its genetic material and builds proteins from the same set of building blocks, known as amino acids. Beyond such basic similarities, however, life on Earth is remarkably diverse.

For most of human history, people generally assumed that all living organisms belonged to either the *plant kingdom* or the *animal kingdom*. But in the mid-20th century, as scientists began to study microbial life in greater detail, we learned that most microbes do not belong to either of those groups. Today, biologists can identify relationships among living species by comparing the sequences of bases in their DNA. For example, two organisms whose DNA sequences differ in only one place for a particular gene are probably more closely related than two organisms whose gene sequences differ in five places. By making many such DNA comparisons, biologists have begun to assemble a *tree of life* that depicts the relationships among all living species. **Figure 15.1** shows key features of the tree of life as it is known today. Notice that life on Earth is divided into three major domains (Bacteria, Archaea, and Eukarya), and that all plants and animals represent only two small and fairly closely related branches of just one domain.

The tree of life offers several key lessons that can help us search for life beyond Earth. First and foremost, it tells us that the plant and animal life with which we are most familiar actually is *not* typical of most life on Earth. Instead, most life on Earth is microscopic, and this microscopic life exhibits far greater diversity in its genetic material than we find among plants and animals. This suggests that we are much more likely to find microscopic life on other worlds than large life forms such as plants or animals.

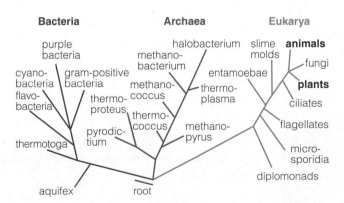

Figure 15.1 | Key features of the tree of life, showing relationships among species determined by comparison of DNA sequences in different organisms. Just two small branches represent *all* plant and animal species on Earth.

Figure 15.2 | This photograph shows a black smoker—a volcanic vent on the ocean floor that spews out hot, mineral-rich water. DNA studies indicate that the microbes living near these vents are evolutionarily older than nearly all other living organisms, suggesting that life may have first arisen in similar environments.

The Copernican revolution brought about a major change in human perspective by teaching us that Earth is not the center of the physical universe. Recent discoveries in biology offer a similar lesson: Despite our apparent dominance over other plants and animals, microbes dominate Earth's overall ecology. In that sense, we are no more the "center" of Earth's biological universe than Earth is the center of the physical universe. Do you think this discovery has or should change our human perspective? Defend your opinion.

A second important lesson is that, despite the great diversity that we see in the tree of life, all life on Earth appears to be related. This suggests that all life evolved from a common ancestor. Organisms on branches located closer to the "root" of the tree of life must contain DNA that has changed less over time, suggesting that these organisms more closely resemble the organisms that lived early in Earth's history. Many of the modern-day organisms that lie closest to this root are microbes that live in very hot water around seafloor volcanic vents called *black smokers* (because of the dark, mineral-rich water that flows out of them) in the deep oceans (**Figure 15.2**). Unlike most life at Earth's surface, which depends on sunlight, these organisms obtain energy from chemical reactions in water heated volcanically by Earth itself. We conclude that sunlight is *not* a necessity for life, a fact that opens possibilities for life even on worlds where sunlight is very weak.

Other key lessons come from more detailed study of the diverse organisms in the tree of life and the conditions under which they live. Most of these organisms live under conditions that seem "extreme" to animals like ourselves. For example, while we humans require oxygen to breathe, most microbes do not. Moreover, many microbes live in conditions that would be almost instantly lethal to us. Organisms near the black smokers thrive in water that is hotter than the normal boiling point; the high pressure of the deep sea keeps this very hot water liquid, but it would still kill any plant or animal quickly. In the freezing cold and dry valleys of Antarctica, the surface appears barren, but microbes have been found living *inside* rocks, surviving on tiny droplets of liquid water and energy from sunlight. Other microscopic organisms have been found living in water-infiltrated rock several kilometers underground, and in conditions of extreme acidity and saltiness. We have even found microbes that can survive high doses of radiation, making it conceivable that they could endure decades of travel through the radiation-filled environment of space. The microbes that live in extreme environments are collectively known as *extremophiles*, and their discovery shows that life can survive under a much wider range of conditions than we would have imagined just a few decades ago.

Requirements for Habitability We can use our understanding of the nature of life on Earth to determine the kinds of places that qualify as **habitable worlds**—worlds on which at least some type of Earth-like life could survive. Note that a habitable world need not necessarily have life, only conditions that could support life. So what exactly does Earth life need to survive?

If we compare all the different forms of life on Earth, including the extremophiles, we find that all share three basic requirements:

- A source of nutrients (elements and molecules) from which to build living cells

- Energy to fuel the activities of life, either from sunlight, from chemical reactions, or from the heat of Earth itself

- Liquid water, which serves several critical functions including the transport of nutrients into living cells and of waste products out of them

Interestingly, only the requirement of liquid water poses much of a constraint. Organic molecules that could serve as nutrients are present on almost all worlds, even on meteorites and comets. Many worlds are large enough to retain internal heat that could provide energy for life, and all worlds in our solar system have sunlight bathing their surfaces, although the inverse square law for light [Section 8.2] means that sunlight provides less energy to worlds farther from the Sun. Nutrients and energy should therefore be available to at least some degree on almost every planet and moon. In contrast, liquid water seems to be relatively rare. And although it is conceivable that other liquids—such as liquid methane or ethane—could substitute for liquid water on some worlds, these liquids are also rare. The search for habitable worlds and life therefore begins with a search for liquids, especially liquid water.

Could there be life elsewhere in our solar system?

The requirement for liquid water or some other liquid seems to rule out the possibility of life on most worlds in our solar system. Mercury and the Moon are barren and dry. Venus's surface is far too hot for liquid water. Pluto, Eris, and other dwarf planets are probably too small and too cold to harbor liquids of any kind. The same is probably also true for most of the other small objects in the solar system, including asteroids, comets, and the small jovian moons. The jovian planets provide a more interesting case. They may have droplets of liquid water and other liquids in their clouds, which has led to speculation about life that might float in their atmospheres. However, these planets also have strong vertical winds that would quickly carry liquid droplets downward to levels at which heat would evaporate them or upward to levels at which they would freeze, making it seem unlikely that they could sustain life. Even after ruling out so many worlds, however, we still find several potentially habitable words in our own solar system.

Mars   Mars is probably the best candidate in our solar system for life beyond Earth. Extensive evidence suggests that Mars had liquid water on its surface in the distant past [Section 5.2], in which case it would have been habitable at that time. If the period of habitability lasted long enough, it is conceivable that life could have arisen and taken hold on Mars. Even if life did not arise on Mars indigenously, it's possible that meteorites ejected from Earth by impacts might have carried microbes from Earth to Mars. If the microbes survived the journey and landed on Mars at a time when it was habitable, then they might have thrived upon arrival.

It is even possible that Mars remains habitable today, though not on its surface. The low atmospheric pressure ensures that liquid water cannot remain stable on the surface, so we do not expect any life to exist there. However, Mars still has plenty of ice and some internal heat, which means that there could be pockets of liquid water underground or water-infiltrated rock. If there was life on Mars in the distant past, then some microbes might still live in such underground environments, much like the microbes that live in underground rock on Earth.

The potential for past or present life on Mars has made it a prime target of interplanetary spacecraft. Our first attempt to search for life on Mars came with the two *Viking* landers that touched down on Mars in 1976. Each was equipped with a robotic arm for scooping up soil samples and feeding them into on-board, robotically controlled experiments. Three *Viking* experiments were designed expressly to look for signs of life. None of these experiments could actually "see" life; rather, they looked for chemical changes that could be attributed to living organisms. Although all three experiments gave results that initially seemed consistent with life, further study suggested that chemical reactions could have produced the same results. Moreover, a fourth experiment, which analyzed the content of Martian soil, found no measurable level of organic molecules—the opposite of what we would expect if life were present. As a result, most scientists have concluded that no life was present at the locations where *Viking* sampled the soil.

Recent missions to Mars have taken a more gradual approach to the search for life (**Figure 15.3**). These missions are designed to help us better understand Martian conditions, because we are more likely to be successful in a search for

Figure 15.3 | The *Spirit* rover studies a rock on Mars; the photo was taken by a camera aboard the rover. Such studies are helping scientists learn if and when Mars may have been habitable.

life if we understand when and where Mars may have been habitable. NASA eventually hopes to launch a mission to Mars that will bring back surface samples for study in laboratories on Earth. Later, we may send more advanced robots or humans to Mars, where they could search for fossils or living organisms in deep canyons, in ancient valley bottoms and dried-up lake beds, or in underground pockets of water near not-quite-dead volcanoes. The search will not be easy, but we should ultimately learn whether life has ever existed on Mars.

 The next mission scheduled to land on Mars is the *Mars Science Laboratory*, which should launch in 2011 and arrive on Mars in 2012. Look up the mission to find its current status. What are its major mission objectives?

Europa, Ganymede, and Callisto   After Mars, the next most likely candidate for life in our solar system is probably Jupiter's moon Europa. Recall that strong evidence suggests that Europa contains a deep ocean of liquid water beneath its icy crust [Section 6.1]. The ice and rock from which Europa formed undoubtedly included the chemical ingredients necessary for life, and Europa's internal heat (due primarily to tidal heating) should be strong enough to power volcanic vents on the seafloor. It's fairly easy to imagine places on Europa's ocean floor that look like black smokers on Earth. Given that life thrives around these undersea volcanic vents on Earth, it seems reasonable to imagine similar life on Europa. And if life got started near these vents, it could have spread farther in Europa's oceans from there.

The possibility of life on Europa is especially interesting because, unlike potential life on Mars, it would not necessarily have to be microscopic. After all, the several kilometers of surface ice that hide Europa's ocean (if it exists) could also hide large creatures swimming within it. However, the potential energy sources for life on Europa are far more limited than the energy sources for life on Earth, mainly because sunlight could not fuel photosynthesis in the subsurface ocean. As a result, most scientists suspect that any life that might exist on Europa would probably be small and primitive.

As we discussed in Chapter 6, some evidence suggests that Jupiter's moons Ganymede and Callisto may also have subsurface oceans. However, these moons would have even less energy for life than Europa. If they have life at all, it is almost certainly small and primitive. Nevertheless, Europa, Ganymede, and Callisto offer the astonishing possibility that Jupiter alone could be orbited by more worlds with life than we find in all the rest of the solar system.

Titan, Enceladus, and Beyond   Saturn's moon Titan offers another enticing place to look for life. Its surface is far too cold for liquid water, but it appears to have lakes and rivers of liquid methane or ethane [Section 6.1]. Although many biologists think it unlikely, it is possible that these liquids could support life much as does water on Earth. Moreover, recent data suggest that Titan may also have a subsurface ocean of liquid water. If so, the rich mix of organic compounds produced by interactions with sunlight on Titan's surface might find its way down into the ocean, in which case there might be far more nutrients and energy available for life on Titan than on Europa.

The discovery of ice fountains on Saturn's moon Enceladus (see Figure 6.17) suggests that it, too, could be habitable if these fountains are driven by subsurface liquids. Although this remains a topic of considerable scientific debate, some scientists think it likely that Enceladus has subsurface lakes or oceans. If this proves to be the case, it would open up the possibility that similar liquid regions—and potentially life—could exist on other solar system bodies, including Neptune's moon Triton and some of the moons of Uranus.

Overall, we have found at least six potentially habitable worlds in our solar system: Mars, Europa, Ganymede, Callisto, Titan, and Enceladus. Although we cannot yet know whether any of these worlds will prove to have life, the possibility certainly exists.

15.2 The Search for Life Among the Stars

We already know of planets orbiting hundreds of stars besides our Sun, and it's likely that billions of planetary systems inhabit our galaxy. These numbers make prospects for life elsewhere seem quite good, but numbers alone don't tell the whole story. In this section, we'll consider the prospects for life on worlds orbiting other stars.

Before we begin, it's important to distinguish between *surface* life like that on Earth and *subsurface* life like that we envision as a possibility on Mars or Europa. While large telescopes could in principle detect the presence of surface life on extrasolar planets, no foreseeable technology could detect life that is hidden deep underground in other star systems (unless it has a noticeable effect on the planet's atmosphere). We therefore will focus on the search for life on planets with habitable surfaces—surfaces with temperatures and pressures that could allow liquid water to exist.

Where might we find habitable planets?

We have not yet discovered any extrasolar planets that seem likely to be habitable. Presumably because of the limitations of current detection techniques [Section 7.1], most of the extrasolar planets found to date are probably jovian

Tools of Science: Planetary Spacecraft

The search for life in our solar system differs from the search for life elsewhere not only because our telescopes can see more detail on nearby worlds, but more importantly because we are able to send spacecraft to visit these worlds. In contrast, the stars are so far away that it would take our current spacecraft tens of thousands of years to reach even the nearest of them. Our ability to send spacecraft to the planets in our own solar system means we can get close-up images and detailed spectra, and by landing on these worlds we can in principle conduct a direct search for life.

With the exception of the *Apollo* astronaut visits to the Moon between 1969 and 1972, all spacecraft sent to other worlds have been robotic. These spacecraft have computers to control their major components, power sources such as solar cells or nuclear batteries for energy, and scientific instruments such as cameras, spectrometers, and devices to measure magnetic fields. Spacecraft that fly through atmospheres or land on surfaces typically also carry instruments to measure winds, temperatures, and atmospheric or soil composition. The spacecraft all carry radio antennas for communication, so that they can receive operating instructions from Earth and radio their data back to Earth.

Some spacecraft simply fly past a world, taking pictures and collecting data during the short time when they are close by. Others go into orbit around a world for longer-term study, while others land on a surface or probe an atmosphere. To date, we have sent spacecraft on flybys of all the planets, as well as several asteroids and comets; the *New Horizons* spacecraft is currently on its way to fly past the dwarf planet Pluto (in 2015) and perhaps a few other objects in the Kuiper belt. We've sent orbiters to the Moon, Venus, Mars, Jupiter, and Saturn; the *MESSENGER* spacecraft will enter orbit of Mercury in 2011. Spacecraft have successfully landed on the Moon, Venus, Mars, Titan, and the asteroid Eros, while the *Galileo* probe entered Jupiter's atmosphere.

Current spacecraft study of potentially habitable worlds focuses on Mars. Every 26 months, Earth and Mars come into a favorable alignment for spacecraft missions, and scientists hope to continue sending missions to Mars at each of these launch opportunities. Each mission builds upon previous discoveries, with the eventual goal of learning whether life exists or has ever existed on Mars.

Europa poses a more difficult challenge, since any life would live in oceans buried under many kilometers of solid ice. Current plans call for a *Europa Orbiter* that should be able to determine definitively whether the subsurface ocean really exists. If it does, a subsequent mission might land on Europa to sample the surface in search of organic material—or even remains of life—from the ocean below. More sophisticated study would require sending a robotic submarine to Europa, which would use a nuclear heater to melt its way down into the ocean.

Similar strategies could be applied to Titan and Enceladus, with each offering advantages. Titan has the advantage of surface liquids, so we need not go underground to begin a search for life. However, Titan is extremely cold, and we would need new technologies for a spacecraft to survive Titan's cold for long. The advantage of Enceladus is that it is relatively small and we can see the vents out of which ice fountains are spraying, so it might be relatively easy to send a spacecraft into its interior. Again, however, the great distance to Saturn's moons and the extremely cold temperatures would make such missions expensive and technologically challenging.

The bottom line is that while spacecraft exploration of the planets remains difficult, we are making great progress. If and when we someday discover life on another world, it will almost certainly be thanks to visits by spacecraft.

Figure 15.4 | The approximate habitable zones around our Sun, a star with 1/2 the mass of the Sun (spectral type K), and a star with 1/10 the mass of the Sun (spectral type M), shown to scale. The habitable zone becomes increasingly smaller and closer in for stars of lower mass and luminosity.

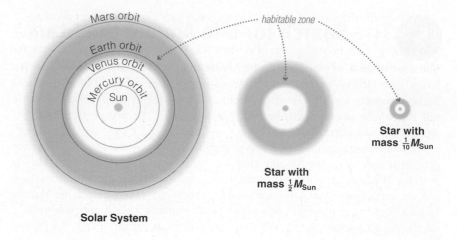

rather than terrestrial in nature. Like the jovian planets in our own solar system, these extrasolar planets are unlikely to have life, though some of them have orbits that could conceivably allow them to have moons with surface water and life.

However, the existence of other jovian planets makes it reasonable to suppose that terrestrial planets are also common around other stars, and that some of these planets are Earth-like in size and orbit. We must therefore consider the questions of where we might expect terrestrial planets to exist and whether they might be habitable.

Constraints on Star Systems Before we consider planets themselves, it's useful to ask how many stars could potentially have planets with life. In other words, which stars would make good "Suns," providing heat and light to the surfaces of terrestrial planets that happen to orbit them?

The first requirement for a star to have life-bearing worlds is that it be old enough that life could have arisen. Stars of greater mass live shorter lives, and the most massive stars live no more than a few million years [Section 9.2]. Given that life on Earth did not arise for hundreds of millions of years after our solar system was born, we can rule out any star with more than a few times the mass of our Sun. However, because low-mass stars are far more common than high-mass stars, the lifetime constraint rules out only about 1% of all stars.

A second requirement is that the star system allow for planets to have stable orbits. About half of all stars are in binary or multiple star systems, in which stable planetary orbits are less likely than around single stars. If life is not possible in such systems, then we can rule out about half the stars in our galaxy as potential homes to life. Of course, the other half—100 billion stars or more—remain possible homes for life. Moreover, under some circumstances, stable planetary orbits are possible in multiple star systems, so we cannot entirely rule out life in such systems.

A third constraint on the likelihood of finding habitable planets is the size of a star's **habitable zone**—the region in which a terrestrial planet of the right size could have a surface temperature that might allow for liquid water and life. **Figure 15.4** shows the approximate sizes, to scale, of the habitable zones around our Sun, around a star with about 1/2 the mass of the Sun (spectral type K), and around a star with about 1/10 the mass of our Sun (spectral type M). Although habitable planets seem possible in all three cases, the smaller size of the habitable zones around the less massive stars makes it less likely that suitable planets would have formed in these regions.

All in all, the vast majority of stars seem capable of having life-bearing planets, and even very conservative assumptions suggest enormous numbers of possibilities. For example, even limiting the search for habitable planets to stars similar to our Sun (spectral type G) would still give us billions of potential other Suns to explore in the Milky Way Galaxy.

Finding Habitable Planets   Finding Earth-size planets is a daunting technological challenge. Recall that looking for an Earth-like planet around a nearby star is like standing on the East Coast of the United States and looking for a pinhead on the West Coast. Nevertheless, new technologies have begun to put Earth-like planets within our telescopic reach [Section 7.1]. The *Kepler* mission, launched in 2009, is searching for transits of Earth-size planets in front of their stars. Scientists hope that *Kepler* will detect dozens of Earth-size planets and that orbital properties measured by *Kepler* will tell us which of these planets lie within their stars' habitable zones.

We will need images or spectra to determine whether the planets really are habitable and whether they host life. Scientists are actively working on technologies that may provide such data, and NASA has preliminary plans for an orbiting telescope that could obtain low-resolution spectra and crude images (a few pixels) of Earth-like planets around nearby stars.

Signatures of Life   Images from future telescopes may tell us whether Earth-size extrasolar planets have continents and oceans like Earth, and studies of seasonal changes could provide hints of life. Spectra should prove even more important to the search for life. Moderate-resolution infrared spectra can reveal the presence and abundance of many atmospheric gases, including carbon dioxide, ozone, methane, and water vapor (**Figure 15.5**). Careful analysis of atmospheric makeup might tell us whether a planet has life. For example, atmospheric oxygen on Earth is a direct result of photosynthetic life. Abundant oxygen in the atmosphere of a distant world might therefore indicate the presence of life, since we know of no nonbiological way to produce an oxygen abundance as high as Earth's. Additional evidence might come from other gases released by life. Scientists are working to improve our understanding of how life influences atmospheric chemistry in hopes that we will be able to recognize particular gas combinations as signatures of life.

Is there intelligent life beyond Earth?

So far, we have focused on search strategies for microbial or other nonintelligent life. However, if intelligent beings and civilizations exist elsewhere, we might be able to find them with a completely different type of search strategy. Instead of searching for hard-to-find spectroscopic signs of life, we simply listen for signals that intelligent beings are sending into interstellar space, either in deliberate attempts to contact other civilizations or as a means of communicating among themselves. The search for signals from other civilizations is generally known as the **search for extraterrestrial intelligence,** or **SETI** for short. But is it reasonable to think that other civilizations really exist and that we might be able to contact them?

The Drake Equation   Given that we do not even know whether microbial life exists anywhere beyond Earth, we certainly don't know whether other civilizations exist. Nevertheless, for the purposes of planning a search for extraterrestrial intelligence, it's helpful to have an organized way of thinking about the number of civilizations that might be out there. In 1961, astronomer

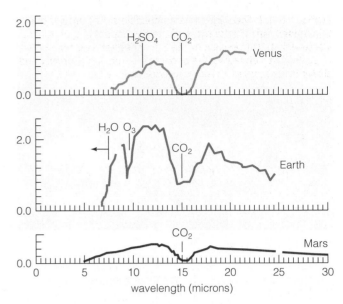

Figure 15.5 | The infrared spectra of Venus, Earth, and Mars, as they might be seen from afar, showing absorption features that point to the presence of carbon dioxide (CO_2), ozone (O_3), and sulfuric acid (H_2SO_4) in their atmospheres. While carbon dioxide is present in all three spectra, only Earth has appreciable oxygen (and hence ozone)—a product of photosynthesis. If we could make similar spectral analyses of distant planets, we might detect atmospheric gases that would indicate life.

Figure 15.6 | Astronomer Frank Drake, with the equation he first wrote in 1961. Some of the scientific ideas behind the equation have changed since that time; therefore, with Dr. Drake's approval, we use a modified form of his original equation in this book.

Frank Drake wrote a simple equation, now known as the **Drake equation,** designed to summarize the factors that would determine the number of civilizations we might contact (**Figure 15.6**). To keep our discussion simple, we'll use the Drake equation to consider only the number of potential civilizations in our own galaxy.

In principle, the Drake equation gives us a simple way to calculate the number of civilizations capable of interstellar communication that are currently sharing the Milky Way Galaxy with us. In a form slightly modified from the original, the Drake equation looks like this:

$$\text{Number of civilizations} = N_{HP} \times f_{life} \times f_{civ} \times f_{now}$$

This equation will make sense once you understand the meaning of each factor:

- N_{HP} is the number of habitable planets in the galaxy—that is, the number of planets that are *potentially* capable of having life.

- f_{life} is the fraction of habitable planets that actually *have* life. For example, $f_{life} = 1$ would mean that all habitable planets have life, and $f_{life} = 1/1,000,000$ would mean that only 1 in a million habitable planets has life. The product $N_{HP} \times f_{life}$ therefore tells us the number of life-bearing planets in the galaxy.

- f_{civ} is the fraction of the life-bearing planets upon which a civilization capable of interstellar communication *has at some time* arisen. For example, $f_{civ} = 1/1000$ would mean that such a civilization has existed on 1 out of 1000 planets with life, while the other 999 out of 1000 have not had a species that learned to build radio transmitters, high-powered lasers, or other devices for interstellar communication. When we multiply this factor by the first two factors to form the product $N_{HP} \times f_{life} \times f_{civ}$, we get the total number of planets upon which intelligent beings have evolved and developed a communicating civilization at some time in the galaxy's history.

- f_{now} is the fraction of these civilization-bearing planets that happen to have a civilization *now*, as opposed to, say, millions or billions of years in the past. This factor is important because we can hope to contact only civilizations that are broadcasting signals we could receive at present (assuming we take into account the light-travel time for signals from other stars).

Because the product of the first three factors tells us the total number of civilizations that have *ever* arisen in the galaxy, multiplying by f_{now} tells us how many civilizations we could potentially make contact with today. Therefore, the result of the Drake equation is the number of civilizations that we might hope to contact.

> **Think about it** Try the following sample numbers in the Drake equation. Suppose that there are 1000 habitable planets in our galaxy, that 1 in 10 habitable planets has life, that 1 in 4 planets with life has at some point had an intelligent civilization, and that 1 in 5 civilizations that have ever existed is in existence now. How many civilizations would exist at present? Explain.

Estimates with the Drake Equation Unfortunately, we don't yet know the value of any of the factors in the Drake equation, so we cannot actually calculate the number of civilizations in our galaxy. Nevertheless, the equation is useful for organizing our thinking about the potential values for each of its factors.

We can make a reasonably educated guess only about the first factor, the number of habitable planets (N_{HP}). Current understanding of solar system formation (see Figure 4.13) suggests that terrestrial planets ought to form fairly easily and, as we discussed earlier, there ought to be 100 billion or more stars in our galaxy with habitable zones large enough to have one or more Earth-like planets. Therefore, while we do not yet have direct evidence to support this idea, it seems reasonable to hypothesize that our galaxy could have as many as 100 billion habitable planets.

The factor f_{life} presents more difficulty. For the moment, we have no reliable way to estimate the fraction of habitable planets upon which life actually arose. The problem is that we cannot generalize when we have only one example—Earth—to study. Still, geological evidence indicates that life arose within a few hundred million years after Earth's formation, and perhaps only tens of millions of years after conditions became conducive for life. This is a short amount of time compared to Earth's history, suggesting that the origin of life was fairly "easy." In that case, we might expect most or all habitable planets to also have life, making the fraction f_{life} close to 1. Of course, until we have solid evidence that life arose anywhere else, such as on Mars, it is also possible that Earth is very lucky to have life and that f_{life} might be so close to zero that life has never arisen on any other planet in our galaxy.

The factor f_{civ}, which represents the fraction of life-bearing planets on which a civilization capable of interstellar communication has arisen, is even more difficult. On Earth, life of some type has flourished for nearly 4 billion years, but our civilization has been capable of sending radio signals to the stars for less than a century. This suggests that even if it was "easy" for life to arise on Earth, it was much harder for that life to evolve into intelligent life and civilization. Some people therefore argue that while life may be common, civilizations will not be. On the other hand, if we assume that Earth is typical, then we might expect that any planet with life would end up with a civilization by the time it reached an age similar to Earth's current age of $4\frac{1}{2}$ billion years. In that case, if habitable planets really are common, as we have guessed, and if life can arise fairly easily, then there might be thousands or millions of civilizations that have arisen at some time in our galaxy.

We can now see the importance of the final factor in the Drake equation, f_{now}. For the sake of argument, let's assume that life and intelligence are likely, and that thousands or millions of civilizations have arisen at some time. Then the number of civilizations that exist now depends on the survivability of civilizations. Consider our own example. In the roughly 12 billion years during which our galaxy has existed, we have been capable of interstellar communication via radio for only about 60 years. If we were to destroy ourselves tomorrow, then other civilizations could have received signals from us during only 60 years out of the galaxy's 12-billion-year existence, equivalent to 1 part in 200 million of the galaxy's history. If such a short technological lifetime is typical of civilizations, then f_{now} would be only 1/200,000,000, and some 200 million civilization-bearing planets would need to have existed at one time or another in the Milky Way in order for us to have a decent chance of finding even one other civilization today.

However, we'd expect f_{now} to be so small only if we are on the brink of self-destruction, because the fraction will grow larger for as long as our civilization survives. Therefore, if civilizations are at all common, survivability is the key factor in whether any are out there now. If most civilizations self-destruct shortly after achieving the technology for interstellar communication, then we are almost certainly alone in the galaxy at present. But if most survive and thrive for thousands or millions of years, the Milky Way may be brimming with civilizations—and most of them would be far more advanced than our own.

Figure 15.7 | The Allen Telescope Array in Hat Creek, California, is used to search for radio signals from extraterrestrial civilizations.

Think about it   Give a few reasons why a civilization capable of interstellar communication would also be capable of self-destruction. Overall, do you believe our civilization can survive for thousands or millions of years? Defend your opinion.

SETI   If there are indeed other civilizations out there, then in principle we ought to be able to make contact with them. Based on our current understanding of physics, it seems likely that even advanced civilizations would communicate much as we do—by encoding signals in radio waves or other forms of light. Most SETI researchers use large radio telescopes to search for alien radio signals (**Figure 15.7**). These telescopes can scan millions of radio frequency bands simultaneously. A few researchers are also beginning to check other parts of the electromagnetic spectrum. For example, some scientists use visible light telescopes to search for communications encoded as laser pulses. Of course, it's also possible that advanced civilizations have invented communication technologies that we cannot even imagine, let alone detect. In that case, we might be capable of detecting communications only from relatively young civilizations like our own, since more advanced civilizations would be using technologies that we do not yet have.

Alien Visitation and Fermi's Paradox   A basic assumption throughout this chapter has been that we do not yet know whether other civilizations exist. However, opinion polls show that many people believe that we have already been visited by aliens in UFOs. Could it be that the evidence for extraterrestrial life is already here?

Claims of UFOs and alien encounters frequently make the news. However, while plenty of evidence has been offered for scientific examination, so far none of it has been enough to convince scientists that the claims of visitation are real. It's possible that more convincing evidence will be found in the future, but for now there is no scientific evidence to back claims of aliens on Earth.

But this lack of scientific evidence for alien life—either through SETI contact or through evidence of visitation to Earth—raises questions of its own. Although interstellar travel remains far beyond our current technology, the rate at which our technology is advancing suggests that we should eventually be capable of such travel. And by modeling the way a civilization might spread among the stars, we find that once a civilization was capable of star travel at even a modest fraction of the speed of light, it would take that civilization no more than a few tens of millions of years to colonize most or all of the galaxy. This time is so short compared to the age of our galaxy that it leads us to an astonishing conclusion: If civilizations are common, then someone else should have been capable of creating a galactic civilization by now. In fact, it should have been done a long time ago.

To see why, let's take some sample numbers. Suppose the factors in the Drake equation tell us that the overall odds of a civilization arising around a star are about the same as your odds of winning the lottery: 1 in a million. Taking a low estimate of 100 billion stars in the Milky Way Galaxy, we come up with some 100,000 civilizations in our galaxy alone. Moreover, current evidence suggests that stars and planetary systems like our own could have formed for at least 5 billion years before our solar system was even born, in which case the first of these 100,000 civilizations would have arisen at least 5 billion years ago. Others would have arisen, on average, about every 50,000 years. Under these assumptions, we would expect the youngest civilization besides ours to be some 50,000 years ahead of us technologically, and most would be millions or billions of years ahead of us.

We thereby encounter a strange paradox: Plausible arguments suggest that a galactic civilization should already exist, yet we have so far found no evidence of such a civilization. This paradox is often called *Fermi's paradox,* after the Nobel Prize–winning physicist Enrico Fermi. During a 1950 conversation with other scientists about the possibility of extraterrestrial intelligence, Fermi responded to speculations by asking "So where is everybody?"

This paradox has many possible solutions, but broadly speaking we can group them into three categories:

1. *We are alone.* There is no galactic civilization because civilizations are extremely rare—so rare that we are the first to have arisen on the galactic scene, perhaps even the first in the universe.

2. *Civilizations are common, but no one has colonized the galaxy.* There are at least three possible reasons why this might be the case. Perhaps interstellar travel is much harder or more expensive than we might guess, and civilizations are unable to venture far from their home worlds. Perhaps the desire to explore is unusual, and other societies either never leave their home star systems or stop exploring before they've colonized much of the galaxy. Most ominously, perhaps many civilizations have arisen, but they have all destroyed themselves before achieving the ability to colonize the stars.

3. *There IS a galactic civilization.* However, it has not yet revealed its existence to us.

We do not know which, if any, of these explanations is the correct solution to Fermi's paradox. However, each category of solution has astonishing implications for our own species.

Consider the first solution—that we are alone. If this is true, then our civilization is a remarkable achievement. It implies that through all of cosmic evolution, among countless star systems, we are the first piece of our galaxy or the universe ever to know that the rest of the universe exists. Through us, the universe has attained self-awareness. Some philosophers and many religions argue that the ultimate purpose of life is to become truly self-aware. If so, and if we are alone, then the destruction of our civilization and the loss of our scientific knowledge would represent an inglorious end to something that took the universe some 14 billion years to achieve. From this point of view, humanity becomes all the more precious, and the collapse of our civilization would be all the more tragic.

The second category of solutions has much more terrifying implications. If thousands of civilizations before us have all failed to achieve interstellar travel on a large scale, what hope do we have? Unless we somehow think differently than all other civilizations, this solution says that we will never go far in space. Because we have always explored when the opportunity arose, this solution almost inevitably leads to the conclusion that failure will come about because we destroy ourselves. We can only hope that this answer is wrong.

The third solution is perhaps the most intriguing. It says that we are newcomers on the scene of a galactic civilization that has existed for millions or billions of years before us. Perhaps this civilization is deliberately leaving us alone for the time being and will someday decide the time is right to invite us to join it.

No matter what the answer turns out to be, learning it is sure to mark a turning point in the brief history of our species. Moreover, this turning point is likely to be reached within the next few decades or centuries. We already have the ability to destroy our civilization. If we do so, then our fate is sealed. But if we survive long enough to develop technology that can take us to the stars, the possibilities seem almost limitless.

15.3 The Theory of Evolution

As you've seen throughout this chapter, the search for life in the universe is built upon our understanding of life on Earth. Today, this understanding is embodied in a single, unifying theory of biology, a theory that was first proposed 150 years ago and that has been refined and strengthened since that time. We are talking about the *theory of evolution*, which is as central to modern biology as the theory of gravity is to astronomy. Because this theory is widely misunderstood, we devote our final case study in the process of science in action to examining the evidence that makes evolution the underpinning of our understanding of life on Earth and beyond.

What is the evidence for evolution?

The word **evolution** simply means "change with time," and a first step in understanding it is to distinguish between *observations* of evolution and the *theory* of evolution. Observations of evolution come primarily from the fossil record. As we examine layers of sediments deposited over millions of years, such as those in the walls of the Grand Canyon, we find that different layers contain fossils of different types. Comparisons of these fossils show that those in older sediment layers are generally more primitive than those in younger layers, and many fossils from the recent geological past represent the extinct ancestors of living species. These observations strongly support the idea that living species have evolved over time. The *theory* of evolution is designed to explain how these observed changes occur. Our modern theory of evolution was first described by Charles Darwin in 1859, though, like other scientific theories, it has been further refined and improved with time. In essence, the fossil record provides strong evidence that evolution *has* occurred, while Darwin's theory of evolution explains *how* it occurs.

Darwin's theory tells us that evolution proceeds through a process called **natural selection.** The many individuals of any species always differ in small ways. If an individual possesses a trait that gives it an advantage in survival and reproduction, then the trait is likely to be passed on to future generations. We say that nature "selects" the advantageous trait, which is why the process is called natural selection. Over time, natural selection can help individuals of a species become better able to compete for scarce resources. If enough new traits accumulate, natural selection can give rise to an entirely new species.

Accumulating Evidence   Darwin compiled a tremendous body of evidence (by studying both fossils and relationships among living species, such as those among animals of the Galapagos Islands) in support of his idea that natural selection is the primary mechanism by which evolution proceeds. Research since Darwin's time has only given further support to his idea, which is why we call it the *theory* of evolution [Section 3.2]. Some of the strongest evidence in favor of evolution has come from the study of DNA, the genetic material of all life on Earth.

In order for evolution to happen as described by Darwin, there must be some way for organisms to pass their genetic traits on to their offspring and for those traits to diversify over time. DNA is what enables that to happen, and its discovery led scientists to understand how evolution occurs on a molecular level. Living organisms reproduce by copying DNA and passing these copies on to their descendants. A molecule of DNA consists of two long strands—somewhat like the interlocking strands of a zipper—wound together in the spiral shape known as a double helix (**Figure 15.8**). The instructions for assembling a living organism are written in the precise order of four chemical bases that make up the interlocking portions of the DNA "zipper." (The four chemical bases are represented in Figure 15.8 by the letters A, T, G, and C, which are the first letters of their chemical names.) These bases pair up in a way that ensures that each of the two strands of a DNA molecule contains the same genetic information. By unwinding and allowing new strands to form alongside the original ones (with the new strands made from chemicals floating around inside a cell), a single DNA molecule can give rise to two identical copies of itself. This is how genetic material is copied and passed on to future generations.

Evolution occurs because the passing of genetic information from one generation to the next is not always perfect. An organism's DNA may be altered by occasional copying errors or by external influences, such as ultraviolet light from the Sun or exposure to toxic or radioactive chemicals. Any change in an organism's DNA is called a **mutation.** Many mutations are lethal, killing the cell in which the mutation occurs. Some, however, may improve a cell's ability to survive and reproduce. The cell then passes on this improvement to its offspring.

Our understanding of the molecular mechanism of natural selection has put the theory of evolution on a stronger foundation than ever. While no theory can ever be proved true beyond all doubt, the theory of evolution today is as solid as any theory in science, including the theory of gravity and the theory of atoms. Biologists routinely witness evolution occurring before their eyes among laboratory microorganisms, or over periods of just a few decades among plants and animals subjected to some kind of environmental stress. Moreover, the theory of evolution has become the underpinning of virtually all modern biology, medicine, and agriculture. For example, agricultural scientists apply the idea of natural selection to develop pest control strategies that reduce the populations of harmful insects without harming the populations of beneficial ones; medical researchers test new drugs on animals that are genetically similar to humans, because the theory of evolution tells us that genetically similar species should have somewhat similar physiological responses; and biologists study the relationships between organisms by comparing their DNA.

To summarize, the theory of evolution is backed by an enormous body of evidence and successfully explains both what we observe in the fossil record and what we observe in the laboratory. The fact that we now understand evolution on a molecular level gives us even more confidence in this model of how life changes through time. And perhaps most important to the topic of this chapter, the idea that life evolves through molecular changes that are passed on to offspring could apply even to organisms that use something besides DNA as their hereditary material. We therefore expect that the theory of evolution will apply to life that we might someday discover on other worlds, and that those discoveries will help us further refine and improve the theory.

The Biological Universe

We are here today not only because of the 4 billion years of biological evolution that have occurred on Earth, but also because of many other processes that have occurred in the universe and that we have studied throughout this book. **Figure 15.9** summarizes some of the critical processes that have made life possible on Earth. If life exists on other worlds, these same processes must have played similar roles. In that sense, the biological universe is intimately connected with all aspects of our physical universe—which is one reason why a *cosmic perspective* is so important to understanding our own lives.

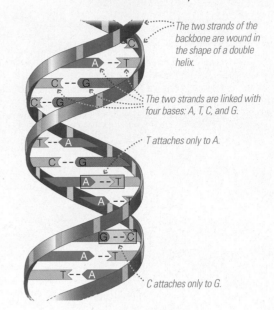

The two strands of the backbone are wound in the shape of a double helix.

The two strands are linked with four bases: A, T, C, and G.

T attaches only to A.

C attaches only to G.

Figure 15.8 | This diagram represents a small piece of a DNA molecule, which looks much like a zipper twisted into a spiral. The important hereditary information is contained in the "teeth" linking the strands. These "teeth" are the DNA bases. Only four DNA bases are used, and they can link the two strands only in specific ways: T attaches only to A, and C attaches only to G. (The color coding is arbitrary and is used only to represent different types of chemical groups; in the backbone, blue and yellow represent sugar and phosphate groups, respectively.)

Throughout this book, we have seen that the history of the universe has proceeded in a way that has made our existence on Earth possible. This figure summarizes some of the key ideas, and leads us to ask: If life arose here, shouldn't it also have arisen on many other worlds? We do not yet know the answer, but scientists are actively seeking to learn whether life is rare or common in the universe.

1 The protons, neutrons, and electrons in the atoms that make up Earth and life were created out of pure energy during the first few moments after the Big Bang, leaving the universe filled with hydrogen and helium gas.

electron

gamma-ray photons

antielectron

Matter can be created from energy: $E = mc^2$.

2 Ripples in the density of the early universe were necessary for life to form later on. Without those ripples, matter would never have collected into galaxies, stars, and planets.

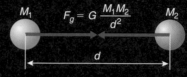

We observe the seeds of structure formation in the cosmic microwave background.

3 The attractive force of gravity pulls together the matter that makes galaxies, stars, and planets.

M_1 $F_g = G \dfrac{M_1 M_2}{d^2}$ M_2

d

Every piece of matter in the universe pulls on every other piece.

4 Our planet and all the life on it is made primarily of elements formed by nuclear fusion in high-mass stars and dispersed into space by supernovae.

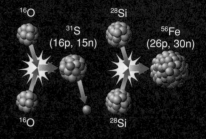

^{16}O ^{28}Si

^{31}S (16p, 15n) ^{56}Fe (26p, 30n)

^{16}O ^{28}Si

High-mass stars have cores hot enough to make elements heavier than carbon.

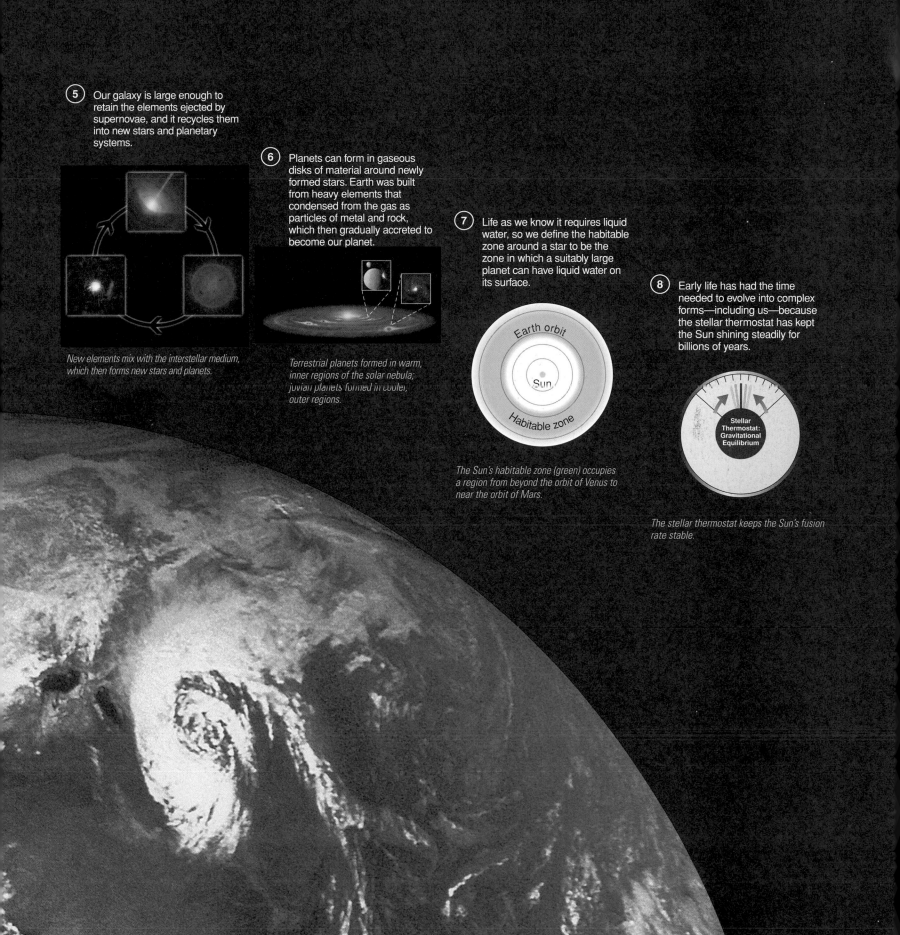

5 Our galaxy is large enough to retain the elements ejected by supernovae, and it recycles them into new stars and planetary systems.

New elements mix with the interstellar medium, which then forms new stars and planets.

6 Planets can form in gaseous disks of material around newly formed stars. Earth was built from heavy elements that condensed from the gas as particles of metal and rock, which then gradually accreted to become our planet.

Terrestrial planets formed in warm, inner regions of the solar nebula; jovian planets formed in cooler, outer regions.

7 Life as we know it requires liquid water, so we define the habitable zone around a star to be the zone in which a suitably large planet can have liquid water on its surface.

Earth orbit

Sun

Habitable zone

The Sun's habitable zone (green) occupies a region from beyond the orbit of Venus to near the orbit of Mars.

8 Early life has had the time needed to evolve into complex forms—including us—because the stellar thermostat has kept the Sun shining steadily for billions of years.

Stellar Thermostat: Gravitational Equilibrium

The stellar thermostat keeps the Sun's fusion rate stable.

Summary of Key Concepts

 15.1 **The Search for Life in the Solar System**

What are the necessities of life on Earth?

Life on Earth is remarkably diverse, but shares a few commonalities that can help us understand the characteristics that would make a **habitable world**—a world on which life is potentially possible. The three main requirements seem to be a source of nutrients, a source of energy, and a liquid such as liquid water.

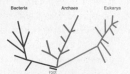

Could there be life elsewhere in our solar system?

At least six worlds in our solar system offer some possibility of harboring life. Mars shows clear evidence of liquid water on its surface in the distant past, suggesting that it was once habitable; liquid water may still exist underground. Europa probably has a subsurface ocean of liquid water, and undersea volcanoes might provide energy for life. Ganymede and Callisto might have oceans as well. Titan has liquid methane or ethane on its surface and may also have a subsurface ocean of liquid water. Enceladus shows evidence of subsurface liquids as well, offering yet another possibility for life.

 15.2 **The Search for Life Among the Stars**

Where might we find habitable planets?

Billions of stars have at least moderate-size **habitable zones** in which life-bearing planets might exist. We do not yet have the technology to search for habitable planets directly, but scientists are working on methods for finding them and looking for signatures of life in their spectra.

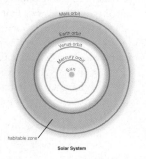

Is there intelligent life beyond Earth?

We don't know, but the **Drake equation** gives us a way to organize our thinking about the question, and scientists are looking for communications from intelligent life through **SETI,** the search for extraterrestrial intelligence. Reasonable assumptions make it seem as if civilizations ought to be common, leading to Fermi's paradox, which asks why we do not yet have any clear evidence of other civilizations.

 ❋ THE PROCESS OF SCIENCE IN ACTION

15.3 **The Evolution of Life on Earth**

What is the evidence for evolution?

Evidence that species have evolved over time comes from the study of fossils and the fossil record. The theory of **evolution** explains *how* these changes have occurred through the mechanism of **natural selection.** Evidence for evolution by natural selection comes from careful observations of relationships among species, and from our modern understanding of the molecular basis for evolution. We expect the theory of evolution to apply to life elsewhere in the universe as well as to life on Earth.

Investigations

Quick Quiz

Choose the best answer to each of the following; answers are in Appendix D. Explain your reasoning with one or more complete sentences.

1. Plants and animals are (a) the two major forms of life on Earth. (b) the only organisms that have DNA. (c) just two small branches of the tree of life on Earth.
2. Careful study of commonalities among different organisms suggests that all life on Earth (a) shares a common ancestor. (b) evolved at a time when Earth had plentiful oxygen in its atmosphere. (c) depends on sunlight for energy.
3. The modern organisms that appear to be most like the earliest life on Earth are (a) simple animals like worms. (b) green algae. (c) microbes living near undersea volcanic vents.
4. The three key requirements for life are a source of nutrients, a source of energy, and (a) H_2O in some form. (b) a liquid such as liquid water. (c) DNA.

5. A habitable world is a world that (a) has life of some kind. (b) is potentially capable of having life. (c) is potentially capable of having an advanced civilization.

6. Which of the following worlds is *not* considered a candidate for harboring life? (a) Europa (b) Mars (c) the Moon

7. If we someday find life on Mars, we expect to find it (a) near the polar ice caps. (b) in dried-up riverbeds. (c) underground.

8. How does the habitable zone around a star of spectral type G compare to that around a star of spectral type M? (a) It is larger. (b) It is hotter. (c) It is closer to its star.

9. In the Drake equation, what would f_{life} = 1/2 mean? (a) Half the stars in the Milky Way Galaxy have a planet with life. (b) Half of all life forms in the universe are intelligent. (c) Half of the habitable worlds in the galaxy have life.

10. According to current scientific understanding, the idea that the Milky Way Galaxy might be home to a civilization millions of years more advanced than ours is (a) a virtual certainty. (b) extremely unlikely. (c) one possible solution to Fermi's paradox.

11. The theory of evolution is (a) a scientific theory backed by extensive evidence. (b) one of several competing scientific models that all seem equally successful in explaining the nature of life on Earth. (c) a speculative hypothesis about how life changes through time.

12. *Natural selection* is the name given to (a) the occasional mutations that occur in DNA. (b) the mechanism by which advantageous traits are passed on from one generation to the next. (c) the idea that species change over time.

Short-Answer/Essay Questions

Explain all answers clearly, with complete sentences and proper essay structure if needed. An asterisk (*) designates a quantitative problem, for which you should show all your work.

13. *Most Likely to Have Life.* Suppose you were asked to vote in a contest to name the world in our solar system (besides Earth) "most likely to have life." Which world would you cast your vote for? Explain and defend your choice in a one-page essay.

14. *Habitable Zones.* Consider the two stars besides the Sun that are shown in Figure 15.4. Briefly describe (a) the size and location of the habitable zone for each star in comparison to the habitable zone of our Sun and (b) the prospects for finding a habitable world around each star. Explain your answers.

15. *Alien Technology.* Suppose that aliens are actually visiting Earth in UFOs. By considering the distances to stars in comparison to the distance to the Moon—which is the farthest distance that humans have so far traveled—describe how the alien technology would have to compare to ours.

16. *Solution to the Fermi Paradox.* Among the various possible solutions to Fermi's question "So where is everybody?" which one do you think is most likely? Write a one- to two-page essay in which you explain why you favor this solution.

17. *What's Wrong with This Picture?* Many science fiction stories have imagined the galaxy divided into a series of empires, each having arisen from a different civilization on a different world, that hold one another at bay because they are all at about the same level of military technology. Is this a realistic scenario? Explain.

18. *Aliens in the Movies.* Choose a science fiction movie or television show that involves an alien species. Do you think aliens like this could really exist? Write a one- to two-page critical review of the movie or show, focusing primarily on the question of whether it portrays the aliens in a scientifically reasonable way.

19. *Extreme Life.* Look for information about a recent discovery of a previously unknown type of extremophile. Describe the organism and the environment in which it lives, and discuss any implications of the finding for the search for life beyond Earth. Summarize your findings in a one-page report.

20. *The Dover Opinion.* Despite the strong scientific evidence for the theory of evolution, the teaching of evolution in public schools has been subject to contentious debate that has often landed in the courts. The most recent major decision on this debate was the Dover opinion, issued by a United States District Court in December 2005. Read the opinion, which is available online, and write a short essay describing its implications for the ongoing public controversy about what should be taught in science classes.

*21. *Time Between Civilizations.* Suppose that there are 100 billion stars in the Milky Way Galaxy, that 1 in 1 million stars gets a civilization at some point, and that these civilizations have arisen at random times over the past 5 billion years. On average, how much time separates the rise of one civilization in the galaxy from the rise of the next? Based on your answer, should we expect to make radio contact with another civilization that is either at or below our own current level of technological development?

*22. *SETI Search.* Suppose there are 10,000 civilizations broadcasting radio signals in the Milky Way Galaxy right now. On average, how many stars would we have to search before we would expect to hear a signal? Assume there are 400 billion stars in the galaxy. How does your answer change if there are only 100 civilizations instead of 10,000?

Appendixes

A Useful Numbers

Astronomical Distances

1 AU $\approx 1.496 \times 10^8$ km $= 1.496 \times 10^{11}$ m

1 light-year $\approx 9.46 \times 10^{12}$ km $= 9.46 \times 10^{15}$ m

1 parsec (pc) $\approx 3.09 \times 10^{13}$ km ≈ 3.26 light-years

1 kiloparsec (kpc) $= 1000$ pc $\approx 3.26 \times 10^3$ light-years

1 megaparsec (Mpc) $= 10^6$ pc $\approx 3.26 \times 10^6$ light-years

Universal Constants

Speed of light: $c = 3.00 \times 10^5$ km/s $= 3 \times 10^8$ m/s

Gravitational constant: $G = 6.67 \times 10^{-11} \dfrac{\text{m}^3}{\text{kg} \times \text{s}^2}$

Planck's constant: $h = 6.63 \times 10^{-34}$ joule $\times$ s

Stefan–Boltzmann constant:

$$\sigma = 5.67 \times 10^{-8} \dfrac{\text{watt}}{\text{m}^2 \times \text{Kelvin}^4}$$

Mass of a proton: $m_\text{p} = 1.67 \times 10^{-27}$ kg

Mass of an electron: $m_\text{e} = 9.11 \times 10^{-31}$ kg

Useful Sun and Earth Reference Values

Mass of the Sun: $1 M_\text{Sun} \approx 2 \times 10^{30}$ kg

Radius of the Sun: $1 R_\text{Sun} \approx 696{,}000$ km

Luminosity of the Sun: $1 L_\text{Sun} \approx 3.8 \times 10^{26}$ watts

Mass of Earth: $1 M_\text{Earth} \approx 5.97 \times 10^{24}$ kg

Radius (equatorial) of Earth: $1 R_\text{Earth} \approx 6378$ km

Acceleration of gravity on Earth: $g = 9.8$ m/s^2

Escape velocity from surface of Earth:

$$v_\text{escape} = 11 \text{ km/s} = 11{,}000 \text{ m/s}$$

Astronomical Times

1 solar day (average) $= 24^\text{h}$

1 sidereal day $\approx 23^\text{h}56^\text{m}4.09^\text{s}$

1 synodic month (average) ≈ 29.53 solar days

1 sidereal month (average) ≈ 27.32 solar days

1 tropical year ≈ 365.242 solar days

1 sidereal year ≈ 365.256 solar days

Energy and Power Units

Basic unit of energy: 1 joule $= 1 \dfrac{\text{kg} \times \text{m}^2}{\text{s}^2}$

Basic unit of power: 1 watt $= 1$ joule/s

Electron-volt: 1 eV $= 1.60 \times 10^{-19}$ joule

B Useful Formulas

Universal law of gravitation for the force between objects of mass M_1 and M_2, with distance d between their centers:

$$F = G\frac{M_1 M_2}{d^2}$$

Newton's version of Kepler's third law (applies to any pair of orbiting objects, such as a star and planet, a planet and moon, or two stars in a binary system; p is the orbital period, a is the distance between the centers of the orbiting objects, and M_1 and M_2 are the object masses):

$$p^2 = \frac{4\pi^2}{G(M_1 + M_2)} a^3$$

Escape velocity at distance R from center of object of mass M:

$$v_{escape} = \sqrt{\frac{2GM}{R}}$$

Relationship between a photon's wavelength (λ), frequency (f), and the speed of light (c):

$$\lambda \times f = c$$

Energy of a photon of wavelength λ or frequency f:

$$E = hf = \frac{hc}{\lambda}$$

Stefan–Boltzmann law for thermal radiation at temperature T (in Kelvin):

$$\text{emitted power per unit area} = \sigma T^4$$

Wien's law for the peak wavelength (λ_{max}) thermal radiation at temperature T (in Kelvin):

$$\lambda_{max} = \frac{2{,}900{,}000}{T} \text{ nm}$$

Doppler shift (radial velocity is positive if the object is moving away from us and negative if it is moving toward us):

$$\frac{\text{radial velocity}}{\text{speed of light}} = \frac{\text{shifted wavelength} - \text{rest wavelength}}{\text{rest wavelength}}$$

Angular separation (α) of two points with an actual separation s, viewed from a distance d (assuming d is much larger than s):

$$\alpha = \frac{s}{2\pi d} \times 360°$$

Inverse square law for light (d is the distance to the object):

$$\text{apparent brightness} = \frac{\text{luminosity}}{4\pi d^2}$$

Parallax formula (distance d to a star with parallax angle p in arcseconds):

$$d \text{ (in parsecs)} = \frac{1}{p \text{ (in arcseconds)}}$$

$$\text{or } d \text{ (in light-years)} = 3.26 \times \frac{1}{p \text{ (in arcseconds)}}$$

The orbital velocity law, to find the mass M_r contained within the circular orbit of radius r for an object moving at speed v:

$$M_r = \frac{r \times v^2}{G}$$

C A Few Mathematical Skills

This appendix reviews the following mathematical skills: powers of 10, scientific notation, working with units, finding a ratio, the metric system, and temperature scale conversions. You should refer to this appendix as needed while studying the textbook.

C.1 Powers of 10

Powers of 10 simply indicate how many times to multiply 10 by itself. For example:

$$10^2 = 10 \times 10 = 100$$

$$10^6 = 10 \times 10 \times 10 \times 10 \times 10 \times 10 = 1,000,000$$

Negative powers are the reciprocals of the corresponding positive powers. For example:

$$10^{-2} = \frac{1}{10^2} = \frac{1}{100} = 0.01$$

$$10^{-6} = \frac{1}{10^6} = \frac{1}{1,000,000} = 0.000001$$

Table C.1 lists powers of 10 from 10^{-12} to 10^{12}. Note that powers of 10 follow two basic rules:

1. A positive exponent tells how many zeros follow the 1. For example, 10^0 is a 1 followed by no zeros, and 10^8 is a 1 followed by eight zeros.

2. A negative exponent tells how many places are to the right of the decimal point, including the 1. For example, $10^{-1} = 0.1$ has one place to the right of the decimal point; $10^{-6} = 0.000001$ has six places to the right of the decimal point.

Multiplying and Dividing Powers of 10

Multiplying powers of 10 simply requires adding exponents, as the following examples show:

$$10^4 \times 10^7 = \underbrace{10,000}_{10^4} \times \underbrace{10,000,000}_{10^7} = \underbrace{100,000,000,000}_{10^{4+7} = 10^{11}} = 10^{11}$$

$$10^5 \times 10^{-3} = \underbrace{100,000}_{10^5} \times \underbrace{0.001}_{10^{-3}} = \underbrace{100}_{10^{5+(-3)} = 10^2} = 10^2$$

$$10^{-8} \times 10^{-5} = \underbrace{0.00000001}_{10^{-8}} \times \underbrace{0.00001}_{10^{-5}} = \underbrace{0.0000000000001}_{10^{-8+(-5)} = 10^{-13}} = 10^{-13}$$

Dividing powers of 10 requires subtracting exponents, as in the following examples:

$$\frac{10^5}{10^3} = \underbrace{100,000}_{10^5} \div \underbrace{1000}_{10^3} = \underbrace{100}_{10^{5-3} = 10^2} = 10^2$$

Table C.1 Powers of 10

Zero and Positive Powers			Negative Powers		
Power	**Value**	**Name**	**Power**	**Value**	**Name**
10^0	1	One			
10^1	10	Ten	10^{-1}	0.1	Tenth
10^2	100	Hundred	10^{-2}	0.01	Hundredth
10^3	1000	Thousand	10^{-3}	0.001	Thousandth
10^4	10,000	Ten thousand	10^{-4}	0.0001	Ten-thousandth
10^5	100,000	Hundred thousand	10^{-5}	0.00001	Hundred-thousandth
10^6	1,000,000	Million	10^{-6}	0.000001	Millionth
10^7	10,000,000	Ten million	10^{-7}	0.0000001	Ten-millionth
10^8	100,000,000	Hundred million	10^{-8}	0.00000001	Hundred-millionth
10^9	1,000,000,000	Billion	10^{-9}	0.000000001	Billionth
10^{10}	10,000,000,000	Ten billion	10^{-10}	0.0000000001	Ten-billionth
10^{11}	100,000,000,000	Hundred billion	10^{-11}	0.00000000001	Hundred-billionth
10^{12}	1,000,000,000,000	Trillion	10^{-12}	0.000000000001	Trillionth

$$\frac{10^3}{10^7} = \underbrace{1000}_{10^3} \div \underbrace{10,000,000}_{10^7} = \underbrace{0.0001}_{10^{3-7} = 10^{-4}} = 10^{-4}$$

$$\frac{10^{-4}}{10^{-6}} = \underbrace{0.0001}_{10^{-4}} \div \underbrace{0.000001}_{10^{-6}} = \underbrace{100}_{10^{-4-(-6)} = 10^2} = 10^2$$

Powers of Powers of 10

We can use the multiplication and division rules to raise powers of 10 to other powers or to take roots. For example:

$$(10^4)^3 = 10^4 \times 10^4 \times 10^4 = 10^{4+4+4} = 10^{12}$$

Note that we can get the same end result by simply multiplying the two powers:

$$(10^4)^3 = 10^{4 \times 3} = 10^{12}$$

Because taking a root is the same as raising to a fractional power (e.g., the square root is the same as the $\frac{1}{2}$ power, the cube root is the same as the $\frac{1}{3}$ power, etc.), we can use the same procedure for roots, as in the following example:

$$\sqrt{10^4} = (10^4)^{1/2} = 10^{4 \times (1/2)} = 10^2$$

Adding and Subtracting Powers of 10

Unlike for multiplying and dividing powers of 10, there is no shortcut for adding or subtracting powers of 10. The values must be written in longhand notation. For example:

$$10^6 + 10^2 = 1,000,000 + 100 = 1,000,100$$

$$10^8 + 10^{-3} = 100,000,000 + 0.001 = 100,000,000.001$$

$$10^7 - 10^3 = 10,000,000 - 1000 = 9,999,000$$

Summary

We can summarize our findings using n and m to represent any numbers:

- To *multiply* powers of 10, *add* exponents: $10^n \times 10^m = 10^{n+m}$

- To *divide* powers of 10, *subtract* exponents: $\dfrac{10^n}{10^m} = 10^{n-m}$

- To *raise* powers of 10 to other powers, multiply exponents: $(10^n)^m = 10^{n \times m}$

C.2   Scientific Notation

When we are dealing with large or small numbers, it's generally easier to write them with powers of 10. For example, it's much easier to write the number 6,000,000,000,000 as 6×10^{12}. This format, in which a number *between* 1 and 10 is multiplied by a power of 10, is called **scientific notation**.

Converting a Number to Scientific Notation

We can convert numbers written in ordinary notation to scientific notation with a simple two-step process:

1. Move the decimal point to come after the *first* nonzero digit.

2. The number of places the decimal point moves tells you the power of 10; the power is *positive* if the decimal point moves to the left and *negative* if it moves to the right.

Examples:

$$3042 \xrightarrow[\text{3 places to left}]{\text{decimal needs to move}} 3.042 \times 10^3$$

$$0.00012 \xrightarrow[\text{4 places to right}]{\text{decimal needs to move}} 1.2 \times 10^{-4}$$

$$226 \times 10^2 \xrightarrow[\text{2 places to left}]{\text{decimal needs to move}} (2.26 \times 10^2) \times 10^2 = 2.26 \times 10^4$$

Converting a Number from Scientific Notation

We can convert numbers written in scientific notation to ordinary notation by the reverse process:

1. The power of 10 indicates how many places to move the decimal point; move it to the *right* if the power of 10 is positive and to the *left* if it is negative.

2. If moving the decimal point creates any open places, fill them with zeros.

Examples:

$$4.01 \times 10^2 \xrightarrow[\text{2 places to right}]{\text{move decimal}} 401$$

$$3.6 \times 10^6 \xrightarrow[\text{6 places to right}]{\text{move decimal}} 3,600,000$$

$$5.7 \times 10^{-3} \xrightarrow[\text{3 places to left}]{\text{move decimal}} 0.0057$$

Multiplying or Dividing Numbers in Scientific Notation

Multiplying or dividing numbers in scientific notation simply requires operating on the powers of 10 and the other parts of the number separately.

Examples:

$$(6 \times 10^2) \times (4 \times 10^5) = (6 \times 4) \times (10^2 \times 10^5) = 24 \times 10^7 = (2.4 \times 10^1) \times 10^7 = 2.4 \times 10^8$$

$$\frac{4.2 \times 10^{-2}}{8.4 \times 10^{-5}} = \frac{4.2}{8.4} \times \frac{10^{-2}}{10^{-5}} = 0.5 \times 10^{-2-(-5)} = 0.5 \times 10^3 = (5 \times 10^{-1}) \times 10^3 = 5 \times 10^2$$

Note that, in both these examples, we first found an answer in which the number multiplied by a power of 10 was *not* between 1 and 10. We therefore followed the procedure for converting the final answer to scientific notation.

Addition and Subtraction with Scientific Notation

In general, we must write numbers in ordinary notation before adding or subtracting.

Examples:

$$(3 \times 10^6) + (5 \times 10^2) = 3,000,000 + 500 = 3,000,500 = 3.0005 \times 10^6$$

$$(4.6 \times 10^9) - (5 \times 10^8) = 4,600,000,000 - 500,000,000 = 4,100,000,000 = 4.1 \times 10^9$$

When both numbers have the *same* power of 10, we can factor out the power of 10 first.

Examples:

$$(7 \times 10^{10}) + (4 \times 10^{10}) = (7 + 4) \times 10^{10} = 11 \times 10^{10} = 1.1 \times 10^{11}$$

$$(2.3 \times 10^{-22}) - (1.6 \times 10^{-22}) = (2.3 - 1.6) \times 10^{-22} = 0.7 \times 10^{-22} = 7.0 \times 10^{-23}$$

C.3 Working with Units

Showing the units of a problem as you solve it usually makes the work much easier and also provides a useful way of checking your work. If an answer does not come out with the units you expect, you probably did something wrong. In general, working with units is very similar to working with numbers, as the following guidelines and examples show.

Five Guidelines for Working with Units

Before you begin any problem, think ahead and identify the units you expect for the final answer. Then operate on the units along with the numbers as you solve the problem. The following five guidelines may be helpful when you are working with units:

1. Mathematically, it doesn't matter whether a unit is singular (e.g., meter) or plural (e.g., meters); we can use the same abbreviation (e.g., m) for both.

2. You cannot add or subtract numbers unless they have the *same* units. For example, 5 apples + 3 apples = 8 apples, but the expression 5 apples + 3 oranges cannot be simplified further.

3. You *can* multiply units, divide units, or raise units to powers. Look for key words that tell you what to do.

   - *Per* suggests division. For example, we write a speed of 100 kilometers per hour as

$$100 \, \frac{km}{hr} \quad \text{or} \quad \frac{100 \, km}{1 \, hr}$$

- *Of* suggests multiplication. For example, if you launch a 50-kg space probe at a launch cost *of* $10,000 per kilogram, the total cost is

$$50 \, \text{kg} \times \frac{\$10,000}{\text{kg}} = \$500,000$$

- *Square* suggests raising to the second power. For example, we write an area of 75 square meters as $75 \, \text{m}^2$.

- *Cube* suggests raising to the third power. For example, we write a volume of 12 cubic centimeters as $12 \, \text{cm}^3$.

4. Often the number you are given is not in the units you wish to work with. For example, you may be given that the speed of light is 300,000 km/s but need it in units of m/s for a particular problem. To convert the units, simply multiply the given number by a *conversion factor:* a fraction in which the numerator (top of the fraction) and denominator (bottom of the fraction) are equal, so that the value of the fraction is 1; the number in the denominator must have the units that you wish to change. In the case of changing the speed of light from units of km/s to m/s, you need a conversion factor for kilometers to meters. Thus, the conversion factor is

$$\frac{1000 \, \text{m}}{1 \, \text{km}}$$

Note that this conversion factor is equal to 1, since 1000 meters and 1 kilometer are equal, and that the units to be changed (km) appear in the denominator. We can now convert the speed of light from units of km/s to m/s simply by multiplying by this conversion factor:

$$\underbrace{300,000 \, \frac{\text{km}}{\text{s}}}_{\substack{\text{speed of light} \\ \text{in km/s}}} \times \underbrace{\frac{1000 \, \text{m}}{1 \, \text{km}}}_{\substack{\text{conversion from} \\ \text{km to m}}} = \underbrace{3 \times 10^8 \, \frac{\text{m}}{\text{s}}}_{\substack{\text{speed of light} \\ \text{in m/s}}}$$

Note that the units of km cancel, leaving the answer in units of m/s.

5. It's easier to work with units if you replace division with multiplication by the reciprocal. For example, suppose you want to know how many minutes are represented by 300 seconds. We can find the answer by dividing 300 seconds by 60 seconds per minute:

$$300 \, \text{s} \div 60 \, \frac{\text{s}}{\text{min}}$$

However, it is easier to see the unit cancellations if we rewrite this expression by replacing the division with multiplication by the reciprocal (this process is easy to remember as "invert and multiply"):

$$300 \, \text{s} \div 60 \frac{\text{s}}{\text{min}} = 300 \, \text{s} \times \underbrace{\frac{1 \, \text{min}}{60 \, \text{s}}}_{\text{invert}} = 5 \, \text{min}$$

$$\underbrace{\phantom{300 \, \text{s} \div 60 \frac{\text{s}}{\text{min}} = 300 \, \text{s} \times \frac{1 \, \text{min}}{60 \, \text{s}} = 5 \, \text{min}}}_{\text{and multiply}}$$

We now see that the units of seconds (s) cancel in the numerator of the first term and the denominator of the second term, leaving the answer in units of minutes.

More Examples of Working with Units

Example 1. How many seconds are there in 1 day?

Solution: We can answer the question by setting up a *chain* of unit conversions in which we start with 1 *day* and end up with *seconds*. We use the facts that there are 24 hours per day (24 hr/day), 60 minutes per hour (60 min/hr), and 60 seconds per minute (60 s/min):

$$1 \, \underbrace{\cancel{\text{day}}}_{\substack{\text{Starting} \\ \text{value}}} \times \underbrace{\frac{24 \, \cancel{\text{hr}}}{\cancel{\text{day}}}}_{\substack{\text{conversion} \\ \text{from} \\ \text{day to hr}}} \times \underbrace{\frac{60 \, \cancel{\text{min}}}{\cancel{\text{hr}}}}_{\substack{\text{conversion} \\ \text{from} \\ \text{hr to min}}} \times \underbrace{\frac{60 \, \text{s}}{\cancel{\text{min}}}}_{\substack{\text{conversion} \\ \text{from} \\ \text{min to s}}} = 86{,}400 \, \text{s}$$

Note that all the units cancel except *seconds,* which is what we want for the answer. There are 86,400 seconds in 1 day.

Example 2. Convert a distance of 10^8 cm to km.

Solution: The easiest way to make this conversion is in two steps, since we know that there are 100 centimeters per meter (100 cm/m) and 1000 meters per kilometer (1000 m/km):

$$\underbrace{10^8 \, \text{cm}}_{\substack{\text{starting} \\ \text{value}}} \times \underbrace{\frac{1 \, \text{m}}{100 \, \text{cm}}}_{\substack{\text{conversion} \\ \text{from} \\ \text{cm to m}}} \times \underbrace{\frac{1 \, \text{km}}{1000 \, \text{m}}}_{\substack{\text{conversion} \\ \text{from} \\ \text{m to km}}} = 10^8 \, \cancel{\text{cm}} \times \frac{1 \, \cancel{\text{m}}}{10^2 \, \cancel{\text{cm}}} \times \frac{1 \, \text{km}}{10^3 \, \cancel{\text{m}}} = 10^3 \, \text{km}$$

Alternatively, if we recognize that the number of kilometers should be smaller than the number of centimeters (because kilometers are larger), we might decide to do this conversion by dividing as follows:

$$10^8 \, \text{cm} \; : \; \frac{100 \, \text{cm}}{\text{m}} \; : \; \frac{1000 \, \text{m}}{\text{km}}$$

In this case, before carrying out the calculation, we replace each division with multiplication by the reciprocal:

$$10^8 \, \text{cm} \div \frac{100 \, \text{cm}}{\text{m}} \div \frac{1000 \, \text{m}}{\text{km}} = 10^8 \, \text{cm} \times \frac{1 \, \text{m}}{100 \, \text{cm}} \times \frac{1 \, \text{km}}{1000 \, \text{m}}$$

$$= 10^8 \, \cancel{\text{cm}} \times \frac{1 \, \cancel{\text{m}}}{10^2 \, \cancel{\text{cm}}} \times \frac{1 \, \text{km}}{10^3 \, \cancel{\text{m}}}$$

$$= 10^3 \, \text{km}$$

Note that we again get the answer that 10^8 cm is the same as 10^3 km, or 1000 km.

Example 3. Suppose you accelerate at 9.8 m/s^2 for 4 seconds, starting from rest. How fast will you be going?

Solution: The question asked "how fast?" so we expect to end up with a speed. Therefore, we multiply the acceleration by the amount of time you accelerated:

$$9.8 \, \frac{\text{m}}{\text{s}^2} \times 4 \, \text{s} = (9.8 \times 4) \, \frac{\text{m} \times \cancel{\text{s}}}{\text{s}^{\cancel{2}}} = 39.2 \, \frac{\text{m}}{\text{s}}$$

Note that the units end up as a speed, showing that you will be traveling 39.2 m/s after 4 seconds of acceleration at 9.8 m/s^2.

Example 4. A reservoir is 2 km long and 3 km wide. Calculate its area, in both square kilometers and square meters.

Solution: We find its area by multiplying its length and width:

$$2 \, \text{km} \times 3 \, \text{km} = 6 \, \text{km}^2$$

Next we need to convert this area of 6 km² to square meters, using the fact that there are 1000 meters per kilometer (1000 m/km). Note that we must square the term 1000 m/km when converting from km² to m²:

$$6 \text{ km}^2 \times \left(1000 \frac{\text{m}}{\text{km}}\right)^2 = 6 \text{ km}^2 \times 1000^2 \frac{\text{m}^2}{\text{km}^2} = 6 \cancel{\text{ km}^2} \times 1{,}000{,}000 \frac{\text{m}^2}{\cancel{\text{km}^2}}$$

$$= 6{,}000{,}000 \text{ m}^2$$

The reservoir area is 6 km², which is the same as 6 million m².

C.4 Finding a Ratio

Suppose you want to compare two quantities, such as the average density of Earth and the average density of Jupiter. The way we do such a comparison is by dividing, which tells us the *ratio* of the two quantities. In this case, Earth's average density is 5.52 g/cm³ and Jupiter's average density is 1.33 g/cm³, so the ratio is

$$\frac{\text{average density of Earth}}{\text{average density of Jupiter}} = \frac{5.52 \text{ g/cm}^3}{1.33 \text{ g/cm}^3} = 4.15$$

Notice how the units cancel on both the top and the bottom of the fraction. We can state our result in two equivalent ways:

- The ratio of Earth's average density to Jupiter's average density is 4.15.

- Earth's average density is 4.15 times Jupiter's average density.

Sometimes, the quantities that you want to compare may each involve an equation. In such cases, you could, of course, find the ratio by first calculating each of the two quantities individually and then dividing. However, it is much easier if you first express the ratio as a fraction, putting the equation for one quantity on top and the other on the bottom. Some of the terms in the equation may then cancel out, making any calculations much easier.

Example 1. Compare the kinetic energy of a car traveling at 100 km/hr to that of the same car traveling at 50 km/hr.

Solution: We do the comparison by finding the ratio of the two kinetic energies, recalling that the formula for kinetic energy is $\frac{1}{2} mv^2$. Since we are not told the mass of the car, you might at first think that we don't have enough information to find the ratio. However, notice what happens when we put the equations for each kinetic energy into the ratio, calling the two speeds v_1 and v_2:

$$\frac{\text{K.E. car at } v_1}{\text{K.E. car at } v_2} = \frac{\frac{1}{2} \cancel{m}_{\text{car}} v_1^2}{\frac{1}{2} \cancel{m}_{\text{car}} v_2^2} = \frac{v_1^2}{v_2^2} = \left(\frac{v_1}{v_2}\right)^2$$

All the terms cancel except those with the two speeds, leaving us with a very simple formula for the ratio. Now we put in 100 km/hr for v_1 and 50 km/hr for v_2:

$$\frac{\text{K.E. car at 100 km/hr}}{\text{K.E. car at 50 km/hr}} = \left(\frac{100 \cancel{\text{ km/hr}}}{50 \cancel{\text{ km/hr}}}\right)^2 = 2^2 = 4$$

The ratio of the car's kinetic energies at 100 km/hr and 50 km/hr is 4. That is, the car has four times as much kinetic energy at 100 km/hr as it has at 50 km/hr.

Example 2. Compare the strength of gravity between Earth and the Sun to the strength of gravity between Earth and the Moon.

Solution: We do the comparison by taking the ratio of the Earth–Sun gravity to the Earth–Moon gravity. In this case, each quantity is found from the equation of Newton's law of gravity. (See Section 3.3.) Thus, the ratio is

$$\frac{\text{Earth--Sun gravity}}{\text{Earth--Moon gravity}} = \frac{\mathcal{G}\dfrac{M_{\text{Earth}} M_{\text{Sun}}}{(d_{\text{Earth-Sun}})^2}}{\mathcal{G}\dfrac{M_{\text{Earth}} M_{\text{Moon}}}{(d_{\text{Earth-Moon}})^2}} = \frac{M_{\text{Sun}}}{(d_{\text{Earth-Sun}})^2} \times \frac{(d_{\text{Earth-Moon}})^2}{M_{\text{Moon}}}$$

Note how all but four of the terms cancel; the last step comes from replacing the division with multiplication by the reciprocal (the "invert and multiply" rule for division). We can simplify the work further by rearranging the terms so that we have the masses and distances together:

$$\frac{\text{Earth--Sun gravity}}{\text{Earth--Moon gravity}} = \frac{M_{\text{Sun}}}{M_{\text{Moon}}} \times \frac{(d_{\text{Earth-Moon}})^2}{(d_{\text{Earth-Sun}})^2}$$

Now it is just a matter of looking up the numbers and calculating:

$$\frac{\text{Earth--Sun gravity}}{\text{Earth--Moon gravity}} = \frac{1.99 \times 10^{30} \text{ kg}}{7.35 \times 10^{22} \text{ kg}} \times \frac{(384.4 \times 10^3 \text{ km})^2}{(149.6 \times 10^6 \text{ km})^2} = 179$$

In other words, the Earth--Sun gravity is 179 times as strong as the Earth--Moon gravity.

C.5 The Metric System (SI)

The modern version of the metric system, known as *Système Internationale d'Unites* (French for "International System of Units") or **SI**, was formally established in 1960. Today, it is the primary measurement system in nearly every country in the world with the exception of the United States. Even in the United States, it is the system of choice for science and international commerce.

The basic units of length, mass, and time in the SI are

- The **meter** for length, abbreviated m
- The **kilogram** for mass, abbreviated kg
- The **second** for time, abbreviated s

Multiples of metric units are formed by powers of 10, using a prefix to indicate the power. For example, *kilo* means 10^3 (1000), so a kilometer is 1000 meters; a microgram is 0.000001 gram, because *micro* means 10^{-6}, or one millionth. Some of the more common prefixes are listed in Table C.2.

Table C.2 SI (Metric) Prefixes

	Small Values			Large Values	
Prefix	Abbreviation	Value	Prefix	Abbreviation	Value
Deci	d	10^{-1}	Deca	da	10^1
Centi	c	10^{-2}	Hecto	h	10^2
Milli	m	10^{-3}	Kilo	k	10^3
Micro	μ	10^{-6}	Mega	M	10^6
Nano	n	10^{-9}	Giga	G	10^9
Pico	p	10^{-12}	Tera	T	10^{12}

Metric Conversions

Table C.3 lists conversions between metric units and units used commonly in the United States. Note that the conversions between kilograms and pounds are valid only on Earth, because they depend on the strength of gravity.

Example 1. International athletic competitions generally use metric distances. Compare the length of a 100-meter race to that of a 100-yard race.

Table C.3 Metric Conversions

To Metric	From Metric
1 inch = 2.540 cm	1 cm = 0.3937 inch
1 foot = 0.3048 m	1 m = 3.28 feet
1 yard = 0.9144 m	1 m = 1.094 yards
1 mile = 1.6093 km	1 km = 0.6214 mile
1 pound = 0.4536 kg	1 kg = 2.205 pounds

Solution: Table C.3 shows that 1 m = 1.094 yd, so 100 m is 109.4 yd. Note that 100 meters is almost 110 yards; a good "rule of thumb" to remember is that distances in meters are about 10% longer than the corresponding number of yards.

Example 2. How many square kilometers are in 1 square mile?

Solution: We use the square of the miles-to-kilometers conversion factor:

$$(1 \text{ mi}^2) \times \left(\frac{1.6093 \text{ km}}{1 \text{ mi}}\right)^2 = (1 \text{ mi}^2) \times \left(1.6093^2 \frac{\text{km}^2}{\text{mi}^2}\right) = 2.5898 \text{ km}^2$$

Therefore, 1 square mile is 2.5898 square kilometers.

C.6   Temperature Scales

In the United States, we usually use the Fahrenheit temperature scale in our daily lives. But most of the rest of the world uses the Celsius scale, and in science we often use the Kelvin scale.

Figure 1 shows the relationship among the three temperature scales. Notice that the Fahrenheit scale is defined so that under ordinary conditions at sea level on Earth, water freezes at 32°F and boils at 212°F. Most of the rest of the world uses the Celsius scale, on which water freezes at 0°C and boils at 100°C. The Kelvin scale is very similar to the Celsius scale, because the boiling point of water is exactly 100 K higher than the freezing point. However, the zero point of the Kelvin scale is defined to be the coldest possible temperature, or *absolute zero*. This puts the freezing point of water at 273.15 K and the boiling point at 375.15 K.

You can use these relationships to convert among the three scales. Because the only difference between the Kelvin scale and the Celsius scale is the starting point, we can convert from Kelvin to Celsius and from Celsius to Fahrenheit with the following approximate relationships:

$$T_{\text{Celsius}} \approx T_{\text{Kelvin}} - 273$$
$$T_{\text{Fahrenheit}} = (1.8 \times T_{\text{Celsius}}) + 32$$

Example 1. Convert a temperature of 300 K to Celsius and Fahrenheit.

Solution: We use the first formula to convert to Celsius:

$$T_{\text{Celsius}} \approx T_{\text{Kelvin}} - 273 = 300 - 273 = 27°C$$

We use the second formula to convert to Fahrenheit:

$$T_{\text{Fahrenheit}} = (1.8 \times T_{\text{Celsius}}) + 32 = (1.8 \times 27) + 32 \approx 81°F$$

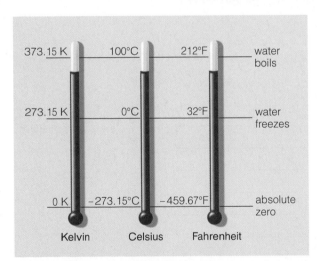

Figure 1 | Three common temperature scales: Kelvin, Celsius, and Fahrenheit. (The degree symbol (°) is not usually used with the Kelvin scale.)

 **D** # Answers to Quick Quiz Questions

Chapter 1

1 a 2 b 3 a 4 c 5 b 6 b 7 c 8 b 9 c 10 a 11 a 12 c

Chapter 2

1 a 2 a 3 c 4 a 5 c 6 a 7 b 8 b 9 b 10 a 11 b 12 b

Chapter 3

1 a 2 b 3 a 4 c 5 c 6 b 7 b 8 c 9 b 10 c 11 c 12 b

Chapter 4

1 b 2 b 3 b 4 c 5 a 6 c 7 b 8 c 9 a 10 c 11 b 12 b

Chapter 5

1 b 2 b 3 a 4 b 5 c 6 b 7 a 8 b 9 c 10 c 11 c 12 c

Chapter 6

1 c 2 c 3 b 4 c 5 a 6 b 7 b 8 b 9 b 10 c 11 b 12 b

Chapter 7

1 c 2 b 3 b 4 b 5 c 6 a 7 b 8 b 9 c 10 b 11 b 12 b

Chapter 8

1 c 2 b 3 c 4 a 5 a 6 a 7 c 8 a 9 c 10 a 11 c 12 c

Chapter 9

1 c 2 a 3 b 4 a 5 b 6 c 7 a 8 b 9 c 10 a 11 c 12 b

Chapter 10

1 a 2 a 3 b 4 c 5 b 6 a 7 c 8 b 9 a 10 c 11 a 12 a

Chapter 11

1 c 2 a 3 c 4 a 5 c 6 c 7 b 8 b 9 c 10 a 11 c 12 c

Chapter 12

1 a 2 b 3 a 4 c 5 c 6 a 7 c 8 c 9 c 10 c 11 c 12 c

Chapter 13

1 b 2 a 3 c 4 b 5 c 6 c 7 a 8 b 9 b 10 a 11 b 12 c

Chapter 14

1 a 2 c 3 c 4 c 5 b 6 a 7 a 8 b 9 a 10 a 11 c 12 b

Chapter 15

1 c 2 a 3 c 4 b 5 b 6 c 7 c 8 c 9 c 10 c 11 a 12 b

Credits

Glossary

absolute magnitude A measure of an object's luminosity; defined to be the apparent magnitude the object would have if it were located exactly 10 parsecs away.

absolute zero The coldest possible temperature, which is 0 K.

absorption (of light) The process by which matter absorbs radiative energy.

absorption line spectrum A spectrum that contains absorption lines.

accelerating universe A universe in which a repulsive force (*see* cosmological constant) causes the expansion of the universe to accelerate with time. Its galaxies will recede from one another increasingly faster, and it will become cold and dark more quickly than a coasting universe.

acceleration The rate at which an object's velocity changes. Its standard units are m/s^2.

acceleration of gravity The acceleration of a falling object. On Earth, the acceleration of gravity, designated by *g*, is $9.8 \ m/s^2$.

accretion The process by which small objects gather together to make larger objects.

accretion disk A rapidly rotating disk of material that gradually falls inward as it orbits a starlike object (e.g., white dwarf, neutron star, or black hole).

active galactic nuclei The unusually luminous centers of some galaxies, thought to be powered by accretion onto supermassive black holes. Quasars are the brightest type of active galactic nuclei; radio galaxies also contain active galactic nuclei.

active galaxy A term sometimes used to describe a galaxy that contains an *active galactic nucleus*.

adaptive optics A technique in which telescope mirrors flex rapidly to compensate for the bending of starlight caused by atmospheric turbulence.

albedo A technical name for *reflectivity; see* reflectivity.

Algol paradox A paradox concerning the binary star Algol, which contains a subgiant star that is less massive than its main-sequence companion.

altitude (above horizon) The angular distance between the horizon and an object in the sky.

amino acids The building blocks of proteins.

analemma The figure 8 path traced by the Sun over the course of a year when viewed at the same place and the same time each day; it represents the discrepancies between apparent and mean solar time.

Andromeda Galaxy (M31; the Great Galaxy in Andromeda) The nearest large spiral galaxy to the Milky Way.

angular momentum Momentum attributable to rotation or revolution. The angular momentum of an object moving in a circle of radius *r* is the product $m \times v \times r$.

angular resolution (of a telescope) The smallest angular separation that two pointlike objects can have and still be seen as distinct points of light (rather than as a single point of light).

angular size (or **angular distance**) A measure of the angle formed by extending imaginary lines outward from our eyes to span an object (or the space between two objects).

annihilation *See* matter–antimatter annihilation.

annular solar eclipse A solar eclipse during which the Moon is directly in front of the Sun but its angular size is not large enough to fully block the Sun; thus, a ring (or *annulus*) of sunlight is still visible around the Moon's disk.

Antarctic Circle The circle on Earth with latitude 66.5°S.

antielectron The antimatter equivalent of an electron. It is identical to an electron in virtually all respects, except it has a positive rather than a negative electrical charge.

antimatter Any particle with the same mass as a particle of ordinary matter but whose other basic properties, such as electrical charge, are precisely opposite.

aphelion The point at which an object orbiting the Sun is farthest from the Sun.

apogee The point at which an object orbiting Earth is farthest from Earth.

apparent brightness The amount of light reaching us *per unit area* from a luminous object; often measured in units of $watts/m^2$.

apparent magnitude A measure of the apparent brightness of an object in the sky, based on the ancient system developed by Hipparchus.

apparent retrograde motion The apparent motion of a planet, as viewed from Earth, during the period of a few weeks or months when it moves westward relative to the stars in our sky.

apparent solar time Time measured by the actual position of the Sun in your local sky, defined so that noon is when the Sun is *on* the meridian.

arcminute (or **minute of arc**) 1/60 of 1°.

arcsecond (or **second of arc**) 1/60 of an arcminute, or 1/3600 of 1°.

Arctic Circle The circle on Earth with latitude 66.5°N.

asteroid A relatively small and rocky object that orbits a star; asteroids are officially considered part of a category known as "small solar system bodies."

asteroid belt The region of our solar system between the orbits of Mars and Jupiter in which asteroids are heavily concentrated.

astrobiology The study of life on Earth and beyond; it emphasizes research into questions of the origin of life, the conditions under which life can survive, and the search for life beyond Earth.

astrometric technique The detection of extrasolar planets through the side-to-side motion of a star caused by gravitational tugs from the planet.

astronomical unit (AU) The average distance (semimajor axis) of Earth from the Sun, which is about 150 million km.

atmosphere A layer of gas that surrounds a planet or moon, usually very thin compared to the size of the object.

atmospheric pressure The surface pressure resulting from the overlying weight of an atmosphere.

atmospheric structure The layering of a planetary atmosphere due to variations in temperature with altitude. For example, Earth's atmospheric structure from the ground up consists of the troposphere, stratosphere, thermosphere, and exosphere.

atomic hydrogen gas Gas composed mostly of hydrogen atoms, though in space it is generally mixed with helium and small amounts of other elements as well; it is the most common form of interstellar gas.

atomic mass number The combined number of protons and neutrons in an atom.

atomic number The number of protons in an atom.

atoms Consist of a nucleus made from protons and neutrons, surrounded by a cloud of electrons.

aurora Dancing lights in the sky caused by charged particles entering our atmosphere; called the *aurora borealis* in the Northern Hemisphere and the *aurora australis* in the Southern Hemisphere.

axis tilt (of a planet in our solar system) The amount by which a planet's axis is tilted with respect to a line perpendicular to the ecliptic plane.

azimuth (usually called **direction** in this book) Direction around the horizon from due north, measured clockwise in degrees. For example, the azimuth of due north is 0°, due east is 90°, due south is 180°, and due west is 270°.

bar The standard unit of pressure, approximately equal to Earth's atmospheric pressure at sea level.

barred spiral galaxies Spiral galaxies that have a straight bar of stars cutting across their centers.

baryonic matter Ordinary matter made from atoms (so called because the nuclei of atoms contain protons and neutrons, which are both baryons).

baryons Particles, including protons and neutrons, that are made from three quarks.

basalt A type of dark, high-density volcanic rock that is rich in iron and magnesium-based silicate minerals; it forms a runny (easy flowing) lava when molten.

belts (on a jovian planet) Dark bands of sinking air that encircle a jovian planet at a particular set of latitudes.

Big Bang The name given to the event thought to mark the birth of the universe.

Big Bang theory The scientific theory of the universe's earliest moments, stating that all the matter in our observable universe came into being at a single moment in time as an extremely hot, dense mixture of subatomic particles and radiation.

Big Crunch The name given to the event that would presumably end the universe if gravity ever reverses the universal expansion and the universe someday begins to collapse.

binary star system A star system that contains two stars.

biosphere The "layer" of life on Earth.

blackbody radiation *See* thermal radiation.

black hole A bottomless pit in spacetime. Nothing can escape from within a black hole, and we can never again detect or observe an object that falls into a black hole.

black smokers Structures around seafloor volcanic vents that support a wide variety of life.

BL Lac objects A class of active galactic nuclei that probably represent the centers of radio galaxies whose jets happen to be pointed directly at us.

blowout Ejection of the hot, gaseous contents of a superbubble when it grows so large that it bursts out of the cooler layer of gas filling the galaxy's disk.

blueshift A Doppler shift in which spectral features are shifted to shorter wavelengths, observed when an object is moving toward the observer.

bosons Particles, such as photons, to which the exclusion principle does not apply.

bound orbits Orbits on which an object travels repeatedly around another object; bound orbits are elliptical in shape.

brown dwarf An object too small to become an ordinary star because electron degeneracy pressure halts its gravitational collapse before fusion becomes self-sustaining; brown dwarfs have mass less than $0.08 M_{Sun}$.

bubble (interstellar) An expanding shell of hot, ionized gas driven by stellar winds or supernovae, with very hot and very low density gas inside.

bulge (of a spiral galaxy) The central portion of a spiral galaxy that is roughly spherical (or football shaped) and bulges above and below the plane of the galactic disk.

Cambrian explosion The dramatic diversification of life on Earth that occurred between about 540 and 500 million years ago.

carbonate rock A carbon-rich rock, such as limestone, that forms underwater from chemical reactions between sediments and carbon dioxide. On Earth, most of the outgassed carbon dioxide currently resides in carbonate rocks.

carbon dioxide cycle (CO₂ cycle) The process that cycles carbon dioxide between Earth's atmosphere and surface rocks.

carbon stars Stars whose atmospheres are especially carbon-rich, thought to be near the ends of their lives; carbon stars are the primary sources of carbon in the universe.

Cassini division A large, dark gap in Saturn's rings, visible through small telescopes on Earth.

CCD (charge coupled device) A type of electronic light detector that has largely replaced photographic film in astronomical research.

celestial coordinates The coordinates of right ascension and declination that fix an object's position on the celestial sphere.

celestial equator (CE) The extension of Earth's equator onto the celestial sphere.

celestial navigation Navigation on the surface of the Earth accomplished by observations of the Sun and stars.

celestial sphere The imaginary sphere on which objects in the sky appear to reside when observed from Earth.

Celsius (temperature scale) The temperature scale commonly used in daily activity internationally, defined so that, on Earth's surface, water freezes at 0°C and boils at 100°C.

center of mass (of orbiting objects) The point at which two or more orbiting objects would balance if they were somehow connected; it is the point around which the orbiting objects actually orbit.

central dominant galaxy A giant elliptical galaxy found at the center of a dense cluster of galaxies, apparently formed by the merger of several individual galaxies.

Cepheid *See* Cepheid variable stars.

Cepheid variable stars A particularly luminous type of pulsating variable star that follows a period–luminosity relation and hence is very useful for measuring cosmic distances.

Chandrasekhar limit *See* white dwarf limit.

charged particle belts Zones in which ions and electrons accumulate and encircle a planet.

chemical enrichment The process by which the abundance of heavy elements (heavier than helium) in the interstellar medium gradually increases over time as these elements are produced by stars and released into space.

chemical potential energy Potential energy that can be released through chemical reactions; for example, food contains chemical potential energy that your body can convert to other forms of energy.

chondrites Another name for primitive meteorites. The name comes from the round chondrules within them. *Achondrites*, meaning "without chondrules," is another name for processed meteorites.

chromosphere The layer of the Sun's atmosphere below the corona; most of the Sun's ultraviolet light is emitted from this region, in which the temperature is about 10,000 K.

circulation cells (or **Hadley cells**) Large-scale cells (similar to convection cells) in a planet's atmosphere that transport heat between the equator and the poles.

circumpolar star A star that always remains above the horizon for a particular latitude.

climate The long-term average of weather.

close binary A binary star system in which the two stars are very close together.

closed universe A universe in which spacetime curves back on itself to the point where its overall shape is analogous to that of the surface of a sphere.

cluster of galaxies A collection of a few dozen or more galaxies bound together by gravity; smaller collections of galaxies are simply called *groups*.

cluster of stars A group of anywhere from several hundred to a million or so stars; star clusters come in two types—open clusters and globular clusters.

CNO cycle The cycle of reactions by which intermediate- and high-mass stars fuse hydrogen into helium.

coasting universe A universe that will keep expanding forever with little change in its rate of expansion; in the absence of a repulsive force (*see* cosmological constant), a coasting universe is one in which the actual mass density is *smaller* than the critical density.

coma (of a comet) The dusty atmosphere of a comet, created by sublimation of ices in the nucleus when the comet is near the Sun.

comet A relatively small, icy object that orbits a star. Like asteroids, comets are officially considered part of a category known as "small solar system bodies."

comparative planetology The study of the solar system by examining and understanding the similarities and differences among worlds.

compound (chemical) A substance made from molecules consisting of two or more atoms with different atomic numbers.

condensates Solid or liquid particles that condense from a cloud of gas.

condensation The formation of solid or liquid particles from a cloud of gas.

conduction (of energy) The process by which thermal energy is transferred by direct contact from warm material to cooler material.

conjunction (of a planet with the Sun) An event in which a planet and the Sun line up in our sky.

conservation of angular momentum (law of) The principle that, in the absence of net torque (twisting force), the total angular momentum of a system remains constant.

conservation of energy (law of) The principle that energy (including mass-energy) can be neither created nor destroyed, but can only change from one form to another.

conservation of momentum (law of) The principle that, in the absence of net force, the total momentum of a system remains constant.

constellation A region of the sky; 88 official constellations cover the celestial sphere.

continental crust The thicker lower-density crust that makes up Earth's continents. It is made when remelting of seafloor crust allows lower-density rock to separate and erupt to the surface. Continental crust ranges in age from very young to as old as about 4 billion years (or more).

continuous spectrum A spectrum (of light) that spans a broad range of wavelengths without interruption by emission or absorption lines.

convection The energy transport process in which warm material expands and rises while cooler material contracts and falls.

convection cell An individual small region of convecting material.

convection zone (of a star) A region in which energy is transported outward by convection.

Copernican revolution The dramatic change, initiated by Copernicus, that occurred when we learned that Earth is a planet orbiting the Sun rather than the center of the universe.

core (of a planet) The dense central region of a planet that has undergone differentiation.

core (of a star) The central region of a star, in which nuclear fusion can occur.

Coriolis effect The effect due to rotation that causes air or objects on a rotating surface or planet to deviate from straight-line trajectories.

corona (solar) The tenuous uppermost layer of the Sun's atmosphere; most of the Sun's X rays are emitted from this region, in which the temperature is about 1 million K.

coronal holes Regions of the corona that barely show up in X-ray images because they are nearly devoid of hot coronal gas.

coronal mass ejections Bursts of charged particles from the Sun's corona that travel outward into space.

cosmic microwave background The remnant radiation from the Big Bang, which we detect using radio telescopes sensitive to microwaves (which are short-wavelength radio waves).

cosmic rays Particles such as electrons, protons, and atomic nuclei that zip through interstellar space at close to the speed of light.

cosmological constant The name given to a term in Einstein's equations of general relativity. If it is not zero, then it represents a repulsive force or a type of energy (sometimes called *dark energy* or *quintessence*) that might cause the expansion of the universe to accelerate with time.

cosmological horizon The boundary of our observable universe, which is where the lookback time is equal to the age of the universe. Beyond this boundary in spacetime, we cannot see anything at all.

Cosmological Principle The idea that matter is distributed uniformly throughout the universe on very large scales, meaning that the universe has neither a center nor an edge.

cosmological redshift The redshift we see from distant galaxies, caused by the fact that expansion of the universe stretches all the photons within it to longer, redder wavelengths.

cosmology The study of the overall structure and evolution of the universe.

cosmos An alternative name for the universe.

crescent (phase) The phase of the Moon (or of a planet) in which just a small portion (less than half) of the visible face is illuminated by sunlight.

critical density The precise average density for the entire universe that marks the dividing line between a recollapsing universe and one that will expand forever.

critical universe A universe that will never collapse, but that expands more and more slowly as time progresses; in the absence of a repulsive force (*see* cosmological constant), a critical universe is one in which the average mass density *equals* the critical density.

crust (of a planet) The low-density surface layer of a planet that has undergone differentiation.

curvature of spacetime A change in the geometry of space that is produced in the vicinity of a massive object and is responsible for the force we call gravity. The overall geometry of the universe may also be curved, depending on its overall mass-energy content.

cycles per second Units of frequency for a wave; describes the number of peaks (or troughs) of a wave that pass by a given point each second. Equivalent to *hertz*.

dark energy Name sometimes given to energy that could be causing the expansion of the universe to accelerate. *See* cosmological constant.

dark matter Matter that we infer to exist from its gravitational effects but from which we have not detected any light; dark matter apparently dominates the total mass of the universe.

daylight saving time Standard time plus 1 hour, so that the Sun appears on the meridian around 1 p.m. rather than around noon.

decay (radioactive) *See* radioactive decay.

declination (dec) Analogous to latitude, but on the celestial sphere; it is the angular north-south distance between the celestial equator and a location on the celestial sphere.

deferent The large circle upon which a planet follows its circle-upon-circle path around Earth in the (Earth-centered) Ptolemaic model of the universe. *See also* epicycle.

degeneracy pressure A type of pressure unrelated to an object's temperature, which arises when electrons (electron degeneracy pressure) or neutrons (neutron degeneracy pressure) are packed so tightly that the exclusion and uncertainty principles come into play.

degenerate object An object, such as a brown dwarf, white dwarf, or neutron star, in which degeneracy pressure is the primary pressure pushing back against gravity.

deuterium A form of hydrogen in which the nucleus contains a proton and a neutron, rather than only a proton (as is the case for most hydrogen nuclei).

differential rotation Rotation in which the equator of an object rotates at a different rate than the poles.

differentiation The process by which gravity separates materials according to density, with high-density materials sinking and low-density materials rising.

diffraction grating A finely etched surface that can split light into a spectrum.

diffraction limit The angular resolution that a telescope could achieve if it were limited only by the interference of light waves; it is smaller (i.e., better angular resolution) for larger telescopes.

dimension (mathematical) Describes the number of independent directions in which movement is possible; for example, the surface of Earth is two-dimensional because only two independent directions of motion are possible (north-south and east-west).

direction (in local sky) One of the two coordinates (the other is altitude) needed to pinpoint an object in the local sky. It is the direction, such as north, south, east, or west, in which you must face to see the object. *See also* azimuth.

disk (of a galaxy) The portion of a spiral galaxy that looks like a disk and contains an interstellar medium with cool gas and dust; stars of many ages are found in the disk.

disk population The stars that orbit within the disk of a spiral galaxy; sometimes called *Population I*.

DNA (deoxyribonucleic acid) The molecule that represents the genetic material of life on Earth.

Doppler effect (shift) The effect that shifts the wavelengths of spectral features in objects that are moving toward or away from the observer.

Doppler technique The detection of extrasolar planets through the motion of a star toward and away from the observer caused by gravitational tugs from the planet.

double shell–burning star A star that is fusing helium into carbon in a shell around an inert carbon core and is fusing hydrogen into helium in a shell at the top of the helium layer.

down quark One of the two quark types (the other is the up quark) found in ordinary protons and neutrons. It has a charge of $-\frac{1}{3}$.

Drake equation An equation that lays out the factors that play a role in determining the number of communicating civilizations in our galaxy.

dust (or **dust grains**) Tiny solid flecks of material; in astronomy, we often discuss interplanetary dust (found within a star system) or interstellar dust (found between the stars in a galaxy).

dust tail (of a comet) One of two tails seen when a comet passes near the Sun (the other is the plasma tail). It is composed of small solid particles pushed away from the Sun by the radiation pressure of sunlight.

dwarf elliptical galaxy A small elliptical galaxy with less than about a billion stars.

dwarf galaxies Relatively small galaxies, consisting of less than about 10 billion stars.

dwarf planet An object that orbits the Sun and is massive enough for its gravity to have made it nearly round in shape, but that does not qualify as an official planet because it has not cleared its orbital neighborhood. The dwarf planets of our solar system include the asteroid Ceres and the Kuiper belt objects Pluto, Eris, Haumea, and Makemake.

Earth-orbiters (spacecraft) Spacecraft designed to study Earth or the universe from Earth orbit.

eccentricity A measure of how much an ellipse deviates from a perfect circle; defined as the center-to-focus distance divided by the length of the semimajor axis.

eclipse An event in which one astronomical object casts a shadow on another or crosses our line of sight to the other object.

eclipse seasons Periods during which lunar and solar eclipses can occur because the nodes of the Moon's orbit are aligned with Earth and Sun.

eclipsing binary A binary star system in which the two stars happen to be orbiting in the plane of our line of sight, so that each star will periodically eclipse the other.

ecliptic The Sun's apparent annual path among the constellations.

ecliptic plane The plane of Earth's orbit around the Sun.

ejecta (from an impact) Debris ejected by the blast of an impact.

electrical charge A fundamental property of matter that is described by its amount and as either positive or negative; more technically, a measure of how a particle responds to the electromagnetic force.

electromagnetic field An abstract concept used to describe how a charged particle would affect other charged particles at a distance.

electromagnetic radiation Another name for light of all types, from radio waves through gamma rays.

electromagnetic spectrum The complete spectrum of light, including radio waves, infrared light, visible light, ultraviolet light, X rays, and gamma rays.

electromagnetic wave A synonym for *light*, which consists of waves of electric and magnetic fields.

electromagnetism (or **electromagnetic force**) One of the four fundamental forces; it is the force that dominates atomic and molecular interactions.

electron degeneracy pressure Degeneracy pressure exerted by electrons, as in brown dwarfs and white dwarfs.

electrons Fundamental particles with negative electric charge; the distribution of electrons in an atom gives the atom its size.

electron-volt (eV) A unit of energy equivalent to 1.60×10^{-19} joule.

electroweak era The era of the universe during which only three forces operated (gravity, strong force, and electroweak force), lasting from 10^{-38} second to 10^{-10} second after the Big Bang.

electroweak force The force that exists at high energies when the electromagnetic force and the weak force exist as a single force.

element (chemical) A substance made from individual atoms of a particular atomic number.

ellipse A type of oval that happens to be the shape of bound orbits. An ellipse can be drawn by moving a pencil along a string whose ends are tied to two tacks; the locations of the tacks are the *foci* (singular: *focus*) of the ellipse.

elliptical galaxies Galaxies that appear rounded in shape, often longer in one direction, like a football. They have no disks and contain very little cool gas and dust compared to spiral galaxies, though they often contain very hot, ionized gas.

elongation (greatest) For Mercury or Venus, the point at which it appears farthest from the Sun in our sky.

emission (of light) The process by which matter emits energy in the form of light.

emission line spectrum A spectrum that contains emission lines.

emission nebula Another name for an ionization nebula. *See* ionization nebula.

energy Broadly speaking, what can make matter move. The three basic types of energy are kinetic, potential, and radiative.

energy balance (in a star) The balance between the rate at which fusion releases energy in the star's core and the rate at which the star's surface radiates this energy into space.

epicycle The small circle upon which a planet moves while simultaneously going around a larger circle (the *deferent*) around Earth in the (Earth-centered) Ptolemaic model of the universe.

equation of time An equation describing the discrepancies between apparent and mean solar time.

equinox *See* fall equinox *and* spring equinox.

equivalence principle The fundamental starting point for general relativity, which states that the effects of gravity are exactly equivalent to the effects of acceleration.

era of atoms The era of the universe lasting from about 500,000 years to about 1 billion years after the Big Bang, during which it was cool enough for neutral atoms to form.

era of galaxies The present era of the universe, which began with the formation of galaxies when the universe was about 1 billion years old.

era of nuclei The era of the universe lasting from about 3 minutes to about 380,000 years after the Big Bang, during which matter in the universe was fully ionized and opaque to light. The cosmic background radiation was released at the end of this era.

era of nucleosynthesis The era of the universe lasting from about 0.001 second to about 3 minutes after the Big Bang, by the end of which virtually all of the neutrons and about one-seventh of the protons in the universe had fused into helium.

erosion The wearing down or building up of geological features by wind, water, ice, and other phenomena of planetary weather.

eruption The process of releasing hot lava on the planet's surface.

escape velocity The speed necessary for an object to completely escape the gravity of a large body such as a moon, planet, or star.

evaporation The process by which atoms or molecules escape into the gas phase from a liquid.

event Any particular point along a worldline; all observers will agree on the reality of an event but may disagree about its time and location.

event horizon The boundary that marks the "point of no return" between a black hole and the outside universe; events that occur within the event horizon can have no influence on our observable universe.

evolution (biological) The gradual change in populations of living organisms responsible for transforming life on Earth from its primitive origins to the great diversity of life today.

exchange particle A type of subatomic particle that transmits one of the four fundamental forces; according to the standard model of physics, these particles are always exchanged whenever two objects interact through a force.

excited state (of an atom) Any arrangement of electrons in an atom that has more energy than the ground state.

exclusion principle The law of quantum mechanics that states that two fermions cannot occupy the same quantum state at the same time.

exosphere The hot, outer layer of an atmosphere, where the atmosphere "fades away" to space.

expansion (of the universe) The idea that the space between galaxies or clusters of galaxies is growing with time.

exposure time The amount of time during which light is collected to make a single image.

extrasolar planet A planet orbiting a star other than our Sun.

extremophiles Living organisms that are adapted to conditions that are "extreme" by human standards, such as very high or low temperature or a high level of salinity or radiation.

Fahrenheit (temperature scale) The temperature scale commonly used in daily activity in the United States; defined so that, on Earth's surface, water freezes at 32°F and boils at 212°F.

fall (September) equinox Refers both to the point in Virgo on the celestial sphere where the ecliptic crosses the celestial equator and to the moment in time when the Sun appears at that point each year (around September 21).

false-color image An image displayed in colors that are *not* the true, visible-light colors of an object.

fault (geological) A place where rocks slip sideways relative to one another.

feedback processes Processes in which a small change in some property (such as temperature) leads to changes in other properties that either amplify or diminish the original small change.

fermions Particles, such as electrons, neutrons, and protons, that obey the exclusion principle.

Fermi's paradox The question posed by Enrico Fermi about extraterrestrial intelligence—"So where is everybody?"—which asks why we have not observed other civilizations even though simple arguments would suggest that some ought to have spread throughout the galaxy by now.

field An abstract concept used to describe how a particle would interact with a force. For example, the idea of a *gravitational field* describes how a particle would react to the local strength of gravity, and the idea of an *electromagnetic field* describes how a charged particle would respond to forces from other charged particles.

filter (for light) A material that transmits only particular wavelengths of light.

fireball A particularly bright meteor.

first-quarter (phase) The phase of the Moon that occurs one-quarter of the way through each cycle of phases, in which precisely half of the visible face is illuminated by sunlight.

fission The process by which one atomic nucleus breaks into two smaller nuclei. It releases energy if the two smaller nuclei together are less massive than the original nucleus.

flare star A small, spectral type M star that displays particularly strong flares on its surface.

flat (or **Euclidean**) **geometry** The type of geometry in which the rules of geometry for a flat plane hold, such as that the shortest distance between two points is a straight line and that the sum of the angles in a triangle is 180°.

flat universe A universe in which the overall geometry of spacetime is flat (Euclidean), as would be the case if the density of the universe was equal to the critical density.

flybys (spacecraft) Spacecraft that fly past a target object (such as a planet), usually just once, as opposed to entering a bound orbit of the object.

focal plane The place where an image created by a lens or mirror is in focus.

foci Plural of *focus*.

focus (of a lens or mirror) The point at which rays of light that were initially parallel (such as those from a distant star) converge.

focus (of an ellipse) One of two special points within an ellipse that lie along the major axis; these are the points around which we could stretch a pencil and string to draw an ellipse. When one object orbits a second object, the second object lies at one focus of the orbit.

force Anything that can cause a change in momentum.

formation properties (of planets) In this book, for the purpose of understanding geological processes, planets are defined to be born with four formation properties: size (mass and radius), distance from the Sun, composition, and rotation rate.

fossil Any relic of an organism that lived and died long ago.

frame of reference *See* reference frame.

free-fall The condition in which an object is falling without resistance; objects are weightless when in free-fall.

free-float frame A frame of reference in which all objects are weightless and hence float freely.

frequency The rate at which peaks of a wave pass by a point, measured in units of 1/s, often called *cycles per second* or *hertz*.

frost line The boundary in the solar nebula beyond which ices could condense; only metals and rocks could condense within the frost line.

fundamental forces There are four known fundamental forces in nature: gravity, the electromagnetic force, the strong force, and the weak force.

fundamental particles Subatomic particles that cannot be divided into anything smaller.

fusion The process by which two atomic nuclei fuse together to make a single, more massive nucleus. It releases energy if the final nucleus is less massive than the two nuclei that went into the reaction.

galactic cannibalism The term sometimes used to describe the process by which large galaxies merge with other galaxies in collisions. *Central dominant galaxies* are products of galactic cannibalism.

galactic fountain A model for the cycling of gas in the Milky Way Galaxy in which fountains of hot, ionized gas rise from the disk into the halo and then cool and form clouds as they sink back into the disk.

galactic wind A wind of low-density but extremely hot gas flowing out from a starburst galaxy, created by the combined energy of many supernovae.

galaxy A huge collection of anywhere from a few hundred million to more than a trillion stars, all bound together by gravity.

galaxy cluster *See* cluster of galaxies.

galaxy evolution The formation and development of galaxies.

Galilean moons The four moons of Jupiter that were discovered by Galileo: Io, Europa, Ganymede, and Callisto.

gamma-ray burst A sudden burst of gamma rays from deep space; such bursts apparently come from distant galaxies, but their precise mechanism is unknown.

gamma rays Light with very short wavelengths (and hence high frequencies)—shorter than those of X rays.

gap moons Tiny moons located within a gap in a planet's ring system. The gravity of a gap moon helps clear the gap.

gas phase The phase of matter in which atoms or molecules can move essentially independently of one another.

gas pressure The force (per unit area) pushing on any object due to surrounding gas. *See also* pressure.

general theory of relativity Einstein's generalization of his special theory of relativity so that the theory also applies when we consider effects of gravity or acceleration.

genetic code The "language" that living cells use to read the instructions chemically encoded in DNA.

geocentric model Any of the ancient Greek models that were used to predict planetary positions under the assumption that Earth lay in the center of the universe.

geocentric universe (ancient belief in) The idea that the Earth is the center of the entire universe.

geological activity Processes that change a planet's surface long after formation, such as volcanism, tectonics, and erosion.

geological processes The four basic geological processes are impact cratering, volcanism, tectonics, and erosion.

geological time scale The time scale used by scientists to describe major eras in Earth's past.

geology The study of surface features (on a moon, planet, or asteroid) and the processes that create them.

geostationary satellite A satellite that appears to stay stationary in the sky as viewed from Earth's surface, because it orbits in the same time it takes Earth to rotate and orbits in Earth's equatorial plane.

geosynchronous satellite A satellite that orbits Earth in the same time it takes Earth to rotate (one sidereal day).

giant galaxies Galaxies that are unusually large, typically containing a trillion or more stars. Most giant galaxies are elliptical, and many contain multiple nuclei near their centers.

giant impact A collision between a forming planet and a very large planetesimal, such as is thought to have formed our Moon.

giant molecular cloud A very large cloud of cold, dense interstellar gas, typically containing up to a million solar masses worth of material. *See also* molecular clouds.

giants (luminosity class III) Stars that appear just below the supergiants on the H-R diagram because they are somewhat smaller in radius and lower in luminosity.

gibbous (phase) The phase of the Moon (or of a planet) in which more than half but less than all of the visible face is illuminated by sunlight.

global positioning system (GPS) A system of navigation by satellites orbiting Earth.

global warming An expected increase in Earth's global average temperature caused by human input of carbon dioxide and other greenhouse gases into the atmosphere.

global wind patterns (or **global circulation**) Wind patterns that remain fixed on a global scale, determined by the combination of surface heating and the planet's rotation.

globular cluster A spherically shaped cluster of up to a million or more stars; globular clusters are found primarily in the halos of galaxies and contain only very old stars.

gluons The exchange particles for the strong force.

grand unified theory (GUT) A theory that unifies three of the four fundamental forces—the strong force, the weak force, and the electromagnetic force (but not gravity)—in a single model.

granulation (on the Sun) The bubbling pattern visible in the photosphere, produced by the underlying convection.

gravitation (law of) *See* universal law of gravitation.

gravitational constant The experimentally measured constant G that appears in the law of universal gravitation:

$$G = 6.67 \times 10^{-11} \frac{m^3}{kg \times s^2}$$

gravitational contraction The process in which gravity causes an object to contract, thereby converting gravitational potential energy into thermal energy.

gravitational encounter An encounter in which two (or more) objects pass near enough so that each can feel the effects of the other's gravity and they can therefore exchange energy.

gravitational equilibrium A state of balance in which the force of gravity pulling inward is precisely counteracted by pressure pushing outward.

gravitational lensing The magnification or distortion (into arcs, rings, or multiple images) of an image caused by light bending through a gravitational field, as predicted by Einstein's general theory of relativity.

gravitationally bound system Any system of objects, such as a star system or a galaxy, that is held together by gravity.

gravitational potential energy Energy that an object has by virtue of its position in a gravitational field; an object has more gravitational potential energy when it has a greater distance that it can potentially fall.

gravitational redshift A redshift caused by the fact that time runs slowly in gravitational fields.

gravitational time dilation The slowing of time that occurs in a gravitational field, as predicted by Einstein's general theory of relativity.

gravitational waves Waves, predicted by Einstein's general theory of relativity, that travel at the speed of light and transmit distortions of space through the universe. Although they have not yet been observed directly, we have strong indirect evidence that they exist.

gravitons The exchange particles for the force of gravity.

gravity One of the four fundamental forces; it is the force that dominates on large scales.

grazing incidence (in telescopes) Reflections in which light grazes a mirror surface and is deflected at a small angle; commonly used to focus high-energy ultraviolet light and X rays.

great circle A circle on the surface of a sphere whose center is at the center of the sphere.

greatest elongation *See* elongation (greatest).

Great Red Spot A large, high-pressure storm on Jupiter.

greenhouse effect The process by which greenhouse gases in an atmosphere make a planet's surface temperature warmer than it would be in the absence of an atmosphere.

greenhouse gases Gases, such as carbon dioxide, water vapor, and methane, that are particularly good absorbers of infrared light but are transparent to visible light.

Gregorian calendar Our modern calendar, introduced by Pope Gregory in 1582.

ground state (of an atom) The lowest possible energy state of the electrons in an atom.

group (of galaxies) A few to a few dozen galaxies bound together by gravity. *See also* cluster of galaxies.

GUT era The era of the universe during which only two forces operated (gravity and the grand-unified-theory, or GUT, force), lasting from 10^{-43} second to 10^{-38} second after the Big Bang.

GUT force The proposed force that exists at very high energies when the strong force, the weak force, and the electromagnetic force (but not gravity) all act as one.

H II region Another name for an ionization nebula. *See* ionization nebula.

habitable world A world with environmental conditions under which life could *potentially* arise or survive.

habitable zone The region around a star in which planets could potentially have surface temperatures at which liquid water could exist.

Hadley cells *See* circulation cells.

half-life The time it takes for half of the nuclei in a given quantity of a radioactive substance to decay.

halo (of a galaxy) The spherical region surrounding the disk of a spiral galaxy.

Hawking radiation Radiation predicted to arise from the evaporation of black holes.

heavy bombardment The period in the first few hundred million years after the solar system formed during which the tail end of planetary accretion created most of the craters found on ancient planetary surfaces.

heavy elements In astronomy, generally all elements *except* hydrogen and helium.

helium-burning star A star that is currently fusing helium into carbon in its core.

helium-capture reactions Fusion reactions that fuse a helium nucleus into some other nucleus; such reactions can fuse carbon into oxygen, oxygen into neon, neon into magnesium, and so on.

helium flash The event that marks the sudden onset of helium fusion in the previously inert helium core of a low-mass star.

helium fusion The fusion of three helium nuclei into one carbon nucleus; also called the *triple-alpha reaction*.

hertz (Hz) The standard unit of frequency for light waves; equivalent to units of 1/s.

Hertzsprung-Russell (H-R) diagram A graph plotting individual stars as points, with stellar luminosity on the vertical axis and spectral type (or surface temperature) on the horizontal axis.

high-mass stars Stars born with masses above about $8M_{Sun}$; these stars will end their lives by exploding as supernovae.

horizon A boundary that divides what we can see from what we cannot see.

horizontal branch The horizontal line of stars that represents helium-burning stars on an H-R diagram for a cluster of stars.

horoscope A predictive chart made by an astrologer; in scientific studies, horoscopes have never been found to have any validity as predictive tools.

hot Jupiter A class of planet that is Jupiter-like in size but orbits very close to it star, causing it to have a very high surface temperature.

hot spot (geological) A place within a plate of the lithosphere where a localized plume of hot mantle material rises.

hour angle (HA) The angle or time (measured in hours) since an object was last on the meridian in the local sky; defined to be 0 hours for objects that are on the meridian.

Hubble's constant A number that expresses the current rate of expansion of the universe; designated H_0, it is usually stated in units of km/s/Mpc. The reciprocal of Hubble's constant is the age the universe would have *if* the expansion rate had never changed.

Hubble's law Mathematical expression of the idea that more distant galaxies move away from us faster: $v = H_0 \times d$, where v is a galaxy's speed away from us, d is its distance, and H_0 is Hubble's constant.

hydrogen compounds Compounds that contain hydrogen and were common in the solar nebula, such as water (H_2O), ammonia (NH_3), and methane (CH_4).

hydrogen shell burning Hydrogen fusion that occurs in a shell surrounding a stellar core.

hydrosphere The "layer" of water on Earth consisting of oceans, lakes, rivers, ice caps, and other liquid water and ice.

hydrostatic equilibrium *See* gravitational equilibrium.

hyperbola The precise mathematical shape of one type of unbound orbit (the other is a parabola) allowed under the force of gravity; at great distances from the attracting object, a hyperbolic path looks like a straight line.

hypernova A term sometimes used to describe a supernova (explosion) of a star so massive that it leaves a black hole behind.

hyperspace Any space with more than three dimensions.

hypothesis A tentative model proposed to explain some set of observed facts, but which has not yet been rigorously tested and confirmed.

ice ages Periods of global cooling during which the polar caps, glaciers, and snow cover extend closer to the equator.

ices (in solar system theory) Materials that are solid only at low temperatures, such as the hydrogen compounds water, ammonia, and methane.

ideal gas law The law relating the pressure, temperature, and number density of particles in an ideal gas.

image A picture of an object made by focusing light.

imaging (in astronomical research) The process of obtaining pictures of astronomical objects.

impact The collision of a small body (such as an asteroid or comet) with a larger object (such as a planet or moon).

impact basin A very large impact crater, often filled by a lava flow.

impact crater A bowl-shaped depression left by the impact of an object that strikes a planetary surface (as opposed to burning up in the atmosphere).

impact cratering The excavation of bowl-shaped depressions (*impact craters*) by asteroids or comets striking a planet's surface.

impactor The object responsible for an impact.

inflation (of the universe) A sudden and dramatic expansion of the universe thought to have occurred at the end of the GUT era.

infrared light Light with wavelengths that fall in the portion of the electromagnetic spectrum between radio waves and visible light.

inner solar system Generally considered to encompass the region of our solar system out to about the orbit of Mars.

intensity (of light) A measure of the amount of energy coming from light of specific wavelength in the spectrum of an object.

interferometry A telescopic technique in which two or more telescopes are used in tandem to produce much better angular resolution than the telescopes could achieve individually.

intermediate-mass stars Stars born with masses between about $2M_{Sun}$ and $8 M_{Sun}$; these stars end their lives by ejecting a planetary nebula and becoming a white dwarf.

interstellar cloud A cloud of gas and dust between the stars.

interstellar dust grains Tiny solid flecks of carbon and silicon minerals found in cool interstellar clouds; they resemble particles of smoke and form in the winds of red giant stars.

interstellar medium The gas and dust that fills the space between stars in a galaxy.

interstellar ramjet A hypothesized type of spaceship that uses a giant scoop to sweep up interstellar gas for use in a nuclear fusion engine.

interstellar reddening The change in the color of starlight as it passes through dusty gas. The light appears redder because dust grains absorb and scatter blue light more effectively than red light.

intracluster medium Hot, X-ray-emitting gas found between the galaxies within a cluster of galaxies.

inverse square law A law followed by any quantity that decreases with the square of the distance between two objects.

inverse square law for light The law stating that an object's apparent brightness depends on its actual luminosity and the inverse square of its distance from the observer:

$$\text{apparent brightness} = \frac{\text{luminosity}}{4\pi \times (\text{distance})^2}$$

inversion (atmospheric) A local weather condition in which air is colder near the surface than higher up in the troposphere—the opposite of the usual condition, in which the troposphere is warmer at the bottom.

ionization The process of stripping an electron from an atom.

ionization nebula A colorful, wispy cloud of gas that glows because neighboring hot stars irradiate it with ultraviolet photons that can ionize hydrogen atoms.

ionosphere A portion of the thermosphere in which ions are particularly common (because of ionization by X rays from the Sun).

ions Atoms with a positive or negative electrical charge.

Io torus A donut-shaped charged-particle belt around Jupiter that approximately traces Io's orbit.

irregular galaxies Galaxies that look neither spiral nor elliptical.

isotopes Forms of an element that have the same number of protons but different numbers of neutrons.

jets High-speed streams of gas ejected from an object into space.

joule The international unit of energy, equivalent to about 1/4000 of a Calorie.

jovian nebulae The clouds of gas that swirled around the jovian planets, from which the moons formed.

jovian planets Giant gaseous planets similar in overall composition to Jupiter.

Julian calendar The calendar introduced in 46 B.C. by Julius Caesar and used until the Gregorian calendar replaced it.

Kelvin (temperature scale) The most commonly used temperature scale in science, defined such that absolute zero is 0 K and water freezes at 273.15 K.

Kepler's first law Law stating that the orbit of each planet about the Sun is an ellipse with the Sun at one focus.

Kepler's laws of planetary motion Three laws discovered by Kepler that describe the motion of the planets around the Sun.

Kepler's second law The principle that, as a planet moves around its orbit, it sweeps out equal areas in equal times. This tells us that a planet moves faster when it is closer to the Sun (near perihelion) than when it is farther from the Sun (near aphelion) in its orbit.

Kepler's third law The principle that the square of a planet's orbital period is proportional to the cube of its average distance from the Sun (semimajor axis), which tells us that more distant planets move more slowly in their orbits; in its original form, written $p^2 = a^3$. *See also* Newton's version of Kepler's third law.

kinetic energy Energy of motion, given by the formula $\frac{1}{2}mv^2$.

Kirchhoff's laws A set of rules that summarizes the conditions under which objects produce thermal, absorption line, or emission line spectra. In brief: (1) An opaque object produces thermal radiation. (2) An absorption line spectrum occurs when thermal radiation passes through a thin gas that is cooler than the object emitting the thermal radiation. (3) An emission line spectrum occurs when we view a cloud of gas that is warmer than any background source of light.

Kirkwood gaps On a plot of asteroid semimajor axes, regions with few asteroids as a result of orbital resonances with Jupiter.

K–T event (or **impact**) The collision of an asteroid or comet 65 million years ago that caused the mass extinction best known for wiping out the dinosaurs. K and T stand for the geological layers above and below the event.

Kuiper belt The comet-rich region of our solar system that spans distances of about 30–100 AU from the Sun. Kuiper belt comets have orbits that lie fairly close to the plane of planetary orbits and travel around the Sun in the same direction as the planets.

Kuiper belt object Any object orbiting the Sun within the region of the Kuiper belt, although the term is most often used for relatively large objects. For example, Pluto and Eris are considered large Kuiper belt objects.

Large Magellanic Cloud One of two small, irregular galaxies (the other is the Small Magellanic Cloud) located about 150,000 light-years away; it probably orbits the Milky Way Galaxy.

large-scale structure (of the universe) Generally refers to the structure of the universe on size scales larger than that of clusters of galaxies.

latitude The angular north-south distance between Earth's equator and a location on Earth's surface.

leap year A calendar year with 366 rather than 365 days. Our current calendar (the Gregorian calendar) incorporates a leap year every 4 years (by adding February 29) except in century years that are not divisible by 400.

length contraction The effect in which you observe lengths to be shortened in reference frames moving relative to you.

lens (gravitational) *See* gravitational lensing.

lenticular galaxies Galaxies that look lens-shaped when seen edge-on, resembling spiral galaxies without arms. They tend to have less cool gas than normal spiral galaxies but more gas than elliptical galaxies.

leptons Fermions *not* made from quarks, such as electrons and neutrinos.

life track A track drawn on an H-R diagram to represent the changes in a star's surface temperature and luminosity during its life; also called an *evolutionary track*.

light-collecting area (of a telescope) The area of the primary mirror or lens that collects light in a telescope.

light curve A graph of an object's intensity against time.

light gases (in solar system theory) Hydrogen and helium, which never condense under solar nebula conditions.

light pollution Human-made light that hinders astronomical observations.

light-year (ly) The distance that light can travel in 1 year, which is 9.46 trillion km.

liquid phase The phase of matter in which atoms or molecules are held together but move relatively freely.

lithosphere The relatively rigid outer layer of a planet; generally encompasses the crust and the uppermost portion of the mantle.

Local Bubble (interstellar) The bubble of hot gas in which our Sun and other nearby stars apparently reside. *See also* bubble (interstellar).

Local Group The group of about 40 galaxies to which the Milky Way Galaxy belongs.

local sidereal time (LST) Sidereal time for a particular location, defined according to the position of the spring equinox in the local sky. More formally, the local sidereal time at any moment is defined to be the hour angle of the spring equinox.

local sky The sky as viewed from a particular location on Earth (or another solid object). Objects in the local sky are pinpointed by the coordinates of *altitude* and *direction* (or *azimuth*).

local solar neighborhood The portion of the Milky Way Galaxy that is located relatively close (within a few hundred to a couple thousand light-years) to our Sun.

Local Supercluster The supercluster of galaxies to which the Local Group belongs.

longitude The angular east-west distance between the prime meridian (which passes through Greenwich) and a location on Earth's surface.

lookback time The amount of time since the light we see from a distant object was emitted. If an object has a lookback time of 400 million years, we are seeing it as it looked 400 million years ago.

low-mass stars Stars born with masses less than about $2M_{Sun}$; these stars end their lives by ejecting a planetary nebula and becoming a white dwarf.

luminosity The total power output of an object, usually measured in watts or in units of solar luminosities ($L_{Sun} = 3.8 \times 10^{26}$ watts).

luminosity class A category describing the region of the H-R diagram in which a star falls. Luminosity class I represents supergiants, III represents giants, and V represents main-sequence stars; luminosity classes II and IV are intermediate to the others.

lunar eclipse An event that occurs when the Moon passes through Earth's shadow, which can occur only at full moon. A lunar eclipse may be total, partial, or penumbral.

lunar maria The regions of the Moon that look smooth from Earth and actually are impact basins.

lunar month *See* synodic month.

lunar phase *See* phase (of the Moon or a planet).

MACHOs One possible form of dark matter in which the dark objects are relatively large, like planets or brown dwarfs; stands for *massive compact halo objects*.

magma Underground molten rock.

magnetic braking The process by which a star's rotation slows as its magnetic field transfers its angular momentum to the surrounding nebula.

magnetic field The region surrounding a magnet in which it can affect other magnets or charged particles.

magnetic field lines Lines that represent how the needles on a series of compasses would point if they were laid out in a magnetic field.

magnetosphere The region surrounding a planet in which charged particles are trapped by the planet's magnetic field.

magnitude system A system for describing stellar brightness by using numbers, called *magnitudes*, based on an ancient Greek way of describing the brightnesses of stars in the sky. This system uses *apparent magnitude* to describe a star's apparent brightness and *absolute magnitude* to describe a star's luminosity.

main sequence The prominent line of points (representing *main-sequence stars*) running from the upper left to the lower right on an H-R diagram.

main-sequence fitting A method for measuring the distance to a cluster of stars by comparing the apparent brightness of the cluster's main sequence with that of the standard main sequence.

main-sequence lifetime The length of time for which a star of a particular mass can shine by fusing hydrogen into helium in its core.

main-sequence stars (luminosity class V) Stars whose temperature and luminosity place them on the main sequence of the H-R diagram. Main-sequence stars are all releasing energy by fusing hydrogen into helium in their cores.

main-sequence turnoff point The point on a cluster's H-R diagram where its stars turn off from the main sequence; the age of the cluster is equal to the main-sequence lifetime of stars at the main-sequence turnoff point.

mantle (of a planet) The rocky layer that lies between a planet's core and crust.

Martian meteorites Meteorites found on Earth that are thought to have originated on Mars.

mass A measure of the amount of matter in an object.

mass-energy The potential energy of mass, which has an amount $E = mc^2$.

mass exchange (in close binary star systems) The process in which tidal forces cause matter to spill from one star to a companion star in a close binary system.

mass extinction An event in which a large fraction of the species living on Earth go extinct, such as the event in which the dinosaurs died out about 65 million years ago.

mass increase (in relativity) The effect in which an object moving past you seems to have a mass greater than its rest mass.

massive star supernova A supernova that occurs when a massive star dies, initiated by the catastrophic collapse of its iron core; often called a *Type II supernova*.

mass-to-light ratio The mass of an object divided by its luminosity, usually stated in units of solar masses per solar luminosity. Objects with high mass-to-light ratios must contain substantial quantities of dark matter.

matter–antimatter annihilation An event that occurs when a particle of matter and a particle of antimatter meet and convert all of their mass-energy to photons.

mean solar time Time measured by the average position of the Sun in your local sky over the course of the year.

meridian A half-circle extending from your horizon (altitude 0°) due south, through your zenith, to your horizon due north.

metallic hydrogen Hydrogen that is so compressed that the hydrogen atoms all share electrons and thereby take on properties of metals, such as conducting electricity. It occurs only under very high-pressure conditions, such as those found deep within Jupiter.

metals (in solar system theory) Elements, such as nickel, iron, and aluminum, that condense at fairly high temperatures.

meteor A flash of light caused when a particle from space burns up in our atmosphere.

meteorite A rock from space that lands on Earth.

meteor shower A period during which many more meteors than usual can be seen.

Metonic cycle The 19-year period, discovered by the Babylonian astronomer Meton, over which the lunar phases occur on the same dates.

microwaves Light with wavelengths in the range of micrometers to millimeters. Microwaves are generally considered to be a subset of the radio wave portion of the electromagnetic spectrum.

mid-ocean ridges Long ridges of undersea volcanoes on Earth, along which mantle material erupts onto the ocean floor and pushes apart the existing seafloor on either side. These ridges are essentially the source of new seafloor crust, which then makes its way along the ocean bottom for millions of years before returning to the mantle at a subduction zone.

Milankovitch cycles The cyclical changes in Earth's axis tilt and orbit that can change the climate and cause ice ages.

Milky Way Used both as the name of our galaxy and to refer to the band of light we see in the sky when we look into the plane of the Milky Way Galaxy.

millisecond pulsars Pulsars with rotation periods of a few thousandths of a second.

minor planets An alternative name for *asteroids*.

model (scientific) A representation of some aspect of nature that can be used to explain and predict real phenomena without invoking myth, magic, or the supernatural.

molecular bands The tightly bunched lines in an object's spectrum that are produced by molecules.

molecular cloud fragments (or **molecular cloud cores**) The densest regions of molecular clouds, which usually go on to form stars.

molecular clouds Cool, dense interstellar clouds in which the low temperatures allow hydrogen atoms to pair up into hydrogen molecules (H_2).

molecular dissociation The process by which a molecule splits into its component atoms.

molecule Technically, the smallest unit of a chemical element or compound; in this text, the term refers only to combinations of two or more atoms held together by chemical bonds.

momentum The product of an object's mass and velocity.

moon An object that orbits a planet.

moonlets Very small moons that orbit within the ring systems of jovian planets.

mutations Errors in the copying process when a living cell replicates itself.

natural selection The process by which mutations that make an organism better able to survive get passed on to future generations.

neap tides The lower-than-average tides on Earth that occur at first- and third-quarter moon, when the tidal forces from the Sun and Moon oppose each another.

nebula A cloud of gas in space, usually one that is glowing.

nebular capture The process by which icy planetesimals capture hydrogen and helium gas to form jovian planets.

nebular theory The detailed theory that describes how our solar system formed from a cloud of interstellar gas and dust.

net force The overall force to which an object responds; the net force is equal to the rate of change in the object's momentum, or equivalently to the object's mass × acceleration.

neutrino A type of fundamental particle that has extremely low mass and responds only to the weak force; neutrinos are leptons and come in three types—electron neutrinos, mu neutrinos, and tau neutrinos.

neutron degeneracy pressure Degeneracy pressure exerted by neutrons, as in neutron stars.

neutrons Particles with no electrical charge found in atomic nuclei, built from three quarks.

neutron star The compact corpse of a high-mass star left over after a supernova; it typically contains a mass comparable to the mass of the Sun in a volume just a few kilometers in radius.

newton The standard unit of force in the metric system:

$$1 \text{ newton} = 1 \frac{\text{kg} \times \text{m}}{\text{s}^2}$$

Newton's first law of motion Principle that, in the absence of a net force, an object moves with constant velocity.

Newton's laws of motion Three basic laws that describe how objects respond to forces.

Newton's second law of motion Law stating how a net force affects an object's motion. Specifically, force = rate of change in momentum, or force = mass × acceleration.

Newton's third law of motion Principle that, for any force, there is always an equal and opposite reaction force.

Newton's universal law of gravitation *See* universal law of gravitation.

Newton's version of Kepler's third law A generalization of Kepler's third law used to calculate the masses of orbiting objects from measurements of orbital period and distance; usually written as

$$p^2 = \frac{4\pi^2}{G(M_1 + M_2)} a^3$$

nodes (of Moon's orbit) The two points in the Moon's orbit where it crosses the ecliptic plane.

nonbaryonic matter Matter that is not part of the normal composition of atoms, such as neutrinos or the hypothetical WIMPs. (More technically, particles that are not made from three quarks.)

nonscience As defined in this book, any way of searching for knowledge that makes no claim to follow the scientific method, such as seeking knowledge through intuition, tradition, or faith.

north celestial pole (NCP) The point on the celestial sphere directly above Earth's North Pole.

nova The dramatic brightening of a star that lasts for a few weeks and then subsides; it occurs when a burst of hydrogen fusion ignites in a shell on the surface of an accreting white dwarf in a binary star system.

nuclear fission The process in which a larger nucleus splits into two (or more) smaller particles.

nuclear fusion The process in which two (or more) smaller nuclei slam together and make one larger nucleus.

nucleus (of a comet) The solid portion of a comet—the only portion that exists when the comet is far from the Sun.

nucleus (of an atom) The compact center of an atom made from protons and neutrons.

observable universe The portion of the entire universe that, at least in principle, can be seen from Earth.

Occam's razor A principle often used in science, holding that scientists should prefer the simpler of two models that agree equally well with observations; named after the medieval scholar William of Occam (1285–1349).

Olbers' paradox A paradox pointing out that if the universe were infinite in both age and size (with stars found throughout the universe), then the sky would not be dark at night.

Oort cloud A huge, spherical region centered on the Sun, extending perhaps halfway to the nearest stars, in which trillions of comets orbit the Sun with random inclinations, orbital directions, and eccentricities.

opacity A measure of how much light a material absorbs compared to how much it transmits; materials with higher opacity absorb more light.

opaque Describes a material that absorbs light.

open cluster A cluster of up to several thousand stars; open clusters are found only in the disks of galaxies and often contain young stars.

open universe A universe in which spacetime has an overall shape analogous to the surface of a saddle.

opposition The point at which a planet appears opposite the Sun in our sky.

optical quality The ability of a lens, mirror, or telescope to obtain clear and properly focused images.

orbit The path followed by a celestial body because of gravity; an orbit may be *bound* (elliptical) or *unbound* (parabolic or hyperbolic).

orbital energy The sum of an orbiting object's kinetic and gravitational potential energies.

orbital resonance A situation in which one object's orbital period is a simple ratio of another object's period, such as 1/2, 1/4, or 5/3. In such cases, the two objects periodically line up with each other, and the extra gravitational attractions at these times can affect the objects' orbits.

orbital velocity law A variation on Newton's version of Kepler's third law that allows us to use a star's orbital speed and distance from the galactic center to determine the total mass of the galaxy contained *within* the star's orbit; mathematically,

$$M_r = \frac{r \times v^2}{G}$$

where M_r is the mass contained within the star's orbit, r is the star's distance from the galactic center, v is the star's orbital velocity, and G is the gravitational constant.

orbiters (of other worlds) Spacecraft that go into orbit of another world for long-term study.

outer solar system Generally considered to encompass the region of our solar system beginning at about the orbit of Jupiter.

outgassing The process of releasing gases from a planetary interior, usually through volcanic eruptions.

oxidation Chemical reactions, often with rocks on the surface of a planet, that remove oxygen from the atmosphere.

ozone The molecule O_3, which is a particularly good absorber of ultraviolet light.

ozone depletion The decline in levels of atmospheric ozone found worldwide on Earth, especially in Antarctica, in recent years.

ozone hole A place where the concentration of ozone in the stratosphere is dramatically lower than is the norm.

pair production The process in which a concentration of energy spontaneously turns into a particle and its antiparticle.

parabola The precise mathematical shape of a special type of unbound orbit allowed under the force of gravity. If an object in a parabolic orbit loses only a tiny amount of energy, it will become bound.

paradigm (in science) A general pattern of thought that tends to shape scientific study during a particular time period.

paradox A situation that, at least at first, seems to violate common sense or contradict itself. Resolving paradoxes often leads to deeper understanding.

parallax The apparent shifting of an object against the background, due to viewing it from different positions. *See also* stellar parallax.

parallax angle Half of a star's annual back-and-forth shift due to stellar parallax; related to the star's distance according to the formula

$$\text{distance in parsecs} = \frac{1}{p}$$

where p is the parallax angle in arcseconds.

parsec (pc) The distance to an object with a parallax angle of 1 arcsecond; approximately equal to 3.26 light-years.

partial lunar eclipse A lunar eclipse during which the Moon becomes only partially covered by Earth's umbral shadow.

partial solar eclipse A solar eclipse during which the Sun becomes only partially blocked by the disk of the Moon.

particle accelerator A machine designed to accelerate subatomic particles to high speeds in order to create new particles or to test fundamental theories of physics.

particle era The era of the universe lasting from 10^{-10} second to 0.001 second after the Big Bang, during which subatomic particles were continually created and destroyed and ending when matter annihilated antimatter.

peculiar velocity (of a galaxy) The component of a galaxy's velocity relative to the Milky Way that deviates from the velocity expected by Hubble's law.

penumbra The lighter, outlying regions of a shadow.

penumbral lunar eclipse A lunar eclipse during which the Moon passes only within Earth's penumbral shadow and does not fall within the umbra.

perigee The point at which an object orbiting Earth is nearest to Earth.

perihelion The point at which an object orbiting the Sun is closest to the Sun.

period–luminosity relation The relation that describes how the luminosity of a Cepheid variable star is related to the period between peaks in its brightness; the longer the period, the more luminous the star.

phase (of matter) The state determined by the way in which atoms or molecules are held together; the common phases are solid, liquid, and gas.

phase (of the Moon or a planet) The state determined by the portion of the visible face of the Moon (or of a planet) that is illuminated by sunlight. For the Moon, the phases cycle through new, waxing crescent, first-quarter, waxing gibbous, full, waning gibbous, third-quarter, waning crescent, and back to new.

photon An individual particle of light, characterized by a wavelength and a frequency.

photosphere The visible surface of the Sun, where the temperature averages just under 6000 K.

pixel An individual "picture element" on a CCD.

Planck era The era of the universe prior to the Planck time.

Planck's constant A universal constant, abbreviated h, with a value of $h = 6.626 \times 10^{-34}$ joule $\times$ s.

Planck time The time when the universe was 10^{-43} second old, before which random energy fluctuations were so large that our current theories are powerless to describe what might have been happening.

planet A moderately large object that orbits a star and shines primarily by reflecting light from its star. More precisely, according to a definition approved in 2006, a planet is an object that (1) orbits a star (but is itself neither a star nor a moon); (2) is massive enough for its own gravity to give it a nearly round shape; and (3) has cleared the neighborhood around its orbit. Objects that meet the first two criteria but not the third, including Ceres, Pluto, and Eris, are designated *dwarf planets*.

planetary geology The extension of the study of Earth's surface and interior to apply to other solid bodies in the solar system, such as terrestrial planets and jovian planet moons.

planetary migration A process through which a planet can move from the orbit on which it is born to a different orbit that is closer to or farther from its star.

planetary nebula The glowing cloud of gas ejected from a low-mass star at the end of its life.

planetesimals The building blocks of planets, formed by accretion in the solar nebula.

plasma A gas consisting of ions and electrons.

plasma tail (of a comet) One of two tails seen when a comet passes near the Sun (the other is the dust tail). It is composed of ionized gas blown away from the Sun by the solar wind.

plates (on a planet) Pieces of a lithosphere that apparently float upon the denser mantle below.

plate tectonics The geological process in which plates are moved around by stresses in a planet's mantle.

polarization (of light) The property of light describing how the electric and magnetic fields of light waves are aligned; light is said to be *polarized* when all of the photons have their electric and magnetic fields aligned in some particular way.

Population I *See* disk population.

Population II *See* spheroidal population.

positron *See* antielectron.

potential energy Energy stored for later conversion into kinetic energy; includes gravitational potential energy, electrical potential energy, and chemical potential energy.

power The rate of energy usage, usually measured in watts (1 watt = 1 joule/s).

precession The gradual wobble of the axis of a rotating object around a vertical line.

precipitation Condensed atmospheric gases that fall to the surface in the form of rain, snow, or hail.

pressure The force (per unit area) pushing on an object. In astronomy, we are generally interested in pressure applied by surrounding gas (or plasma). Ordinarily, such pressure is related to the temperature of the gas (*see* thermal pressure). In objects such as white dwarfs and neutron stars, pressure may arise from a quantum effect (*see* degeneracy pressure). Light can also exert pressure (*see* radiation pressure).

primary mirror The large, light-collecting mirror of a reflecting telescope.

prime focus (of a reflecting telescope) The first point at which light focuses after bouncing off the primary mirror; located in front of the primary mirror.

prime meridian The meridian of longitude that passes through Greenwich, England; defined to be longitude 0°.

primitive meteorites Meteorites that formed at the same time as the solar system itself, about 4.6 billion years ago. Primitive meteorites from the inner asteroid belt are usually stony, and those from the outer belt are usually carbon-rich.

processed meteorites Meteorites that apparently once were part of a larger object that "processed" the original material of the solar nebula into another form. Processed meteorites can be rocky if chipped from the surface or mantle, or metallic if blasted from the core.

proper motion The motion of an object in the plane of the sky, perpendicular to our line of sight.

protogalactic cloud A huge, collapsing cloud of intergalactic gas from which an individual galaxy formed.

proton–proton chain The chain of reactions by which low-mass stars (including the Sun) fuse hydrogen into helium.

protons Particles found in atomic nuclei with positive electrical charge, built from three quarks.

protoplanetary disk A disk of material surrounding a young star (or protostar) that may eventually form planets.

protostar A forming star that has not yet reached the point where sustained fusion can occur in its core.

protostellar disk A disk of material surrounding a protostar; essentially the same as a protoplanetary disk, but may not necessarily lead to planet formation.

protostellar wind The relatively strong wind from a protostar.

protosun The central object in the forming solar system that eventually became the Sun.

pseudoscience Something that purports to be science or may appear to be scientific but that does not adhere to the testing and verification requirements of the scientific method.

Ptolemaic model The geocentric model of the universe developed by Ptolemy in about 150 A.D.

pulsar A neutron star from which we see rapid pulses of radiation as it rotates.

pulsating variable stars Stars that grow alternately brighter and dimmer as their outer layers expand and contract in size.

quantum laws The laws that describe the behavior of particles on a very small scale; *see also* quantum mechanics.

quantum mechanics The branch of physics that deals with the very small, including molecules, atoms, and fundamental particles.

quantum state The complete description of the state of a subatomic particle, including its location, momentum, orbital angular momentum, and spin, to the extent allowed by the uncertainty principle.

quantum tunneling The process in which, thanks to the uncertainty principle, an electron or other subatomic particle appears on the other side of a barrier that it does not have the energy to overcome in a normal way.

quarks The building blocks of protons and neutrons; quarks are one of the two basic types of fermions (leptons are the other).

quasar The brightest type of active galactic nucleus.

radar mapping Imaging of a planet by bouncing radar waves off its surface, especially important for Venus and Titan, where thick clouds mask the surface.

radar ranging A method of measuring distances within the solar system by bouncing radio waves off planets.

radial motion The component of an object's motion directed toward or away from us.

radial velocity The portion of any object's total velocity that is directed toward or away from us. This part of the velocity is the only part that we can measure with the Doppler effect.

radiation pressure Pressure exerted by photons of light.

radiation zone (of a star) A region of the interior in which energy is transported primarily by radiative diffusion.

radiative diffusion The process by which photons gradually migrate from a hot region (such as the solar core) to a cooler region (such as the solar surface).

radiative energy Energy carried by light; the energy of a photon is Planck's constant times its frequency, or $h \times f$.

radioactive decay The spontaneous change of an atom into a different element, in which its nucleus breaks apart or a proton turns into an electron. It releases heat in a planet's interior.

radioactive element (or **radioactive isotope**) A substance whose nucleus tends to fall apart spontaneously.

radio galaxy A galaxy that emits unusually large quantities of radio waves; thought to contain an active galactic nucleus powered by a supermassive black hole.

radio lobes The huge regions of radio emission found on either side of radio galaxies. The lobes apparently contain plasma ejected by powerful jets from the galactic center.

radiometric dating The process of determining the age of a rock (i.e., the time since it solidified) by comparing the present amount of a radioactive substance to the amount of its decay product.

radio waves Light with very long wavelengths (and hence low frequencies)—longer than those of infrared light.

random walk A type of haphazard movement in which a particle or photon moves through a series of bounces, with each bounce sending it in a random direction.

recession velocity (of a galaxy) The speed at which a distant galaxy is moving away from us because of the expansion of the universe.

recollapsing universe A universe in which the collective gravity of all its matter eventually halts and reverses the expansion, causing the galaxies to come crashing back together and the universe to end in a fiery Big Crunch.

red giant A giant star that is red in color.

red-giant winds The relatively dense but slow winds from red giant stars.

redshift (Doppler) A Doppler shift in which spectral features are shifted to longer wavelengths, observed when an object is moving away from the observer.

reference frame (or **frame of reference**) What two people (or objects) share if they are *not* moving relative to one another.

reflecting telescope A telescope that uses mirrors to focus light.

reflection (of light) The process by which matter changes the direction of light.

reflection nebula A nebula that we see as a result of starlight reflected from interstellar dust grains. Reflection nebulae tend to have blue and black tints.

refracting telescope A telescope that uses lenses to focus light.

resonance *See* orbital resonance.

rest wavelength The wavelength of a spectral feature in the absence of any Doppler shift or gravitational redshift.

retrograde motion Motion that is backward compared to the norm. For example, we see Mars in apparent retrograde motion during the periods of time when it moves westward, rather than the more common eastward, relative to the stars.

revolution The orbital motion of one object around another.

right ascension (RA) Analogous to longitude, but on the celestial sphere; the angular east-west distance between the spring equinox and a location on the celestial sphere.

rings (planetary) The collections of numerous small particles orbiting a planet within its Roche zone.

Roche tidal zone The region within two to three planetary radii (of any planet) in which the tidal forces tugging an object apart become comparable to the gravitational forces holding it together; planetary rings are always found within the Roche tidal zone.

rocks (in solar system theory) Materials common on the surface of Earth, such as silicon-based minerals, that are solid at temperatures and pressures found on Earth but typically melt or vaporize at temperatures of 500–1300 K.

rotation The spinning of an object around its axis.

rotation curve A graph that plots rotational (or orbital) velocity against distance from the center for any object or set of objects.

runaway greenhouse effect A positive feedback cycle in which heating caused by the greenhouse effect causes more greenhouse gases to enter the atmosphere, which further enhances the greenhouse effect.

saddle-shaped (or **hyperbolic**) **geometry** The type of geometry in which the rules—such as that two lines that begin parallel eventually diverge—are most easily visualized on a saddle-shaped surface.

Sagittarius Dwarf A small dwarf elliptical galaxy that is currently passing through the disk of the Milky Way Galaxy.

saros cycle The period over which the basic pattern of eclipses repeats, which is about 18 years $11\frac{1}{3}$ days.

satellite Any object orbiting another object.

scattered light Light that is reflected into random directions.

Schwarzschild radius A measure of the size of the event horizon of a black hole.

science The search for knowledge that can be used to explain or predict natural phenomena in a way that can be confirmed by rigorous observations or experiments.

scientific method An organized approach to explaining observed facts through science.

scientific theory A model of some aspect of nature that has been rigorously tested and has passed all tests to date.

seafloor crust On Earth, the thin, dense crust of basalt created by seafloor spreading.

seafloor spreading On Earth, the creation of new seafloor crust at mid-ocean ridges.

search for extraterrestrial intelligence (SETI) The name given to observing projects designed to search for signs of intelligent life beyond Earth.

secondary mirror A small mirror in a reflecting telescope, used to reflect light gathered by the primary mirror toward an eyepiece or instrument.

sedimentary rock A rock that formed from sediments created and deposited by erosional processes.

seismic waves Earthquake-induced vibrations that propagate through a planet.

selection effect (or **selection bias**) A type of bias that arises from the way in which objects of study are selected and that can lead to incorrect conclusions. For example, when you are counting animals in a jungle it is easiest to see brightly colored animals, which could mislead you into thinking that these animals are the most common.

semimajor axis Half the distance across the long axis of an ellipse; in this text, it is usually referred to as the *average* distance of an orbiting object, abbreviated *a* in the formula for Kepler's third law.

Seyfert galaxies The name given to a class of galaxies that are found relatively nearby and that have nuclei much like those of quasars, except that they are less luminous.

shepherd moons Tiny moons within a planet's ring system that help force particles into a narrow ring; a variation on *gap moons*.

shield volcano A shallow-sloped volcano made from the flow of low-viscosity basaltic lava.

shock wave A wave of pressure generated by gas moving faster than the speed of sound.

sidereal day The time of 23 hours 56 minutes 4.09 seconds between successive appearances of any particular star on the meridian; essentially, the true rotation period of the Earth.

sidereal month The time required for the Moon to orbit Earth once (as measured against the stars); about $27\frac{1}{4}$ days.

sidereal period (of a planet) A planet's actual orbital period around the Sun.

sidereal time Time measured according to the position of stars in the sky rather than the position of the Sun in the sky. *See also* local sidereal time.

sidereal year The time required for Earth to complete exactly one orbit as measured against the stars; about 20 minutes longer than the tropical year on which our calendar is based.

silicate rock A silicon-rich rock.

singularity The place at the center of a black hole where, in principle, gravity crushes all matter to an infinitely tiny and dense point.

Small Magellanic Cloud One of two small, irregular galaxies (the other is the Large Magellanic Cloud) located about 150,000 light-years away; it probably orbits the Milky Way Galaxy.

small solar system body An asteroid, comet, or other object that orbits a star but is too small to qualify as a planet or dwarf planet.

snowball Earth Name given to a hypothesis suggesting that, some 600–700 million years ago, Earth experienced a period in which it became cold enough for glaciers to exist worldwide, even in equatorial regions.

solar activity Short-lived phenomena on the Sun, including the emergence and disappearance of individual sunspots, prominences, and flares; sometimes called *solar weather*.

solar circle The Sun's orbital path around the galaxy, which has a radius of about 28,000 light-years.

solar day 24 hours, which is the average time between appearances of the Sun on the meridian.

solar eclipse An event that occurs when the Moon's shadow falls on the Earth, which can occur only at new moon. A solar eclipse may be total, partial, or annular.

solar flares Huge and sudden releases of energy on the solar surface, probably caused when energy stored in magnetic fields is suddenly released.

solar luminosity The luminosity of the Sun, which is approximately 4×10^{26} watts.

solar maximum The time during each sunspot cycle at which the number of sunspots is the greatest.

solar minimum The time during each sunspot cycle at which the number of sunspots is the smallest.

solar nebula The piece of interstellar cloud from which our own solar system formed.

solar neutrino problem The disagreement between the predicted and observed number of neutrinos coming from the Sun.

solar prominences Vaulted loops of hot gas that rise above the Sun's surface and follow magnetic field lines.

solar sail A large, highly reflective (and thin, to minimize mass) piece of material that can "sail" through space using pressure exerted by sunlight.

solar system (or **star system**) A star (sometimes more than one star) and all the objects that orbit it.

solar thermostat *See* stellar thermostat; the solar thermostat is the same idea applied to the Sun.

solar wind A stream of charged particles ejected from the Sun.

solid phase The phase of matter in which atoms or molecules are held rigidly in place.

solstice *See* summer solstice *and* winter solstice.

sound wave A wave of alternately rising and falling pressure.

south celestial pole (SCP) The point on the celestial sphere directly above Earth's South Pole.

spacetime The inseparable, four-dimensional combination of space and time.

spacetime diagram A graph that plots a spatial dimension on one axis and time on another axis.

special theory of relativity Einstein's theory that describes the effects of the fact that all motion is relative and that everyone always measures the same speed of light.

spectral lines Bright or dark lines that appear in an object's spectrum, which we can see when we pass the object's light through a prismlike device that spreads out the light like a rainbow.

spectral resolution The degree of detail that can be seen in a spectrum; the higher the spectral resolution, the more detail we can see.

spectral type A way of classifying a star by the lines that appear in its spectrum; it is related to surface temperature. The basic spectral types are designated by a letter (OBAFGKM, with O for the hottest stars and M for the coolest) and are subdivided with numbers from 0 through 9.

spectrograph An instrument used to record spectra.

spectroscopic binary A binary star system whose binary nature is revealed because we detect the spectral lines of one or both stars alternately becoming blueshifted and redshifted as the stars orbit each other.

spectroscopy (in astronomical research) The process of obtaining spectra from astronomical objects.

spectrum (of light) *See* electromagnetic spectrum.

speed The rate at which an object moves. Its units are distance divided by time, such as m/s or km/hr.

speed of light The speed at which light travels, which is about 300,000 km/s.

spherical geometry The type of geometry in which the rules—such as that lines that begin parallel eventually meet—are those that hold on the surface of a sphere.

spheroidal component (of a galaxy) The portion of any galaxy that is spherical (or football-like) in shape and contains very little cool gas; it generally contains only very old stars. Elliptical galaxies have only a spheroidal component, while spiral galaxies also have a disk component.

spheroidal galaxy Another name for an *elliptical galaxy*.

spheroidal population Stars that orbit within the spheroidal component of a galaxy; sometimes called *Population II*. Elliptical galaxies have only a spheroidal population (they lack a disk population), while spiral galaxies have spheroidal population stars in their bulges and halos.

spin (quantum) *See* spin angular momentum.

spin angular momentum The inherent angular momentum of a fundamental particle; often simply called *spin*.

spiral arms The bright, prominent arms, usually in a spiral pattern, found in most spiral galaxies.

spiral density waves Gravitationally driven waves of enhanced density that move through a spiral galaxy and are responsible for maintaining its spiral arms.

spiral galaxies Galaxies that look like flat white disks with yellowish bulges at their centers. The disks are filled with cool gas and dust, interspersed with hotter ionized gas, and usually display beautiful spiral arms.

spreading centers (geological) Places where hot mantle material rises upward between plates and then spreads sideways, creating new seafloor crust.

spring (March) equinox Refers both to the point in Pisces on the celestial sphere where the ecliptic crosses the celestial equator and to the moment in time when the Sun appears at that point each year (around March 21).

spring tides The higher-than-average tides on Earth that occur at new and full moon, when the tidal forces from the Sun and Moon both act along the same line.

standard candle An object for which we have some means of knowing its true luminosity, so that we can use its apparent brightness to determine its distance with the luminosity–distance formula.

standard model (of physics) The current theoretical model that describes the fundamental particles and forces in nature.

standard time Time measured according to the internationally recognized time zones.

star A large, glowing ball of gas that generates energy through nuclear fusion in its core. The term *star* is sometimes applied to objects that are in the process of becoming true stars (e.g., protostars) and to the remains of stars that have died (e.g., neutron stars).

starburst galaxy A galaxy in which stars are forming at an unusually high rate.

star cluster *See* cluster of stars.

star–gas–star cycle The process of galactic recycling in which stars expel gas into space, where it mixes with the interstellar medium and eventually forms new stars.

star system *See* solar system.

state (quantum) *See* quantum state.

steady state theory A now-discredited theory that held that the universe had no beginning and looks about the same at all times.

Stefan–Boltzmann constant A constant that appears in the laws of thermal radiation, with value

$$\sigma = 5.7 \times 10^{-8} \frac{\text{watt}}{\text{m}^2 \times \text{Kelvin}^4}$$

stellar evolution The formation and development of stars.

stellar parallax The apparent shift in the position of a nearby star (relative to distant objects) that occurs as we view the star from different positions in Earth's orbit of the Sun each year.

stellar thermostat The regulation of a star's core temperature that comes about when a star is in both energy balance (the rate at which fusion releases energy in the star's core is balanced with the rate at which the star's surface radiates energy into space) and gravitational equilibrium.

stellar wind A stream of charged particles ejected from the surface of a star.

stratosphere An intermediate-altitude layer of Earth's atmosphere that is warmed by the absorption of ultraviolet light from the Sun.

stratovolcano A steep-sided volcano made from viscous lavas that can't flow very far before solidifying.

string theory New ideas, not yet well-tested, that attempt to explain all of physics in a much simpler way than current theories.

stromatolites Large bacterial "colonies."

strong force One of the four fundamental forces; it is the force that holds atomic nuclei together.

subduction (of tectonic plates) The process in which one plate slides under another.

subduction zones Places where one plate slides under another.

subgiant A star that is between being a main-sequence star and being a giant; subgiants have inert helium cores and hydrogen-burning shells.

sublimation The process by which atoms or molecules escape into the gas phase from a solid.

summer (June) solstice Refers both to the point on the celestial sphere where the ecliptic is farthest north of the celestial equator and to the moment in time when the Sun appears at that point each year (around June 21).

sunspot cycle The period of about 11 years over which the number of sunspots on the Sun rises and falls.

sunspots Blotches on the surface of the Sun that appear darker than surrounding regions.

superbubble Essentially a giant interstellar bubble, formed when the shock waves of many individual bubbles merge to form a single giant shock wave.

superclusters The largest known structures in the universe, consisting of many clusters of galaxies, groups of galaxies, and individual galaxies.

supergiants The very large and very bright stars (luminosity class I) that appear at the top of an H-R diagram.

supermassive black holes Giant black holes, with masses millions to billions of times that of our Sun, thought to reside in the centers of many galaxies and to power active galactic nuclei.

supernova The explosion of a star.

Supernova 1987A A supernova witnessed on Earth in 1987; it was the nearest supernova seen in nearly 400 years and helped astronomers refine theories of supernovae.

supernova remnant A glowing, expanding cloud of debris from a supernova explosion.

surface area–to–volume ratio The ratio defined by an object's surface area divided by its volume; this ratio is larger for smaller objects (and vice versa).

synchronous rotation The rotation of an object that always shows the same face to an object that it is orbiting because its rotation period and orbital period are equal.

synchrotron radiation A type of radio emission that occurs when electrons moving at nearly the speed of light spiral around magnetic field lines.

synodic month (or **lunar month**) The time required for a complete cycle of lunar phases, which averages about $29\frac{1}{2}$ days.

synodic period (of a planet) The time between successive alignments of a planet and the Sun in our sky; measured from opposition to opposition for a planet beyond Earth's orbit, or from superior conjunction to superior conjunction for Mercury and Venus.

tangential motion The component of an object's motion directed across our line of sight.

tangential velocity The portion of any object's total velocity that is directed across (perpendicular to) our line of sight. This part of the velocity cannot be measured with the Doppler effect. It can be measured only by observing the object's gradual motion across our sky.

tectonics The disruption of a planet's surface by internal stresses.

temperature A measure of the average kinetic energy of particles in a substance.

terrestrial planets Rocky planets similar in overall composition to Earth.

theories of relativity (special and general) Einstein's theories that describe the nature of space, time, and gravity.

theory (in science) *See* scientific theory.

theory of evolution The theory, first advanced by Charles Darwin, that explains how evolution occurs through the process of natural selection.

thermal emitter An object that produces a thermal radiation spectrum; sometimes called a *blackbody*.

thermal energy The collective kinetic energy, as measured by temperature, of the many individual particles moving within a substance.

thermal escape The process in which atoms or molecules in a planet's exosphere move fast enough to escape into space.

thermal pressure The ordinary pressure in a gas arising from motions of particles that can be attributed to the object's temperature.

thermal pulses The predicted upward spikes in the rate of helium fusion, occurring every few thousand years, that occur near the end of a low-mass star's life.

thermal radiation The spectrum of radiation produced by an opaque object that depends only on the object's temperature; sometimes called *blackbody radiation*.

thermosphere A high, hot X-ray-absorbing layer of an atmosphere, just below the exosphere.

third-quarter (phase) The phase of the Moon that occurs three-quarters of the way through each cycle of phases, in which precisely half of the visible face is illuminated by sunlight.

tidal force A force that occurs when the gravity pulling on one side of an object is larger than that on the other side, causing the object to stretch.

tidal friction Friction within an object that is caused by a tidal force.

tidal heating A source of internal heating created by tidal friction. It is particularly important for satellites with eccentric orbits such as Io and Europa.

time dilation The effect in which you observe time running more slowly in reference frames moving relative to you.

timing (in astronomical research) The process of tracking how the light intensity from an astronomical object varies with time.

torque A twisting force that can cause a change in an object's angular momentum.

total apparent brightness *See* apparent brightness. The word "total" is sometimes added to make clear that we are talking about light across all wavelengths, not just visible light.

totality (eclipse) The portion of a total lunar eclipse during which the Moon is fully within Earth's umbral shadow or a total solar eclipse during which the Sun's disk is fully blocked by the Moon.

total luminosity *See* luminosity. The word "total" is sometimes added to make clear that we are talking about light across all wavelengths, not just visible light.

total lunar eclipse A lunar eclipse in which the Moon becomes fully covered by Earth's umbral shadow.

total solar eclipse A solar eclipse during which the Sun becomes fully blocked by the disk of the Moon.

transit An event in which a planet passes in front of a star (or the Sun) as seen from Earth. Only Mercury and Venus can be seen in transit of our Sun. The search for transits of extrasolar planets is an important planet detection strategy.

transmission (of light) The process in which light passes through matter without being absorbed.

transparent Describes a material that transmits light.

tree of life (evolutionary) A diagram that shows relationships between different species as inferred from genetic comparisons.

triple-alpha reaction *See* helium fusion.

Trojan asteroids Asteroids found within two stable zones that share Jupiter's orbit but lie 60° ahead of and behind Jupiter.

tropical year The time from one spring equinox to the next, on which our calendar is based.

tropic of Cancer The circle on Earth with latitude 23.5°N, which marks the northernmost latitude at which the Sun ever passes directly overhead (which it does at noon on the summer solstice).

tropic of Capricorn The circle on Earth with latitude 23.5°S, which marks the southernmost latitude at which the Sun ever passes directly overhead (which it does at noon on the winter solstice).

tropics The region on Earth surrounding the equator and extending from the Tropic of Capricorn (latitude 23.5°S) to the Tropic of Cancer (latitude 23.5°N).

troposphere The lowest atmospheric layer, in which convection and weather occur.

Tully–Fisher relation A relationship among spiral galaxies showing that the faster a spiral galaxy's rotation speed, the more luminous it is. It is important because it allows us to determine the distance to a spiral galaxy once we measure its rotation rate and apply the luminosity–distance formula.

turbulence Rapid and random motion.

21-cm line A spectral line from atomic hydrogen with wavelength 21 cm (in the radio portion of the spectrum).

ultraviolet light Light with wavelengths that fall in the portion of the electromagnetic spectrum between visible light and X rays.

umbra The dark central region of a shadow.

unbound orbits Orbits on which an object comes in toward a large body only once, never to return; unbound orbits may be parabolic or hyperbolic in shape.

uncertainty principle The law of quantum mechanics that states that we can never know both a particle's position and its momentum, or both its energy and the time it has the energy, with absolute precision.

universal law of gravitation The law expressing the force of gravity (F_g) between two objects, given by the formula

$$F_g = G\frac{M_1 M_2}{d^2}$$

$$\left(\text{where } G = 6.67 \times 10^{-11}\frac{\text{m}^3}{\text{kg} \times \text{s}^2}\right)$$

universal time (UT) Standard time in Greenwich (or anywhere on the prime meridian).

universe The sum total of all matter and energy.

up quark One of the two quark types (the other is the down quark) found in ordinary protons and neutrons; has a charge of $+\frac{2}{3}$.

velocity The combination of speed and direction of motion; it can be stated as a speed in a particular direction, such as 100 km/hr due north.

virtual particles Particles that "pop" in and out of existence so rapidly that, according to the uncertainty principle, they cannot be directly detected.

viscosity The thickness of a liquid described in terms of how rapidly it flows; low-viscosity liquids flow quickly (e.g., water), while high-viscosity liquids flow slowly (e.g., molasses).

visible light The light our eyes can see, ranging in wavelength from about 400 to 700 nm.

visual binary A binary star system in which both stars can be resolved through a telescope.

voids Huge volumes of space between superclusters that appear to contain very little matter.

volatiles Substances, such as water, carbon dioxide, and methane, that are usually found as gases, liquids, or surface ices on the terrestrial worlds.

volcanic plains Vast, relatively smooth areas created by the eruption of very runny lava.

volcanism The eruption of molten rock, or lava, from a planet's interior onto its surface.

waning (phases) The set of phases in which less and less of the visible face of the Moon is illuminated; the phases that come after full moon but before new moon.

watt The standard unit of power in science; defined as 1 watt = 1 joule/s.

wavelength The distance between adjacent peaks (or troughs) of a wave.

waxing (phases) The set of phases in which more and more of the visible face of the Moon is becoming illuminated; the phases that come after new moon but before full moon.

weak bosons The exchange particles for the weak force.

weak force One of the four fundamental forces; it is the force that mediates nuclear reactions, and it is the only force besides gravity felt by weakly interacting particles.

weakly interacting particles Particles, such as neutrinos and WIMPs, that respond only to the weak force and gravity; that is, they do not feel the strong force or the electromagnetic force.

weather The ever-varying combination of winds, clouds, temperature, and pressure in a planet's troposphere.

weight The net force that an object applies to its surroundings; in the case of a stationary body on the surface of the Earth, it equals mass × acceleration of gravity.

weightlessness A weight of zero, as occurs during free-fall.

white dwarf limit (or **Chandrasekhar limit**) The maximum possible mass for a white dwarf, which is about $1.4M_{Sun}$.

white dwarfs The hot, compact corpses of low-mass stars, typically with a mass similar to that of the Sun compressed to a volume the size of Earth.

white dwarf supernova A supernova that occurs when an accreting white dwarf reaches the white-dwarf limit, ignites runaway carbon fusion, and explodes like a bomb; often called a *Type Ia supernova*.

WIMPs A possible form of dark matter consisting of subatomic particles that are dark because they do not respond to the electromagnetic force; stands for *weakly interacting massive particles*.

winter (December) solstice Refers both to the point on the celestial sphere where the ecliptic is farthest south of the celestial equator and to the moment in time when the Sun appears at that point each year (around December 21).

worldline A line that represents an object on a spacetime diagram.

wormholes The name given to hypothetical tunnels through hyperspace that might connect two distant places in our universe.

X-ray binary A binary star system that emits substantial amounts of X rays, thought to be from an accretion disk around a neutron star or black hole.

X-ray burster An object that emits a burst of X rays every few hours to every few days; each burst lasts a few seconds and is thought to be caused by helium fusion on the surface of an accreting neutron star in a binary system.

X-ray bursts Burst of X rays coming from sudden ignition of fusion on the surface of an accreting neutron star in an X-ray binary system.

X rays Light with wavelengths that fall in the portion of the electromagnetic spectrum between ultraviolet light and gamma rays.

Zeeman effect The splitting of spectral lines by a magnetic field.

zenith The point directly overhead, which has an altitude of 90°.

zodiac The constellations on the celestial sphere through which the ecliptic passes.

zones (on a jovian planet) Bright bands of rising air that encircle a jovian planet at a particular set of latitudes.

Index